Modern and Ancient
Coal-Forming Environments

Edited by

James C. Cobb
Kentucky Geological Survey
228 MMRB
University of Kentucky
Lexington, Kentucky 40506-0107

and

C. Blaine Cecil
U.S. Geological Survey
956 National Center
Reston, Virginia 22092

SPECIAL PAPER

286

1993

Published by The Geological Society of America, Inc.
3300 Penrose Place, P.O. Box 9140, Boulder, Colorado 80301

Printed in U.S.A.

GSA Books Science Editor Richard A. Hoppin

Library of Congress Cataloging-in-Publication Data
Modern and ancient coal-forming environments / edited by James C. Cobb
  and C. Blaine Cecil.
      p.   cm. — (Special paper / The Geological Society of America
  ; 286)
    Includes bibliographical references and index.
    ISBN 0-8137-2286-1
    1. Peat. 2. Coal—Geology. I. Cobb, James C. II. Cecil, C. B.
III. Series: Special papers (Geological Society of America) ; 286.
TN837.M55   1993
553.2'1—dc20
                                                        93-31787
                                                        CIP

Cover: Computer-enhanced bathymetric patterns in the eastern and central part of the Strait of Malacca between the Malay peninsula of Malaysia and Sumatra, Indonesia. Red areas are intertidal, light blue is shallowest, and maroon is deepest water. Yellow to darkest green illustrates increasing elevation of highly generalized land topography. Restricted sediment supply and deposition within the strait is inferred to be primarily a function of marine rather than fluvial processes because erosion and subsequent fluvial-sediment discharge into the strait are highly restricted by rain forest cover. See "Allogenic and autogenic controls on sedimentation in the central Sumatra basin as an analogue for Pennsylvanian coal-bearing strata in the Appalachian basin" by C. Blaine Cecil, Frank T. Dulong, James C. Cobb, and Supardi, p. 3–22.

10 9 8 7 6 5 4 3 2 1

Contents

Contents

ANCIENT COAL-FORMING ENVIRONMENTS

Geological Society of America
Special Paper 286
1993

Introduction

James C. Cobb
Kentucky Geological Survey, 228 MMRB, University of Kentucky, Lexington, Kentucky 40506-0107
C. Blaine Cecil
U.S. Geological Survey, 956 National Center, Reston, Virginia 22092

Coal is the mainstay of electrical power generation in the United States and the world. It is also vital as a fuel in some industrial applications and in coke manufacture for use in the steel industry. The role of coal in power generation worldwide is growing because less developed nations are using more coal to increase their industrial capacities as well as meet their domestic electrical power requirements. Therefore, the use of coal as a fuel is expected to continue to increase well into the next century. Tens of billions of tons of coal will be produced and burned worldwide in the next 25 years, perhaps as much as 30 billion tons in the United States alone. This increase is occurring at a time when concerns about the environment are also increasing in the United States and the world.

The Clean Air Act Amendments of 1990, which regulate the emissions of sulfur dioxide in the United States, place real monetary penalties on sulfur emissions from coal and therefore on the sulfur content of the nation's coal resources. Restrictions on airborne toxic substances from coal combustion may also be enforced in the near future. Such restrictions cause concerns about the distribution of other chemical constituents in coal in addition to concerns regarding sulfur. Therefore, understanding the geology of coal has never been more important, because geologic interpretations of coal beds can be used to predict chemical and physical characteristics of coal beds for utilization.

For more than 50 years, coal geology has relied heavily on stratigraphic and sedimentologic studies for the interpretation of the environments of coal formation. Although some attempts have been made to model coal formation on the basis of modern peat deposits, most of these deposits have not had the thickness, lateral extent, or purity necessary to produce a coal bed of commercial interest. Knowledge of the origin of commercial-quality coal has been hindered by the lack of study of appropriate analogs. Therefore, a search for more appropriate analogs has been accelerated over the past decade. This volume presents observations, data, and interpretations on analogs of coal formation in equatorial Indonesia that may prove to be important for improving models that can be used to predict the occurrence and quality of coal deposits.

Over the past few years, attention has become focused on the accumulation of peat and associated sediments in equatorial ever-wet climates as an analog for coal formation. Data have been collected on the chemical and physical characteristics of tropical, ombrogenous peat deposits and associated sedimentary environments from field investigations in Indonesia and Malaysia. These new data are important for furthering the understanding of the geology of coal.

One objective of this volume is to present research on those peat deposits that appear to have the necessary characteristics of thickness, areal extent, chemistry, and geology to satisfy the criteria required for the formation of low-ash and low-sulfur coal resources. Thus, Part I of this volume contains seven research papers on modern tropical peat and associated sedimentary environments in Indonesia. The first through sixth papers present data and interpretations on the geology, sedimentology, chemistry, mineralogy, and petrography of tropical, ombrogenous domed peat deposits having characteristics needed to produce low-ash and low-sulfur coal resources. These papers are related to one another in that they deal with different aspects of the same deposits. Therefore, they present very thorough analyses of the chief areas of interest for coal geologists. The seventh paper examines an allochthonous deposit of peat that may be an analog for terrestrial organic matter as a source for petroleum.

The second objective of this volume, presented in Part II, is to address the origin of coal and coal-bearing strata. Four papers in Part II are directed toward this objective. These four papers present data on coal petrology, coal palynology, coal-ball paleobotany, and physical sedimentology. The data are then used to interpret the origin of Pennsylvanian coal deposits and coal-bearing strata.

Doming, as a mechanism of peat formation, may have been an important factor in the origin of certain types of coal deposits. Thus, the use of domed peat swamps as analogs for coal formation has become an important tool in the past decade for explaining the origin of some coal beds. However, data on the chemical, physical, and petrographic character of domed peat deposits and associated sediments have been limited. This volume augments existing information on domed peat formation and should provide much-needed new data for models of coal formation.

Cobb, J. C., and Cecil, C. B., 1993, Introduction, *in* Cobb, J. C., and Cecil, C. B., eds., Modern and Ancient Coal-Forming Environments: Boulder, Colorado, Geological Society of America Special Paper 286.

OVERVIEW

Part I

The paper by C. Blaine Cecil, Frank T. Dulong, James C. Cobb, and Supardi discusses the allogenic and autogenic controls on peat and siliciclastic deposition in an epeiric setting in a humid tropical region.

The paper by Sandra G. Neuzil, Supardi, C. Blaine Cecil, Jean S. Kane, and Kadar Soedjono presents data on the content and distribution of the major inorganic constituents of water and peat in three low-ash peat deposits and discusses possible sources, cycling, and leaching of these inorganic elements.

The regional setting for two domed, ombrogenous peat deposits in Riau, Sumatra, Indonesia, is presented in the paper by Supardi, A. D. Subekty, and Sandra Neuzil. The discussion of the setting includes the lateral extent, cross-sectional geometry, age, and rate of peat accumulation.

The paper by William C. Grady, Cortland F. Eble, and Sandra G. Neuzil contains the results of petrographic analyses of two peat cores at the margin and interior of a peat deposit from Riau, Sumatra, Indonesia. This paper compares the vertical variability of the petrography of the peat to that of coal beds in the Appalachian Basin.

The paper by Leslie F. Ruppert, Sandra G. Neuzil, C. Blaine Cecil, and Jean S. Kane examines the discrete crystalline, biogenic, and other mineral matter in a peat deposit and discusses possible sources and evidence for cycling of mineral matter in the peat.

Analytical methods used in determining the concentrations of major inorganic constituents in peat and the statistical correlations of these elements with ash, which were used as tools to understand inorganic and organic affinities in the peat stage of coal formation, are discussed in the paper by Jean Kane and Sandra G. Neuzil.

Little research has been done on allochthonous terrestrial organic-matter deposits. Such deposits may occur at the interface between certain fluvial and marine environments. The paper by Robert A. Gastaldo, George P. Allen, and Alain Y. Huc describes the sedimentology, plant composition, and geochemistry of a detrital peat deposit in the Mahakan River delta of Eastern Kalimantan, Indonesia (Borneo), from the perspective of its potential as a source for petroleum, not as an analog for commercial coal beds.

Part II

Part II of this volume addresses research on the origin of Pennsylvanian-age coal deposits and associated strata.

The paper by Cortland F. Eble and William C. Grady presents the palynology and petrography of two commercially important Central Appalachian Basin coal beds, and demonstrates that these coal beds were the result of transitions from planar to domed peat swamps.

Vegetational patterns in the Springfield coal bed of the Eastern Interior Basin (Illinois Basin) are described in a paper by Debra A. Willard. The paper describes the vegetational composition in this coal bed on the basis of coal-ball and miospore records. These records give important insights into the paleoecology of the Springfield coal bed, one of the principal producing beds in the United States.

The paper by J. H. Calder examines the lithotype, maceral, and miospore characteristics of a Pennsylvanian coal bed in Nova Scotia, Canada. These data show evidence for the evolution of a groundwater-influenced, peat-forming ecosystem in a piedmont setting.

The paper by Allen W. Archer and Erik P. Kvale focuses on the roof strata above coal beds in the Eastern Interior Basin (Illinois Basin). Detailed observations of the sedimentology of certain shales revealed evidence for tidal origins. Recognition of tidal deposits associated with coal beds is important because of the implications for reconstructing the paleogeography and depositional environments of coal-bearing strata.

ACKNOWLEDGMENTS

We would like to thank all the reviewers for their helpful reviews of the papers in this volume. The following geologists served as reviewers: J. F. Ferm (University of Kentucky), J. M. Coleman (Louisiana State University), J. R. Beerbower (State University of New York), W. A. Dimichele (Smithsonian Institution), E. H. Clifton (Conoco Oil Company), S. F. Greb (Kentucky Geological Survey), D. I. Siegel (Syracuse University), C. F. Eble (Kentucky Geological Survey), R. B. Finkelman (U.S. Geological Survey), Z. S. Altschuler (U.S. Geological Survey), T. Malterer (University of Minnesota), L. Coleman-Wnuk (U.S. Geological Survey), W. F. Outerbridge (U.S. Geological Survey), R. W. Stanton (U.S. Geological Survey), T. A. Moore (Wyoming Geological Survey), J. C. Hower (University of Kentucky), R. D. Harvey (Illinois Geological Survey), P. C. Bennett (University of Texas), R. Raymond, Jr. (Los Alamos National Laboratory), H. E. Francis (Kentucky Geological Survey), R. E. Hunter (U.S. Geological Survey), R. J. Burnham (New Mexico Museum of Natural History), T. L. Phillips (University of Illinois), G. W. Rothwell (Ohio University), R. M. Bustin (University of British Columbia). These reviewers are acknowledged for their timely and thorough reviews.

MANUSCRIPT ACCEPTED BY THE SOCIETY JANUARY 14, 1993

Geological Society of America
Special Paper 286
1993

Allogenic and autogenic controls on sedimentation in the central Sumatra basin as an analogue for Pennsylvanian coal-bearing strata in the Appalachian basin

C. Blaine Cecil and Frank T. Dulong
U.S. Geological Survey, 956 National Center, Reston, Virginia 22092
James C. Cobb
Kentucky Geological Survey, 228 MMRB, University of Kentucky, Lexington, Kentucky 40506-0107
Supardi
Directorate of Mineral Resources, J1. Soekorno-Hatta 444, Bandung, West Java 40254, Indonesia

ABSTRACT

Recent sedimentation patterns in the central Sumatra basin, Republic of Indonesia, may help to explain the cyclic stratigraphy of the Pennsylvanian System of the eastern United States. Modern influx of fluvial siliciclastic sediment to the epeiric seas of the Sunda shelf, including the Strait of Malacca, appears to be highly restricted by rain forest cover within the ever-wet climate belt of equatorial Sumatra. As a result, much of the marine and estuarine environments appear to be erosional or nondepositional except for localized deposition of sediment in slack water areas, such as the down-stream end of islands. Contemporaneously, thick (>13 m), laterally extensive (>70,000 km^2), peat deposits are forming on poorly drained coastal lowlands. Modern peat formation in this study, therefore, is not coeval with aggrading fluvial siliciclastic systems, a situation that commonly is assumed in many depositional models of coal formation. The stratigraphy of Pleistocene and Holocene sediments on the Sunda shelf, as well as those of the Pennsylvanian System, appears to be better explained by the allocyclic controls of climate and sea-level change on sediment flux rather than by depositional models that are based on autocyclic processes. The objective of this paper is to evaluate allocyclic and autocyclic controls on sedimentation in an epeiric setting in a humid (ever-wet) tropical region. Of particular interest are the factors that control peat formation and siliciclastic sediment flux in rivers, estuaries, and open marine environments.

INTRODUCTION

Models of coal formation

Although the origin of coal has been studied and debated for over a century, models that can be used to predict the occurrence, distribution, and quality of coal continue to be wanting. Current models that focus on parameters such as environment of deposition (Ferm and Horne, 1979), or tectonics (Tankard, 1986), and/or eustasy versus tectonics (Klein and Willard, 1989; Klein, 1990) implicitly assume a constant sediment supply. Other recent models recognize that sediment supply is probably not constant (e.g., McCabe, 1987; Belt and Lyons, 1989). To account for variations in sediment supply, McCabe (1987) suggested that peat-forming environments must be far removed from clastic sources, whereas Belt and Lyons (1989) suggested that blind thrust sheets may have diverted fluvial sediments away from peat swamps. None of these models adequately explains the (1) cyclic stratigraphic distribution of coal beds; (2) stratigraphic variation in coal quality, either within or among coal beds (Cecil et al., 1985); (3) extensive lateral continuity of many coal-bed horizons; or (4) chemical and physical stratigraphy of coal-bearing se-

Cecil, C. B., Dulong, F. T., Cobb, J. C., and Supardi, 1993, Allogenic and autogenic controls on sedimentation in the central Sumatra basin as an analogue for Pennsylvanian coal-bearing strata in the Appalachian basin, *in* Cobb, J. C., and Cecil, C. B., eds., Modern and Ancient Coal-Forming Environments: Boulder, Colorado, Geological Society of America Special Paper 286.

quences. Current models do not adequately explain such parameters because they often focus on description of strata associated with coal rather than deciphering the geologic factors that control the origin of coal beds. Most models implicitly assume that sedimentology and stratigraphy (both physical and chemical) are controlled by physical processes such as (1) autocyclic delta switching or stream meandering or (2) the allocyclic processes of tectonic or eustatic sea-level change. As a result, models that fully integrate the allogenic processes of tectonics, eustasy, and climate change (Beerbower, 1964) have not been fully developed, even though they were anticipated by Wanless and Shepard (1936). Integrated models may better explain both physical and chemical processes of sedimentation and, hence, both physical and chemical stratigraphy. Chemical stratigraphy, as used herein, refers to strata that have primary depositional and geochemical signatures such as paleosols, coal beds, carbonates, evaporites, and red beds, among others.

Allogenic controls on sedimentation (Beerbower, 1964) have been qualitatively recognized for many years, but, commonly, only one or two variables are evaluated. For example, Klein and Willard (1989) and Klein (1990) reviewed and summarized the inferred effects of tectonics versus eustasy on the origin of Pennsylvanian coal-bearing strata in the eastern United States without discussing the effects of paleoclimatic change on siliciclastic and chemical sediment flux. As with facies models, the effects of tectonics and eustasy are physical and, therefore, cannot account for chemical stratigraphy in laterally extensive sequences. Chemical stratigraphy can be accounted for, however, through changes in paleoclimate (Cecil, 1990; Cecil et al., 1985). This can be illustrated by comparing the chemical stratigraphy of Pennsylvanian with Mississippian strata in the eastern United States. Tectonics and eustasy were active during both periods as evidenced by regional changes in accommodation space and the occurrence of numerous transgressive-regressive events. However, differences in chemical stratigraphy, such as the paucity of coal and abundance of calcareous paleosols and other highly calcareous nonmarine material in Mississippian strata are in sharp contrast to the occurrence of abundant coal beds, highly leached paleosols, and a paucity of calcareous material of any type in Lower and lower Middle Pennsylvanian strata (Cecil et al., 1985; Cecil, 1990). These chemical differences cannot be explained solely by transgressive-regressive or tectonic cycles but are readily explained by including the allocyclic effects of paleoclimate change; the Mississippian was relatively dry as compared to the Pennsylvanian (White, 1913; White, 1925; Cecil et al., 1985; Cecil, 1990).

Humid tropical paleoclimates prevailed during deposition of Pennsylvanian coal-bearing strata of the eastern United States (Schopf, 1975; Cross, 1975; and Donaldson et al., 1985). Phillips and Peppers (1984) attributed changes in paleo-peat swamp flora and coal occurrence in Pennsylvanian strata to differences in rainfall patterns. In an attempt to account for the chemical stratigraphy of uppermost Mississippian and Pennsylvanian strata in the central Appalachian basin, Cecil et al. (1985) inferred a relatively dry seasonal climate for the Late Mississippian, a tropical ever-wet climate for the Early through the mid-Middle Pennsylvanian, and a return to a drier, more seasonal, climate during the upper Middle and Late Pennsylvanian.

Resolution of allogenic and autogenic process control on sedimentation is partially addressed by sequence stratigraphy. Physical allocyclic and autocyclic controls on sedimentation are described by this technique through recognition of sequence boundaries (attributed to sea-level change, an allocyclic process) and the concept of parasequences (the autocyclic response to sea-level change, summarized in Van Wagoner et al., 1988). Sequence stratigraphy is recognized in both siliciclastic and carbonate sequences (e.g., Van Wagoner et al., 1988). However, as in facies modeling, sequence stratigraphy is based on physical processes (primarily eustatic changes in sea level) and, therefore, does not address or explain chemical stratigraphy. Thus, the occurrence or quality of chemical rocks such as coal or limestone cannot be successfully predicted within stratigraphic sequences.

Although poorly understood, and rarely applied, climate change has long been recognized as a major control on the origin of stratigraphic sequences (e.g., Huntington, 1907; Wanless and Shepard, 1936; Cecil et al., 1985, Perlmutter and Matthews, 1989; Cecil, 1990), and, therefore, climate change should be included along with tectonics and eustasy in comprehensive allocyclic models. It may then be possible to develop comprehensive models that can predict both chemical and physical stratigraphy.

Indonesian analogue for Pennsylvanian coal-bearing sequences

Possible similarities between the genesis of Carboniferous coal-bearing sequences, such as those in the Appalachian basin, and the extensive coastal-plain peat deposits in equatorial Southeast Asia have been recognized for many years (e.g., White, 1913). Furthermore, on the basis of coal petrography and palynology, Smith (1962) suggested that the genesis of the Lower and Middle Yorkshire coal measures in England (Early through early Middle Pennsylvanian) might be analogous to these modern equatorial deposits. Subsequently, a comparison of Carboniferous stratigraphic data and inferences on paleoclimate with information presented by Anderson (1961, 1964, 1983) on modern tropical peat deposits in coastal Malaysia and Indonesia led Neuzil and Cecil (1984) and Cecil and others (1985) to suggest that the domed (convex upper surface), ombrogenous (water source from rainfall), peat deposits of these ever-wet equatorial areas might represent a modern analogue for Lower through mid-Middle Pennsylvanian coal deposits of the eastern United States. By comparing the physical and chemical stratigraphy of Lower and lower Middle Pennsylvanian strata with that of upper Middle and Upper Pennsylvanian strata, Cecil et al. (1985) concluded that the latter were deposited under relatively drier and perhaps more seasonal climatic conditions and that paleoswamps of this interval were planar (horizontal upper surface) and topogenous (water source from ground and surface water). The present study was undertaken, therefore, to (1) compare Pennsylvanian coal-bearing sequences with a specific tropical peat-forming environment and (2) determine if the modern conditions of sedimentation, including peat formation in this area, are analogous to the origin of either Lower and lower Middle Pennsylvanian, or upper Middle and Upper Pennsylvanian strata.

The central Sumatra basin in Riau Province, the Republic of Indonesia, was selected for study (Figs. 1 and 2). The central Sumatra basin is a foreland basin situated in the northwestern part of the epicontinental Sunda shelf (Hamilton, 1979). Much of this epicontinental platform was exposed during the last glacial maximum (Geyh et al., 1979), but now it is mostly flooded by the southern part of the South China Sea, the Java Sea, and the Strait of Malacca (Fig. 1). During this former low stand, the climate of the region was probably drier than today, and tropical rain forests were replaced by a treed savanna throughout the epicontinental shelf (Verstappen, 1975). Extensive modern peat formation in the study area is occurring on a broad, flat coastal plain and on nearby offshore islands. The coastal plain and islands appear to be the result of transgressive-regressive events and subsidence (allogenic processes) in the central Sumatra foreland basin (Hamilton, 1979; Cameron et al., 1982). Modern annual rainfall in this equatorial region is high (≥2.5 m/yr, see Neuzil et al., this volume) and rather uniformly distributed throughout the year, a climatic condition herein referred to as ever wet. This rainfall pattern is a key allogenic control on (1) weathering, leaching, and soil development; (2) vegetative cover that highly restricts erosion and fluvial siliciclastic sediment load; and (3) formation of ombrogenous domed-peat deposits.

A wide variety of environments consisting of upland soil, rivers, estuaries, and open sea form the borders of the peat deposits. Tropical weathering is contributing to soil development in upland areas adjacent to the peat deposits. Furthermore, human disturbance of the peat-swamp forests within the study area is minimal. Because the central Sumatra basin is an active foreland basin, situated in an epicontinental setting that has been subjected to eustatic changes in sea level, and is in a modern tropical ever-wet climate, the area was judged to be an ideal location to study peat formation and associated sedimentological conditions as a potential analogue for Pennsylvanian coal-bearing strata in the eastern United States.

This analogue study was designed to evaluate the allogenic processes of climate, tectonics, and eustasy, and the resulting autogenic responses. Our evaluation included (1) the allogenic Pleistocene and Holocene transgressive-regressive cycles and tectonic evolution of the study area (Cameron et al., 1982), (2) the physical and chemical parameters of the peat deposits (Neuzil et al., this volume; Supardi et al., this volume), and (3) the allogenic and autogenic controls on both chemical and siliciclastic sedimentation that are the subject of this paper.

FIELD INVESTIGATIONS AND OBSERVATIONS

Field investigations included three studies. The initial study, conducted by the Directorate of Mineral Resources, Republic of Indonesia, consisted of topographic surveys, peat-thickness mapping, and resources assessment of two peat deposits, Siak Kanan and Bengkalis Island (Supardi et al., this volume). These peat deposits, hereinafter referred to as Siak and Bengkalis, are bordered by the Minas Hills, Siak River, Kampar River, and the Strait of Malacca on the southwest, northwest, southeast, and northeast respectively (Fig. 2). This initial study was followed by

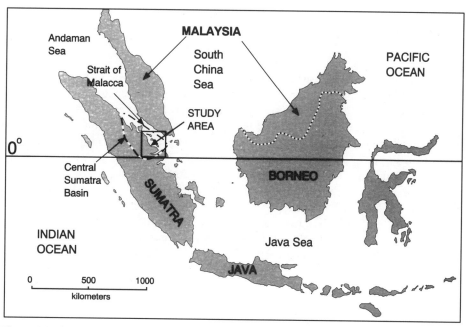

Figure 1. Index map of the study area, which is situated in western Indonesia in the central Sumatra basin. The equator is shown by the line marked 0°.

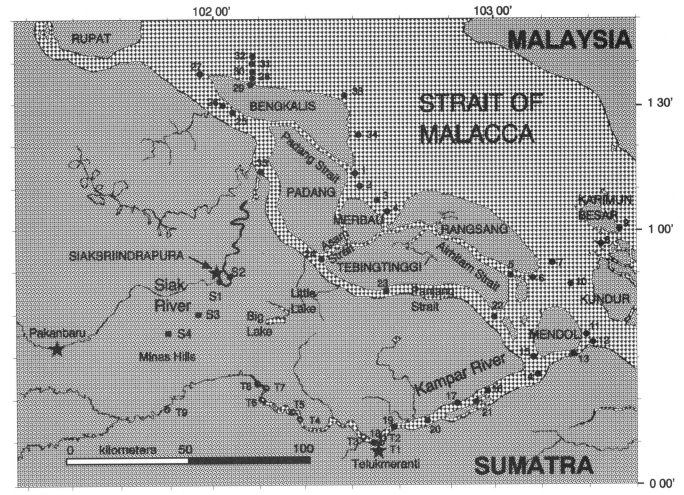

Figure 2. Sample location map for the study area, Riau Province, Indonesia. Symbols used in Figure 2 and subsequent figures are as follows: solid star, towns and villages; solid circle, marine and estuarine sampling stations during first field season; open circle, estuarine and riverine sampling stations during second field season; solid box, nonpeat soil localities.

two additional studies that included peat and water sampling (Neuzil et al., this volume) and an investigation of contemporaneous, sedimentary environments.

Sedimentary regimes in the study area were classified as chemical or siliciclastic. Chemical sedimentary regimes are limited to thick deposits of high-quality, laterally continuous, domed, ombrogenous peat (Supardi et al., this volume; Neuzil et al., this volume). Carbonate deposits, with the exception of minor amounts of siderite, were not found. Evaluation of sedimentary environments included upland soil, river, estuary, levee, flood plain, peat substratum, coastal areas, and the Strait of Malacca and associated lesser straits (sample locations are shown in Fig. 2).

Upland soil

A reconnaissance study of soil structure, mineralogy, and weathering was conducted in upland areas near the margin of the peat deposits (Fig. 2, sites S3 and S4) and in a levee along the Siak River (Fig. 2, sites S1 and S2). The upland soils are forming on the tectonically uplifted marine Minas Formation of Pleistocene age (Cameron et al., 1982). The uplift and regression have exposed the Minas to weathering and subsequent soil formation.

Approximately 2.5 m of the Minas Formation was exposed at site S3. On the basis of textural and X-ray diffraction analyses, the exposure at S3 ranges from sandy clay at the top to clay at the bottom (Soil Survey Staff, 1975); mineralogically, it is composed primarily of quartz with secondary amounts of various clay minerals that are dominated by kaolinite. The lower part of the exposure contains kaolinite clasts as large as 10 cm in diameter. The origin of the clasts is unknown, but they appear to be of primary origin rather than the result of soil-forming processes. Calcareous material, including calcareous peds, was not observed in the field nor were calcareous minerals detected by X-ray diffraction analyses.

Burrows made by marine organisms, which were probably originally cemented with siderite or pyrite (Fig. 3A), weather and

Figure 4. Desiccation structures in clayey soil developed on the Pleistocene Minas Formation (Fig. 2, site S4). The hat in the photograph is ~30 cm across.

Figure 3. A, Trace fossils cemented with siderite and/or pyrite in Pleistocene Minas Formation (Fig. 2, sites S3). B, Weathered trace fossils give red (dark) mottled appearance to immature soils developed on the Pleistocene Minas Formation (Fig. 2, site S4). Penknife in photos is 8 cm in length.

oxidize during soil formation giving the light-gray soils at sites S3 and S4 a red (dark) mottled appearance (Fig. 3B). Cemented burrows are easily recognized below the weathering zone, but, in most of the surface exposures, weathering and soil formation have altered the burrows beyond recognition. As a result of the pedogenic overprint, the oxidized burrows could easily be mistaken for rhizomorphs. The presence of analogous oxidized burrows in the rock record may be useful as a criterion to recognize and interpret paleo–sea-level change. The presence of oxidized burrows stratigraphically bounded by unoxidized burrows (i.e., still containing siderite or pyrite) would suggest subaerial exposure and incipient soil formation during the regressive phase of transgressive-regressive cycles.

Polygonal desiccation structures, approximately 0.5 m in diameter are well developed in clayey soils in some road cuts (Fig. 4). Vertical cracks with openings as much as approximately 5 cm at the surface taper downward for more than 1 m. These structures may have developed in response to unusually dry conditions that prevailed at the time of our study and/or to exposure through devegetation and road construction. We have observed, however, similar structures in Middle Pennsylvanian strata in the Appalachian basin that are suggestive of subaerial exposure and desiccation during periods of relative dryness under what otherwise may have been a relatively wet climate. These vertical polygonal desiccation structures contrast with the cross-cutting "vertic" or pseudoanticlinal structures in Vertisols that develop in response to seasonal wet/dry conditions (Retallack, 1989).

Subaerial exposure and the onset of weathering appears to be relatively recent, because upland soils in the study area are immature (Inceptisols). Weathering is in progress, as indicated by the transformation of siderite and pyrite to iron oxides in the mottled horizons and the presence of poorly crystallized kaolinite as determined by X-ray diffraction. Well-developed laterites (Spodosols or Oxisols) were not observed in the study area, although they are well known in other parts of Indonesia and Malaysia where the duration of subaerial exposure has been longer.

If the upland soils in the study area have been exposed to weathering as long as the peat has been forming (since 5000 yr B.P., Supardi et al., this volume), then weathering over the last five thousand years has not produced a mature soil profile. The development of a mature Latosol profile may require more time than is necessary for the formation of domed peat deposits. The rock record, therefore, may contain immature paleosols in juxtaposition with coal beds.

Cutbank, levee, and flood plain of the Siak River estuary

The terms "cutbank," "levee," and "flood plain," as used in this section are morphologic descriptors of the Siak River estuary (Fig. 2). As defined by tidal range, the estuary extends inland more than 100 km. Tidal range in the Siak River estuary was observed up the estuary as far as Pakanbaru (Fig. 2). Our measurements of suspended and bed load in the estuary and observations of sedimentary features in associated environments, however, were mostly near the village of Siaksriindrapura some 50 km inland from Panjang Strait (Fig. 2).

Profile samples were collected in 20-cm increments from the top of the levee soil to the base of the cutbank (Fig. 2, sites S1, S2). A somewhat sparse diatom flora, having planktonic marine and/or brackish affinity (Victoria Andrle, 1989, personal communication), indicates the sediment into which the Siak estuary is entrenched (Fig. 2, sites S1, S2) is primarily a marine deposit (Minas Formation?) rather than alluvium as suggested by Cameron et al. (1982). Estuarine currents erode the marine Minas Formation(?) as evidenced by undercutting and slumping of cutbanks. Mineralogically, the cutbank profile is dominated by quartz (>90%) and kaolinite (~10%) and texturally, the profile is silt. There is no change in grain size or mineralogy from top to bottom. An immature soil, developed atop the thin veneer of quartzose silt of the levee, is darkened by organic material; primary sedimentary structures have been destroyed by roots and soil-forming processes. The rooted soil profile grades downward into undisturbed planar and ripple-bedded quartzose silt in the cutbank. Uniform mineralogy and grain size, limited aggradation, the absence of any apparent discontinuity, the absence of freshwater diatoms, and the presence of marine and brackish water diatoms indicate that the thin veneer of levee material (<1 m) was derived from the cutbank and deposited by estuarine processes.

The flood plain of the Siak estuary is covered with water from rainfall and runoff from the adjacent peat dome during the two rainiest periods of the year (Supardi, personal communication). The flood plain, which occupies a belt between the levee and the peat dome (Fig. 5A), consists of light gray clay. One meter deep probes taken along a traverses across the flood plain (along A–A', Fig. 5A) indicate that the edge of the peat dome does not receive sediment from water in the flood plain because the edge of the peat is not being buried. Aggradation of siliciclastic material on the flood plain appears to be very limited indicating very low sediment influx from the Siak River/estuary. This observation is consistent with our assessment of the origin of levee material and the low suspended-sediment concentration (generally <60 mg/l) of the Siak River/estuary observed during our study. The dominance of marine- and brackish-water diatoms in the flood-plain clay (Victoria Andrle, 1990, written communication) is a further indication that sediment is only supplied by estuarine processes. Fluvial supply is negligible. The only freshwater diatoms noted are typical of those found in Indonesian domed peat deposits (Victoria Andrle, 1990, written communication) and apparently were transported to the flood plain by run off from the adjacent peat dome.

Peat substratum

Samples of peat substratum were analyzed for texture and mineralogy. The samples that were analyzed are shown in Table 1 and the location of these samples is shown in Figure 6A and B.

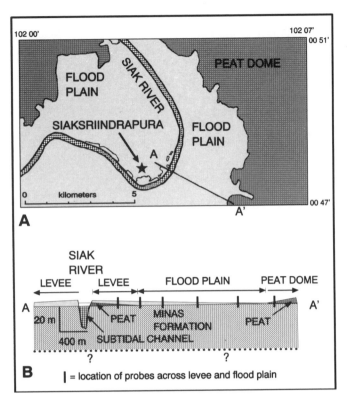

Figure 5. A. Plan view of Siak River estuary in the vicinity of the village of Siaksriindrapura. B. Cross-sectional profile (Fig. 5A, A–A') illustrating the Siak River/estuary, "flood plain," and the edge of the Siak peat dome (Fig. 2, site S2). The pattern in the channel profile denotes water at low tide. The Siak River/estuary channel is devoid of a modern bed load except for an estuary mouth bar. Lines are dotted where contacts are unknown.

Texturally, samples from beneath the Siak peat dome were generally a silty-clay whereas samples from the Bengkalis peat dome were clay (Table 1). Compositionally all bulk samples were dominated by quartz (>50%) but contained significant amounts of kaolinite with minor amounts of mixed-layer illite-smectite (Table 2). The mineralogy of the modern substratum is different from that of Lower and lower Middle Pennsylvanian underclays (Dulong and Cecil, 1989) because the peat substratum contains more quartz and smectite than the Pennsylvanian underclays. Burial diagenesis would probably alter mixed-layer clays and enhance crystallinity; thus we believe that the modern underclays are diagenetically immature as compared to the Pennsylvanian underclays.

The peat-dome substratum has been assumed to consist of alluvial material (Cameron et al., 1982). However, marine- and brackish-water diatoms were found in the samples of the substratum from both the Siak and Bengkalis peat deposits; nonmarine forms were not detected (Victoria Andrle, 1989, written communication).

Additional evidence for a marine depositional environment for the substratum includes marine borings in logs (Fig. 2, site S1)

TABLE 1. SIZE ANALYSIS AND CLASSIFICATION* OF SAMPLES FROM THE PEAT SUBSTRATUM, CENTRAL SUMATRA

Sample Identification Number	Depth* (m)	Sand§ (%)	Silt** (%)	Clay‡ (%)	Soil Classification§§
Siak peat dome					
SK5-C1	0.00 to 0.25	18	62	20	Silt loam
SK5-C2	0.25 to 0.58	16	53	31	Silty clay loam
SK6-C1	0.00 to 0.35	24	49	27	Loam
SK6-C3	0.60 to 0.85	1	44	55	Silty clay
SK7-C1	0.10 to 0.35	18	45	36	Silty clay loam
SK7-C2	0.35 to 0.60	---	48	52	Silty clay
SK8-C1	0.00 to 0.25	5	45	51	Silty clay
SK8-C3	0.50 to 0.85	2	56	43	Silty clay
SKLL3-PC3	0.00 to 0.15	12	48	39	Silty clay loam
SKLL3-PC5	0.30 to 0.46	8	49	43	Silty clay
Bengkalis peat dome					
BK1-C1	0.00 to 0.20	2	68	30	Silty clay loam
BK1-C3	0.45 to 0.95	1	56	43	Silty clay
BK4-C1	0.00 to 0.35	1	35	64	Clay
BK4-C3	1.35 to 1.85	2	32	66	Clay
BK5-C1	0.00 to 0.50	3	39	59	Clay
BK5-C2	1.00 to 2.00	1	34	64	Clay
BK8-C1	0.00 to 0.65	1	40	59	Clay/silty clay
BK8-C2	0.65 to 1.15	1	34	65	Clay

*Soil Survey Staff, 1975. SKLL3 samples are from the bottom of Lake Bawah (Fig. 6) where the substratum is overlain by approximately 0.5 m of organic ooze (Supardi and others, this volume), all other samples were taken below the base of the peat.
†Depth below peat/substrate contact.
§Sand size, >63μm. Organic free.
**Silt size, <63μm and >2μm. Organic free.
‡Clay size, <2μm. Organic free.
§§Soil classification based on percent sand, silt, and clay.

that are now being exhumed by tidal-current erosion in the lower part of the intertidal zone. The logs were carbon dated at 4240 ± 100 yr B.P. Trenches cut parallel to the logs and into the cutbank revealed that the boring occurred prior to burial indicating that the logs may have been deposited either under marine conditions during a previous high stand of sea level that peaked about 5000 yr B.P. (Geyh et al., 1979), or by an earlier esturine tidal system. In either case, the presence of marine fossils and borings by marine organisms suggest that the peat developed directly on the marine Minas Formation. Jansen et al. (1985) suggested that the peat substratum in West Kalimantan, Indonesia, might be of marine origin based on the occurrence of pyrite and some shell material in substrate samples. The data from our study and the data of Jansen et al. (1985) indicate that the peat substratum in both Sumatra and West Kalimantan were deposited by marine processes during a prior high stand of sea level.

Lenticular deposits of thin peat (as much as 200 m long and ~1 meter thick) are exposed in a few areas near the top of the intertidal zone in cutbanks along the Siak estuary (Fig. 5B). These exposures suggest that the river/estuary system might be eroding into the main Siak peat body, which is supratidal. However, evaluation by a series of probes, taken across the back of the levee and the flood plain (Fig. 5B), determined that the peat exposed in the cutbanks of the levee pinches out and is stratigraphically below and, therefore, not connected to the main peat dome (Fig. 5B). The elevation of the top of the lenticular peat in the cutbank is approximately 1 to 2 m below the elevation of the base of the main peat dome. A log, contained within the thin lenticular peat exposed in the cutbank, was dated at 5560 ± 200 yr B.P.; dates at the base of the peat dome are generally slightly younger than 5000 yr B.P. (Supardi et al., this volume). These lenticular peat deposits in the substratum are significant because they may indicate the onset of conditions favorable for the formation of peat. More importantly, they would be incorporated into the rock record as "leader beds" of coal.

Coastal plain

On the basis of the marine fossils in the Siak and Bengkalis peat substratum, radio-carbon dating of the marine-bored buried wood in the Siak estuary, and the exhumation of the paleosol in the intertidal zone along the north coast of Bengkalis Island, the modern coastal plain and offshore islands of Sumatra appear to be the result of complex sedimentation in a marine environment during prior high stands of sea level and subareal exposure during the past 5,000 years. Furthermore, coastal erosion was, by far, the most commonly observed feature of coastal areas. This erosion is occurring in sediment that was deposited approximately 5000 yr B.P. Marine currents appear to be the primary agent of erosion, sediment transport, and deposition in near-shore marine and estuarine environments. Siliciclastic sedimentation and modern coastal progradation, which is primarily deposition of silt and mud, appears to be limited to slack-water areas such as the downstream ends of islands and peninsulas. The present coastal plain appears, therefore, to be mainly the result of sedimentation during prior high stands of sea level and not recent and modern deltaic coastal progradation (e.g., Cameron et al., 1982). Modern peat formation commenced immediately after the last drop in sea level, approximately 5000 yr B.P. (Supardi et al., this volume).

Estuarine sedimentary environments

Two river/estuary systems, the Siak and Kampar Rivers, were evaluated during a reconnaissance study (Fig. 2). Each system was evaluated with respect to tidal influence, bed load, suspended-solids concentration, and dissolved-sediment concentration. Both systems are tidally influenced for more than 100 km (62 mi) inland, as indicated by a visible tidal range. The Kampar River/estuary was visited during two field seasons, 1987 and 1989. The 1987 field season on the Kampar proceeded upstream from the mouth as far as the village of Telukmeranti (Fig. 2, sampling stations at sites 14 through 21). The 1989 field season

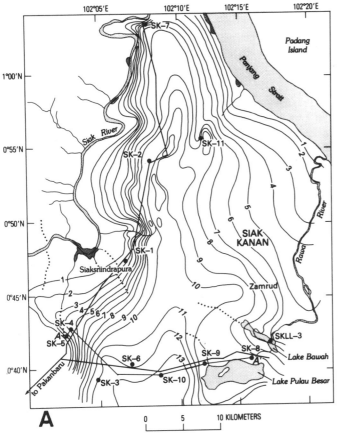

on the Kampar proceeded down stream from just south of Pakan-baru to Telukmeranti (Fig. 7; Fig. 2, sampling stations included sites T1–T9).

Tidal influence and bed load. At its confluence with Pan-jang Strait, the Siak estuary appears to have a micro- to mesotidal range (1.0 m neap and 2.5 m spring). Tidal current dominance in the Siak estuary was noted during our reconnaissance investiga-tion when extremely strong flood-tidal currents were strong enough to inhibit bottom sampling with a Pfleger corer near the village of Siaksriindrapura (Fig. 2, site S1). Weaker currents were observed on the ebb tide.

Although tidal ranges vary within each of the estuaries dur-ing the year, they both appear to be influenced by similarly strong tidal currents and ranges. The Kampar estuary is macrotidal (1.5 m neap and 3.9 m spring) at the confluence with Panjang Strait where the tidal range was more than 4 m. Like the Siak estuary, very strong currents were noted on the flood tide as compared to the ebb tide.

Although tidal ranges appear to be similar in both of the estuaries studied, bottom soundings, recording–depth-finder pro-files, and bottom sampling indicate two very different sedimento-logical systems. The depth of the Siak River can accommodate ocean-going vessels, and on the basis of a cross-sectional profile (Fig. 5B) at site S2 (Fig. 2) and bottom sampling, the Siak ap-pears to be devoid of recent siliciclastic sediment except for an estuary mouth bar. The Siak River estuary appears to be an erosional and/or nondepositional system throughout the study

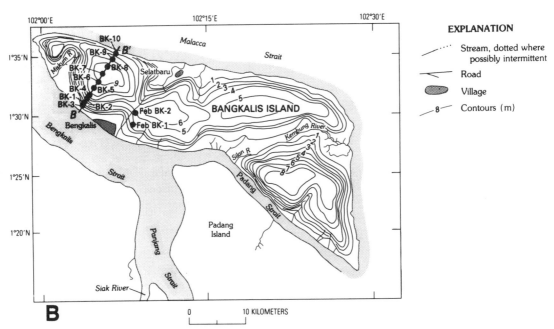

Figure 6. Location map for samples of peat substratum shown in Tables 1 and 2 (adapted from Neuzil et al., this volume). Peat isopach contours are in meters. A, Siak Kanan peat deposit; B, Bengkalis peat deposit.

TABLE 2. SEMIQUANTITATIVE MINERAL ANALYSIS OF THE BULK SAMPLES OF THE PEAT SUBSTRATUM SAMPLES FROM CENTRAL SUMATRA

Sample Identification Number	Quartz	Feldspar	Illite-smectite	Kaolinite
Siak area				
SK5-C1	90			5
SK5-C2	85	tr*		10
SK6-C1	80	tr		10
SK6-C3	65	tr	tr	25
SK7-C1	65	5	10	15
SK7-C2	60	tr	20	10
SK8-C1	60	5	10	20
SK8-C3	70	5	10	15
SKLL3-PC3	75	5	10	10
SKLL3-PC5	70	tr	10	15
Bengkalis area				
BK1-C1	85	5	5	10
BK1-C3	60	5	15	15
BK4-C1	50	tr	15	25
BK4-C3	45	5	20	30
BK5-C1	60	5	15	20
BK5-C2	50	tr	15	15
BK8-C1	55	tr	20	15
BK8-C2	50	5	15	25

*Trace amount <5%.

area; the dominant source of mouth-bar sediment appears to be from Panjang Strait or from tidal-current erosion of the estuary channel. This observation is consistent with sedimentation in other flood-tide dominated estuaries (Dyer, 1986, Chapter 9).

The Kampar estuary appears to be tidally influenced for up to 180 km, and to be flood tide dominated for more than 100 km inland (Fig. 7). In contrast to the scoured bottom of the Siak estuary, the Kampar estuary is choked with coarse silt and fine sand for at least 90 km inland; we refer to this part of the estuarine system as the lower estuary (Fig. 7). Only small boats of shallow draft can navigate the lower estuary of the Kampar because of shallow water depth. As illustrated in Figure 7, most of the main channel of the lower 80 km of the lower estuary is straight. However, depth-finder profiles, and repeated problems in running aground while trying to navigate the lower estuary indicate that flow in the lower estuary is not straight, but rather the flow is channelized in what appear to be anastomosing or meandering subtidal channels. Depth-finder profiles indicate that these channels have subaqueous levees as illustrated in the schematic cross-sectional profile shown in Figure 8 at site T2 (Fig. 2). Flood-tide dominance in the lower Kampar estuary is indicated by the upstream orientation of the butt-end of logs on medial bars and tidal flats and current ripples observed at low tide. Shallow water and strong flood-tide conditions are also indicated by a tidal bore that extends more than 80 km inland from the coast. Ebb-tide currents may slightly modify ripple crests on tidal flats but generally do not appear to be of sufficient strength to reverse current ripples into a downstream orientation, move large pieces of wood debris that are deposited on the flood tide, or erase scour caused by transport of the debris on the flood tide (Fig. 9).

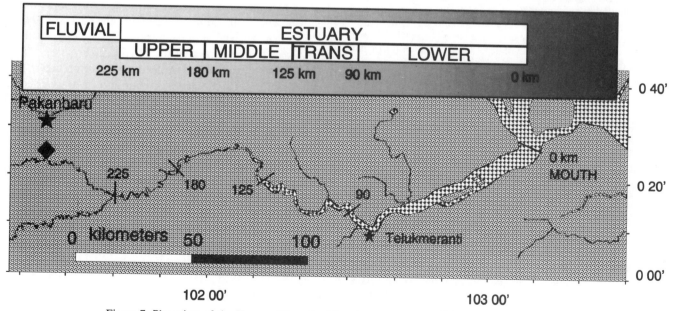

Figure 7. Plan view of the Kampar River fluvial and estuary systems, Riau Province, Sumatra. ◆ indicates the point of embarkation, 1989 field season.

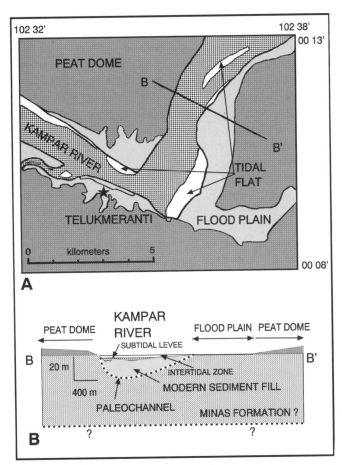

Figure 8. A, Plan view and profile of Kampar River estuary channel in upper part of the lower estuary (Fig. 2, site T2). B, Cross sectional profile along B–B′. Total depth of the paleochannel is inferred on the basis of channel depth in the middle estuary where the channel bottom is erosional or nondepositional. Total depth is probably the result of incision during a previous glacial low stand of sea level. Channel fill at site T2 (Fig. 2) appears to be the result of deposition by tidal processes that developed during transgression.

Figure 9. View upstream of a root-scrape trace left by a tree stump that was dragged by flood-tide currents across a tidal flat in the upper part of the lower estuary of the Kampar River (Fig. 2, site 18). Ebb-tide currents were not adequate to significantly alter the trace or reverse flood tide ripples.

Perennial flood-tide dominance, even during rainier periods and perhaps flooding, is indicated by the permanency of depositional bars and channel fill observed on topographic maps and satellite imagery. Channel fill, which is composed of fine quartzose sand and coarse silt, is constantly being reworked for more than 100 km inland primarily by flood tides. The primary source of the sand and silt in the lower estuary appears to be from bank erosion because (1) the channel is overly wide with respect to the remainder of the river/estuary system, (2) the sandy silt of the banks is highly susceptible to erosion, and (3) the grain size and composition of the channel fill is similar to that of the channel banks. It is also possible that along-shore currents from the Java Sea to the southeast deliver some sediment to the flood-tide dominated system.

In the upper part of the lower estuary (above the village of Telukmeranti, Fig. 7) channel fill in meanders is on the "cutbank" side of the meanders, and the modern channel is located on the inside or where point bars would normally occur (Fig. 10). The modern channel morphology is reversed when compared to typical fluvial or ebb-tide dominated systems. The meander pattern is probably a relic feature from fluvial processes that were active during the last glacial maximum when there was a low stand of sea level. However, the modern channel morphology appears to be the result of the present tidal system, in which sediment is deposited by the flood tide on the cutbank side of meanders, and the channel is located where point bars would normally occur in a fluvial or ebb-tide dominated system (Fig. 10B). In addition to the enigmatic channel fill in meanders, the mouths of tributaries in the upper part of the lower estuary are filled with sand deposited by the flood tide; the tidal bore in this area is known to extend up a major tributary located just west of the village of Telukmeranti near site T3 (Fig. 2).

Mud drapes were well developed by tidal processes at the mouths of some of the tributaries draining into the lower estuary of the Kampar River. Bioturbation of these drapes, however, was rare to nonexistent. As a further indication of flood-tide control on sedimentation, planktonic diatoms of marine and brackish-water origin were found in sediment samples from the lower estuary (Fig. 2, site 17); fresh-water diatoms were not found (Victoria Andrle, 1989, written communication). Megascopic shell material of marine origin was not observed in any part of the Kampar or Siak estuaries. Invertebrate trace fossils also were exceedingly rare. Clearly, interpretation of such environments in the rock record would be hampered by the paucity of megascopic shell material and invertebrate traces.

Proceeding inland from the lower estuary, in the region situated approximately 90–125 km from the coast, channel form and depositional features in the Kampar system are transitional (transition zone) between the shallow water of the lower estuary and the much deeper water of the middle estuary (Fig. 7). Numerous islands have developed in the transition zone. Depth-

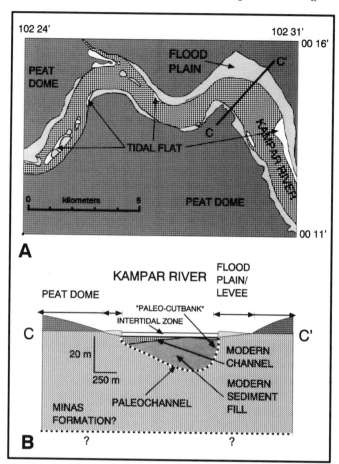

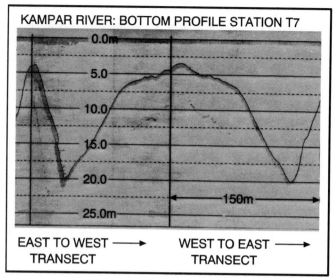

Figure 11. Duplicate cross-sectional profiles of the bottom of the middle estuary at station T7 of Figure 2. Profile 1 is a transect from east to west, and profile 2 is from west to east. The channel is presently erosional or nondepositional as evidenced by what appears to be a relic channel lag and the absence of any apparent fluvially derived bed load as determined by dredge sampling.

Figure 10. A, Plan view of the Kampar River estuary in the transition zone between the lower and middle estuary approximately 3 km (1.86 mi) inland from site T3 in Figure 2. B, Schematic cross section of the paleo- and modern-channel profile. In the modern system, channel fill is on the "cutbank" side of the meander and the channel is on the "point bar" side of the meander. The base of the paleochannel cut is inferred on the basis of channel depths of ~20–25 m at sites T7 and T8 upstream in the middle estuary beyond tidal deposition. Total depth of the paleochannel may be the result of down cutting during a previous glacial low stand of sea level; channel fill from tidal deposition appears to have occurred as sea level rose.

biculidae, suggestive of fresh-water conditions (Lynn Wingard, 1989, personal communication). These samples also contain a lag deposit consisting of angular pebble- to cobble-sized clasts that were coated with algae suggestive of a stable bottom with limited erosion or bed-load transport. Although most of the middle estuary is nondepositional, small dunes having a wave height of ~0.5 m were detected on a depth-finder profile in a limited area of the lower part of the middle estuary (Fig. 2, site T8). The dunes exhibited an upstream orientation indicative of flood-tide current dominance (Fig. 12). These bed forms represent the only observed occurrence of sedimentary structures indicative of a bed load in the middle estuary. Point bars are absent in the lower and middle part of the middle estuary. The deep part of the channel and point-bar development in the upper reaches of the middle estuary are reversed from that of a typical fluvial or ebb-tide dominated system. Minor point-bar development in the transition between the upper part of the middle estuary and the upper estuary is the only indication of fluvial or ebb-tide siliciclastic transport in the middle estuary.

Further inland, the upper estuary extends from approximately 180–220 km (Fig. 7). The upper estuary exhibits a tidal range of as much as 2 m. Channel morphology and deposition are transitional with both the middle estuary downstream and the fluvial part of the system upstream. Stream patterns characteristic of fluvial systems are poorly developed in the upper part of the upper estuary. Meanders with point bars and cutbanks are present, but meander cutoffs are rare when compared to the fluvial part of the system that does not have a tidal range.

The Kampar River fluvial system is transitional with the upper estuary. An arbitrary boundary was drawn 225 km inland

finder profiles of the bottom indicate that channels are deeper next to the islands, as compared to the lower estuary. This region appears to represent a zone in which the flood tide continues to be predominate over the ebb tide. This tidal regime, which consists of slack water twice a day, has also led to the development of the numerous midchannel islands.

The middle estuary of the Kampar River extends from approximately 125 km to approximately 180 km inland (Fig. 7). In marked contrast to the shallow lower estuary, the middle estuary is much deeper and reaches depths of at least 20 m (Fig. 11). Bottom sampling and depth-finder profiles indicate nondeposition, or erosion in most of the middle estuary. Bottom samples from the deepest part of the channel (25 m; Fig. 2, station at site T7) included articulate and disarticulate clams of the family Cor-

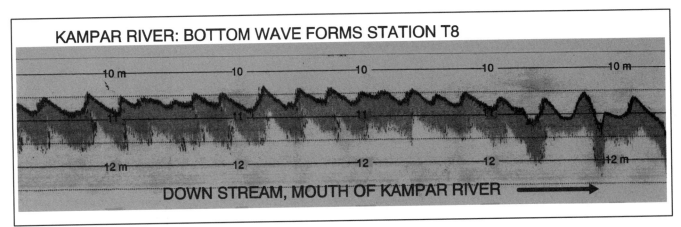

Figure 12. Bottom profile in the middle estuary illustrating 0.5 m high bed forms with steep or avalanche slope oriented upstream. These bed forms are indicative of flood tide transport of bed load.

(Fig. 7) where evidence for a tidal range was no longer observable and channel morphology and depositional features are typical of fluvial systems. On the basis of satellite imagery and field observations, neither the Siak nor Kampar estuaries shows any evidence of the formation of an estuarine delta, presumably as a result of limited fluvial siliciclastic input into the upper estuary.

Suspended solids. Suspended-sediment concentrations in the Siak River estuary were generally low (<60 mg/l) except at the mouth of the estuary (Fig. 2, site 35) where a suspended-sediment concentration of 214 mg/l was measured. Suspended-sediment concentrations in the Kampar estuary ranged from 8 mg/l at site T8 (Fig. 2) to 1,745 mg/l at the mouth (site 15, Fig. 2).

Suspended-sediment concentration decreases away from the mouth of the Kampar in all directions. A profile of suspended-sediment concentration is depicted in Figure 13. These data and the absence of a fluvially derived bed load in the middle, transitional, and lower estuary indicate that fluvial influx of siliciclastic sediment to the mouth of the Kampar estuary is negligible. Instead, the bed load and high suspended concentration at the mouth (Fig. 13) appears to be related to strong turbulence that develops in response to the complex tidal system. Currents from the estuary, Panjang Strait to the northwest, and the Strait of Malacca and Java Sea to the northeast converge and diverge at the confluence of the Kampar with Panjang strait (Fig. 2). The mixing of these three water masses at the confluence of the Kampar estuary with Panjang Strait causes extreme turbulence in this complex macrotidal system. The tidally induced turbulence suspends silt- and sand-size material (>1,700 mg/l of fine sand and coarse silt) where it is constantly reworked throughout the lower 100 km of the estuary and the eastern extremity of Panjang Strait (Fig. 2). On the basis of the suspended sediment data, the absence of a fluvially derived bed load, field observations, and bathymetric data, net sediment transport into the Strait of Malacca by the Kampar system appears to be very limited. The modern sediment in the lower estuary and the part of Panjang Strait proximal to the

mouth of the Kampar may best be classified as an estuarine mouth bar whose sediment is primarily derived from bank widening by tidal processes within the lower Kampar estuary. This interpretation, which is similar to our interpretation of sediment supply and load in the Siak estuary, is in marked contrast to the generally held assumptions that rivers, such as the Kampar and Siak, contribute large volumes of fluvially transported sediment to the Sunda shelf (e.g., Keller and Richards, 1967, p. 105–106).

Dissolved solids. Dissolved-solids concentrations in samples collected in the Siak and Kampar estuaries were determined from conductivity measurements (Hem, 1989). The results are shown in Table 3 along with an analysis of rainwater in the region and data from some other rivers of the world (compiled from various sources by Berner and Berner, 1987, Chapter 5). Dissolved-solids concentrations in both the Siak and Kampar are very low upstream of marine water influence. Concentrations for the Kampar are depicted in Figure 14, which indicates that salt-water influence has not reached as far upstream as site 18. In the middle estuary of the Kampar, dissolved-solids concentrations are approximately equivalent to rain water (Table 3). Beyond the effects of marine water mixing, dissolved-solids concentrations in both the Siak and Kampar are comparable to other rivers in the humid tropics such as the Amazon, its tributaries, and the Orinoco (Table 3). In contrast, rivers draining regions that are drier and more seasonal such as the Mississippi, Colorado, and Ganges are considerably higher in dissolved-solids concentrations (Table 3). The low dissolved-solids concentrations and the pH of the systems would inhibit or preclude deposition and preservation of calcareous sediments, a condition analogous to nonmarine Lower and lower Middle Pennsylvanian strata in the Appalachian basin.

Marine environments

Offshore, a Pfleger corer was used to sample bottom sediment in the Strait of Malacca, associated lesser straits, and a small area of the Java Sea. Sample-station sites are shown in Figure 2 and our observations of bottom samples are shown in Table 4. Much of the marine environment appears to be erosional or

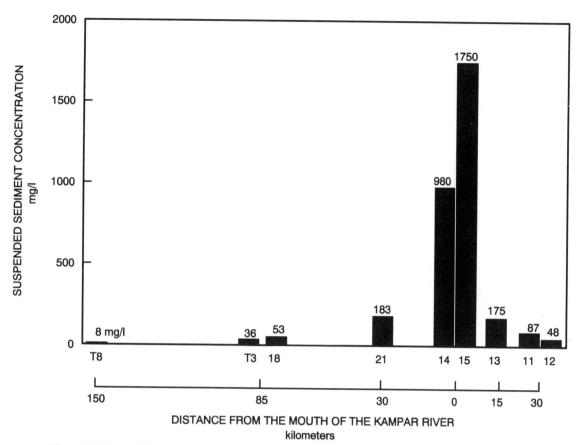

Figure 13. Suspended-sediment profile for the Kampar River estuary. Sample locations are shown in Figure 2.

nondepositional, but, in some areas, modern sediment is accumulating over partially indurated older sediment (herein referred to as substratum). The modern muds are easily distinguished from the older sediment because the modern sediment is soft and easily penetrated with the corer whereas the silty mud substratum is partially indurated and penetration was restricted to a few centimeters; similar conditions were noted by Keller and Richards (1967). In general, much of the study area appears to be either erosional or nondepositional. Exceptions to this generalization are limited to (1) the lee sides of islands (e.g., sites 1, 2, 3, and 27); (2) the mouth bar of the Siak estuary at sites 35 and 3, the estuarine delta associated with the lower estuary of the Kampar River, and the eastern part of Panjang Strait (sites 11 through 22).

Marine erosion and deposition. The substratum is widely exposed to marine bottom currents and is undergoing erosion or nondeposition except where buried by modern sediment. The substratum is generally composed of clayey silt or silty clay, a characteristic that is also consistent with the observations by Keller and Richards (1967). The top of the substratum is a modern ravinement surface that appears to be forming as a result of the ongoing transgression.

In contrast to erosion by marine currents, wave energy does not appear to be a significant agent of coastal erosion or of ravinement surface formation in the study area because surface winds are minimal within this equatorial atmospheric low-pressure belt; waves, therefore, also are minimal. The region also is protected from waves from the Indian Ocean and South China Sea by Sumatra and the Malay Peninsula. Wave-generated beaches were not observed within the study area. Waves, therefore, do not appear to be a significant factor in coastal erosion and long-shore transport.

Sand deposition in coastal areas was only observed in intertidal flats along the north coast of Bengkalis Island and the intertidal zone at site 9. Along the central part of the north coast of Bengkalis Island, in the vicinity of the village of Selatbaru (Fig. 6), thin, isolated, longitudinal bars of fine sand are migrating toward the northwest. Residence time of such bars is unknown. The bars are tens of meters in length, generally less than 10 m wide, and less than 1 m thick. Lateral spacing between the bars was 2 to 3 m. The ravinement surface over which the bars are migrating is a paleosol that is being excavated by strong alongshore currents that move through the Strait of Malacca toward the northwest. The paleosol, which contains in-situ tree stumps and rhizomorphs, is composed of mottled brown and gray clay, silt, and fine sand. The intertidal zone is sediment starved. There is not enough sand in the system to cover this modern ravinement surface, except for the longitudinal bars.

In the same coastal area north of the village of Selatbaru

TABLE 3. DISSOLVED-SOLIDS CONCENTRATIONS IN SIAK AND RIVER/ESTUARY SYSTEMS

Sample Location	pH	Dissolved Solids (mg/l)
Siak estuary		
S1	5.1	41
Rain water		
Near Siak	4.5	8
Mendol Island strait		
11	6.3	22,750
12	6.3	22,610
13	6.4	22,330
Kampar estuary		
14	6.4	20,790
15	6.2	21,840
16	7.0	11,480
18	4.5	18
21	5.7	5,530
T3		10
T8		10
Other rivers of the world (Berner and Berner, 1987)		
Lower Amazon (Brazil)		38
Rio Negro (Amazon tributary)		6
Orinoco (Venezuela)		34
Mississippi (U.S.A., 1905)		216
Colorado (U.S.A.)		703
Ganges (India)		167
Normal sea water (Kennett, 1982)		35,000

*Sample locations are shown in Figure 2. Concentrations near the mouth of the Kampar estuary are approximately two-thirds that of normal sea water. Dissolved-solids concentrations for some other rivers of the world illustrate the similarities between rivers in the humid tropics in contrast to rivers that drain regions that have drier and more seasonal climatic regimes.

(Fig. 6), the substratum of the intertidal flat is separated from the modern vegetated soil by an erosional vertical cliff approximately 1 m high. The height of this cliff increases to more than 3 m to the west and, at site 28, erosion has laterally cut into both the peat substratum and the domed peat. East and west of site 28, spanning a total distance of 10 to 20 km, erosion is undercutting the cliff, and slumping of large blocks of peat into the intertidal zone is common. The domed peat appears to be easily eroded and large amounts of peat debris are being carried away by the strong currents that flow northwest through the Strait of Malacca during each tidal cycle. Organic debris, including large logs that have been buried in the peat dome for as long as 5,000 years (Supardi et al., this volume), is being eroded and floated away.

Tidal-current erosion of the peat dome is important because (1) it demonstrates that certain types of peat are easily eroded, and (2) eroded peat may be a significant source of terrestrial organic matter in marine sediments. Because the fibric peat of these ombrogenous peat domes is highly susceptible to erosion, the concept that all peat deposits are resistent to erosion (e.g., McCabe, 1984) does not apply to the domed peat deposits of Sumatra. The eroded peat, which is being carried northwestward through the Strait of Malacca toward the Andaman Sea, could be a significant source of terrestrial organic matter in the Andaman basin, a relatively deep water (depths of more than 2,500 m), back-arc tectonic setting (Hamilton, 1979).

Offshore of Bengkalis Island in the Strait of Malacca (Fig. 2, site 32), recent bottom sediment appears to consist of isolated ripples of sand on a ravinement surface that is developed on Pleistocene(?) marine clay, probably the Minas Formation. Repeated attempts to obtain cores generally encountered the partially indurated ravinement surface rather than sand ripples. These observations are consistent with those of Keller and Richards (1967) who noted sand waves in the narrow part of the strait that appear to be migrating toward the northwest over older sediment. Deposition in the vicinity of site 32 (Fig. 2) is apparently limited to the isolated ripples, which probably also are migrating toward the northwest. Our data and observations indicate that present-day sediment supply is from marine erosion as inferred by Umbgrove (1949), who suggested that much of the modern Sunda shelf is erosional rather than depositional. All indications are that the modern system is sediment starved with respect to fluvial input in contrast to the interpretations of Keller and Richards (1967) who suggest a large sediment influx from fluvial systems to account for sand waves in the narrow part of the strait. Thus, even though the area is situated in the modern tropical ever-wet climatic belt (Indonesian low-pressure system) where annual rainfall and runoff are high, fluvial sediment input is minor and marine erosion is occurring. Interpretations of analogous ravinement surfaces (even if they could be recognized) in the rock record would typically attribute erosion to wave action during marine transgression rather than other types of marine currents.

In the Panjang Strait, modern sediment thickness decreases toward the northwest from sites 15 and 22 near the mouth of the Kampar estuary to site 23 (distal part of the Kampar estuarine delta). Shallow water in the proximal estuarine delta in the vicinity of the mouth of the Kampar (sites 14, 15, and 22) precludes navigation except by vessels of very shallow draft. Panjang Strait becomes progressively deeper in a northwesterly direction away from the mouth of the Kampar (Table 4; Haslam, 1983), and the bottom is erosional at site 24 beyond the distal edge of the Kampar estuarine delta.

Off the south-central coast of Bengkalis Island at site 25 (Fig. 2) the bottom of Panjang Strait appears to be an erosional- or nondepositional-ravinement surface. At this locality, modern sediment was not encountered, and the ravinement surface consisted of a hardened red-mottled, silty clay. We attribute the mottling to subaerial exposure and weathering that occurred during a previous low stand of sea level. As in the Strait of Malacca, the ravinement surface in Panjang Strait appears to be continuously modified by strong marine currents.

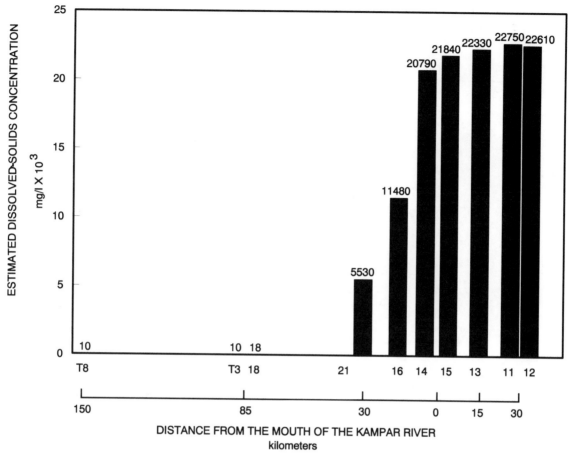

Figure 14. Dissolved-solids concentration profile for the Kampar River/estuary and the strait between Mendol Island and the Sumatra coastal plain. Sample locations are shown in Figure 2.

Suspended- and dissolved-sediment concentrations. Suspended- and dissolved-sediment concentrations in marine environments were measured at sampling stations in the Strait of Malacca and intra-island straits (Fig. 2). A continuous sample of water was taken from the surface to the bottom by lowering a modified "milk-bottle" sampler. The results are included in Table 5. Suspended sediment concentrations are, in general, relatively low except at sites 6, 22, and 35 where the higher concentrations must be related to marine-current erosion and sediment transport.

Dissolved-solids concentrations (salinity) at all sites in marine environments (Table 5) average $23.9^0/_{00}$ which is approximately two-thirds that of normal sea water ($35^0/_{00}$; e.g., Kennett, 1982, p. 237). Salinity in open waters over much of the Sunda shelf is approximately $32^0/_{00}$ throughout the year (Wyrtki, 1971). The data in Table 5 strongly indicate that the reduced salinities in the study area are the result of mixing of river discharge with water from the South China Sea as it moves across the Sunda shelf and through the Strait of Malacca. These data are a further indication that river discharge causes a significant reduction in salinity in Sumatran coastal environments (e.g., Wyrtki, 1971). The reduced salinity may partially explain the paucity of calcareous sediment in near-shore marine environments and, on the basis of the salinity data, we suggest that the entire Sunda shelf is probably not a region of high carbonate production. Conditions such as these may explain the paucity of calcareous material in Lower and lower Middle Pennsylvanian marine strata in the Appalachian basin.

DISCUSSION

Physical sedimentology

Our observations on fluvial siliciclastic-sediment flux are in sharp contrast to most published interpretations of sedimentation in the study area, which assume a large sediment supply from the river/estuary systems to the Strait of Malacca (e.g., Keller and Richards, 1967; Coleman et al., 1970; Cameron et al., 1982). Our data indicate that the estuaries and marine environments are sediment-starved with respect to influx of fluvially derived siliciclastic sediment as suggested by Umbgrove (1949). Keller and Richards (1967) inferred a large influx of fluvial siliciclastic sediments into the Strait of Malacca from rivers in Sumatra such as the Kampar and Siak. Our observations, however, are not consistent with their inference. Fluvial siliciclastic input to the Strait

TABLE 4. OBSERVATIONS ON BOTTOM SAMPLES FROM MARINE ENVIRONMENTS*

Site	Bottom Depth (m)	Core Penetration (cm)	Bottom Age, Texture, and Appearance
1	6.5	42	Recent(?) sand over gray mud
2	12.5	33	Recent(?) organic-rich muddy sand
3	7.5	38	Recent(?)
4	7.5 to 9.0	10	Recent sand over hard-clay substratum
5	14.0	45	Recent(?) gray mud
6	12.0	26	Recent organic-rich mud
7	10.0	61	Recent organic-rich clay over brown mud
8	12.5		Recent coarse sand and gravel
9	Intertidal	Grab sample	Recent quartz sand
10	12.0	56	Recent(?) brown and gray mud
11	9.0	51	Recent(?) muddy sand
12	9.0	41	Recent brown sandy silt with organic debris
13	15.5	31	Recent silt and mud
14	9.5	8	Recent, hard, fine sand
15	7.0	0 (0 for 3)	Recent, hard, fine sand
16	~2.0	0	Recent, rippled, fine sand
17	Intertidal	Grab sample	Recent, mud drape on levee
18	Intertidal	Grab sample	Recent, rippled, fine sand
19	~2.0	0	Recent, hard sand
20	Intertidal		Recent, cuspate ripples in sand
21	4.5		Recent, sand
22	6.5	20	Recent, hard, fine sand
23	~25	58	Recent, gray mud
24	23.0	15 and 22	Substratum, hard clay
25	16.0	17 and 30	Recent, silty clay with shells over gray-green substratum
26	15.0	17	Recent sand and shells over red-mottled substratum
27	8.5	10	Recent clay sand over substratum
28	7.0	40	Substratum, green-gray clay
29	Intertidal	Grab sample	Substratum, hard underclay of peat
30	12.0	12	Recent sand over hard gray clay substratum
31	14.0	5	Recent(?) green-gray clay
32	35 to 40	10	Recent, sand over gray clay substratum
33	3.5	33	Recent, sand over clay substratum
34	19.0	20 and 38	Recent(?) gray silty mud
35	10.0	25 and 47	Recent, bedded sand over organic-rich dark brown clay

*Bottom samples were collected with a Pfleger corer or dredge. Depths were determined by wire-line measurements.

of Malacca appears to be limited to a very small supply of suspended sediment of clay-size material (<60 mg/l total suspended solids). We found no evidence for transport of a fluvially derived bed load through either the Siak or Kampar estuaries as suggested by Keller and Richards (1967).

Although weathering is occurring and rainfall and run off are high under the modern ever-wet climatic conditions of this equatorial region, erosion and fluvial transport appear to be curtailed by the extensive vegetative cover of the tropical rain forest. This is evidenced by very low suspended-sediment concentrations, absence of a fluvially derived bed load, and the absence of fluvial-deltaic sedimentation associated with the extensive estuaries in the Siak and Kampar river/estuary systems. Upstream of

salt-water influence, dissolved-solids concentrations are also minimal because soluble metallic ions have been leached from soils through tropical weathering. High rainfall and runoff also may cause dilution of dissolved load in the rivers, estuaries, and the marine environment.

Our data indicate that siliciclastic sediment movement in marine environments is primarily controlled by marine currents that are eroding and reworking preexisting marine sediments. These currents, which may be the result of atmospheric and tidal effects, move through the Sunda shelf epeiric seaway generally in a westerly direction. These currents may be generated in response to global atmospheric circulation but probably are modified by tidal effects. In the equatorial western Pacific, the Indonesian

TABLE 5. SUSPENDED- AND DISSOLVED-SOLIDS CONCENTRATIONS FOR MARINE ENVIRONMENTS*

Site	Sample Depth (m)	Estimated Dissolved-Solids Concentration (salinity) (ppt)	Suspended Sediment Concentration (mg/l)
1	0 to 6.5	....	29
2	0 to 12.5	24.2	....
3	0 to 7.5	24.1	14
4	0 to 8.5	24.2	85
5	0 to 14.0	23.5	45
6	0 to 12.0	22.5	244
7	0 to 10.0	23.2	76
8	0 to 12.5	23.6	17
10	0 to 12.0	22.9	37
16	~2.0	Brackish to taste	
22	0 to 6.5	21.8	115
23	0 to ~25	24.6	24
24	0 to 23.0	24.6	34
25	0 to 15.0	24.7	21
26	0 to 13.0	....	18
27	0 to 8.5	25.6	4
28	0 to 7.0	....	42
30	0 to 12.0	....	23
31	0 to 14.0	....	13
32	0 to ~35	25.0	3
33	0 to 3.5	....	40
34	0 to 19.0	....	17
35	0 to 10.0	23.6	214

*Dissolved solids concentrations (salinity) are generally less than the 35,000 mg/l (35 ppt) for normal sea water. The low concentrations are probably, in part, the result of dilution of the lower than normal salinity of open marine water (~32 ppm; e.g., Wyrtki, 1971) from the Sunda shelf by runoff from rivers in the region of the study area.

atmospheric low-pressure cell (Ahrens, 1991, Chapter 12) causes the sea surface to be elevated ~170 m (Mörner, 1976) relative to the Indian Ocean. Part of the water from the topographically high, warm, and low-density water mass in the western Pacific (e.g., Ahrens, 1991, Chapter 12) flows westward across the epicontinental shelf and into the Indian Ocean where the sea surface is depressed by an equatorial atmospheric high-pressure cell. The northwestward flow of the pressure-cell–induced currents across the shelf and through the Strait of Malacca is both augmented and retarded by tidal processes. Augmentation accelerates the flow, whereas tidal retardation slows the flow, but, in most areas, the northwestward current does not completely stop or reverse.

The strong marine currents are erosional in high-energy areas such as the banks of estuaries, unprotected coast zones, and straits such as the narrow part of the Strait of Malacca and Banka Strait (Umbgrove, 1949), but the currents are depositional in slack water areas such as the lee side of islands and where the

Strait of Malacca widens in the vicinity of latitude 2°30′N and longitude 101°10′E (Fig. 15). If deposition is occurring where the strait widens, it is probably because transport energy is lost. Sediment transported through the strait toward the northwest appears to be deposited on both the north and south sides of the strait as indicated by the bathymetry depicted in Figure 15. Sediment deposited on the north side of the strait has been described as a high-tide tropical delta of the Klang-Langat Rivers (Coleman et al., 1970). Although Coleman et al. (1970, p. 185), indicate that the Klang-Langat Rivers are the major contributors of sediment, they note that accurate data on river discharge and sediment load were not available to them. They suggest that relatively small quantities of sediment are being transported by these small tropical rivers (Coleman et al., 1970, p. 185).

On the basis of our observations, bathymetry, and the current and sedimentological data of Coleman and others (1970) and Keller and Richards (1967), we suggest that the deposits in the vicinity of the Klang-Langat River mouth on the north and Rokan River mouth on the south sides of the Strait represent either (1) relict deltaic deposits that are currently being reworked by tidal processes or (2) deposition of sediment primarily contributed by currents moving through the Strait of Malacca rather than sediment delivered by the modern Klang-Langat Rivers and Rokan River. These deposits may actually be "deltas" of the Strait of Malacca rather than deltas of the nearby rivers. Shoreline progradation is primarily the result of sediment supply and deposition by marine currents rather than a modern fluvial supply. The Klang-Langat "delta" is often cited in deltaic models of coal formation (e.g., McCabe, 1984). Our observations indicate, however, that peat formation in the region, including that of the Klang-Langat "delta" is neither coeval with, nor in any way controlled by, fluvio-deltaic sedimentation. Instead, peat formation is the result of the ever-wet climate and a poorly drained coastal plain and islands. The physiography of the coastal plain and islands appears to be mainly the result of marine deposition during former higher stands of sea level followed by peat formation and estuarine processes at present.

Chemical sedimentology

Carbonate deposition. Marine and nonmarine carbonates, other than minor amounts of siderite, are not being formed. The lack of nonmarine calcareous sediment is apparently related to low dissolved-solids concentrations and low pH in rivers and estuaries (Table 3). Unavailability of cations as a result of leached soils, dilution of dissolved solids by high rainfall, and the acidic water from peat deposits contribute to the water chemistry of nonmarine environments. The absence of marine carbonate deposition may be related to lowered or fluctuating pH and salinity (Table 5) caused by the discharge of acidic black-water rivers. The unfavorable conditions for carbonate deposition in our study area are consistent with the paucity of marine and nonmarine calcareous carbonates in Lower and lower Middle Pennsylvanian strata, but not consistent with the commonly calcareous strata in

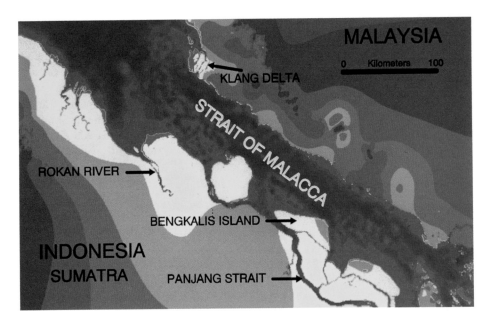

Figure 15. Bathymetric patterns in the eastern and central part of the Strait of Malacca. Red areas are intertidal, light blue is shallowest, and maroon is deepest water. Sediment supply and deposition are inferred to be primarily a function of marine rather than fluvial processes.

Upper Mississippian and upper Middle and Upper Pennsylvanian sequences.

Peat formation. Chemical sedimentation in the study area is mainly limited to extensive bodies of domed peat. Peat formation has been controlled primarily by the allogenic processes of sea-level change and the modern ever-wet climate (Neuzil et al., this volume; Supardi et al., this volume). Autogenic processes, such as delta switching, stream meandering, or barrier-bar migration, which are generally implicit in many facies models of coal formation (e.g., Ferm and Horne, 1979), are of secondary importance as a control on the origin of peat. We postulate that in our study area peat began to form on a very flat, poorly drained, coastal plain when this surface was exposed after a drop in sea level ca. 5000 yr B.P. Peat formation probably progressed from an initial planar, topogenous phase to the domed, ombrogenous peat swamp of the present because of a tropical ever-wet climate, which appears to have become optimum for peat formation over the area ca. 5000 yr B.P. (Verstappen, 1975). All stages of peat formation appear to have been oligotrophic (nutrient poor) including the early topogenous stage (Neuzil et al., this volume). The topogenous oligotrophism appears to have resulted from leaching and flushing of available dissolved inorganic solids under the ever-wet climatic regime, whereas oligotrophism in the modern domed swamps is the result of ombrogenous conditions (ever-wet climate). The modern ever-wet climatic condition also has contributed to an extensive rain-forest cover in both the coastal plain and upland areas, which restricts erosion and, hence, siliciclastic transport and deposition. The conditions of sea level and climate over the past 5,000 years appear to have been ideal for

extensive peat formation and poor for fluvial siliciclastic and inorganic dissolved-sediment transport.

It is generally assumed that peat formation in the region is progressing seaward on a prograding coastal plain or deltas (e.g., Cameron et al., 1982). However, such an interpretation is inconsistent with (1) similar age dates at the base of the peat deposits in the Siak and Bengkalis deposits (Supardi et al., this volume), thus precluding recent progradation; (2) coastal erosion as the dominant coastal process, and (3) the restricted supply of fluvially derived sediment.

SUMMARY AND CONCLUSIONS

Allocyclic controls on sedimentation

Conditions of sedimentation in the study area are controlled by the allogenic processes of climate, eustasy, and tectonics. At present, climate appears to be the dominant factor controlling fluvial siliciclastic influx and peat formation. The tropical ever-wet climate favors the formation of laterally extensive, thick, low-sulfur, low-ash peat deposits rather than terrestrial erosion and siliciclastic aggradation. Erosion and fluvial sediment transport (both siliciclastic and dissolved solids) are highly restricted because of the vegetative cover and leaching in the tropical rainforest environment. Hence, the study area is sediment starved with respect to particulate fluvial input even though rainfall and relief in the catchment basins are high. This is in sharp contrast to the commonly held assumption that the rivers and estuaries of the region carry large fluvial sediment loads to the Strait of Malacca (e.g., Keller and Richards, 1967). As a result, modern siliciclastic sedimentation is controlled almost exclusively by erosion and redeposition of siliciclastic material by estuarine and marine processes.

Chemical sedimentation in the form of domed peat deposits dominates the coastal plain setting of the study area. If the climate should become drier and more seasonal, domed peat formation would cease, and the present tropical rain forest cover would be altered or even lost. Such a condition would favor increased erosion and siliciclatic transport and deposition. If the system should then become aggradational, the estuaries would fill and the domed peat deposits would be buried by the fluvial influx of siliciclastic sediment. Such shifts in climate and vegetational cover have been suggested in the region for the Pleistocene and Holocene (Verstappen, 1975). Similar shifts in climate may also have occurred during the late Paleozoic in what is now North America (e.g., Huntington, 1907; Wanless and Shepard, 1936; Cecil, 1990).

A climate-change mechanism for turning on or off siliciclastic influx and chemical sedimentation has numerous implications for interpreting coal-bearing and other stratigraphic sequences. Such a mechanism represents a viable alternative to the more commonly used autocyclic depositional (or) facies models. Recognition of allocyclic process controls on sedimentation allows a clearer definition of the requirements for low-ash, low-sulfur coal formation, such as proper water budget and chemistry and the almost total absence of siliciclastic and dissolved-sediment input coeval with peat formation.

Implications for depositional modeling

If the present Siak peat dome and Siak estuary channel were to become incorporated into the rock record, application of current facies models to the ensuing rock record would suggest that the peat developed in a flood-plain environment. In contrast to such a facies (depositional) model, the modern flood plain of the Siak system is devoid of peat; instead, peat forms the boundary of the flood plain. Thus, the peat dome is the interfluve between the Siak and Kampar river/estuary systems. Thick, high-quality, laterally continuous peat has continued to form over the past 5,000 years in a system that has been almost totally devoid of fluvial siliciclastic or dissolved sediment influx, that is, at least a 5,000-year period of nonaggradation of siliciclastic material. If existing allocyclic conditions were to change, causing siliciclastic deposition and aggradation in the present Siak channel, application of current facies models would suggest that siliciclastic deposition was coeval with peat formation. Obviously, such an interpretation would be erroneous. Facies analysis of subsequent clastic deposition would likely be misinterpreted as to an apparent coeval timing of chemical and siliciclastic sedimentation. Our observations indicate, therefore, that formation of extensive peat deposits in the geologic past was not necessarily coeval with siliciclastic aggradation as most earlier analyses, based on facies models, assume.

Application of the Sumatrain analogue to Carboniferous coal-bearing strata

The results of the analogue study indicate that the climatic conditions and the resulting sedimentary geochemistry in the study area appear to be analogous to those that existed during deposition of coal-bearing strata, or more specifically, during periods of peat formation in the Early through mid-Middle Pennsylvanian in the Appalachian basin as suggested by Neuzil and Cecil (1984) and Cecil et al. (1985). The data of Supardi et al. (this volume) and Neuzil et al. (this volume) indicate that thickness, lateral continuity, and sulfur content and ash yield of the peat deposits are analogous to Lower through mid-Middle Pennsylvanian coal beds in the Appalachian basin. However, climatic and sedimentary geochemical conditions in the study area are not analogous to those that prevailed during deposition of upper Middle and Upper Pennsylvanian strata when the climate was generally more seasonal, and peat formation was confined to topogenous swamps during pluvial periods (Cecil, 1990).

Many Lower and lower Middle Pennsylvanian coal-bed horizons in the Appalachian basin are stratigraphically bounded by siliciclastic sediments of marine or near-shore marine origin (see Ferm and Horne, 1979). They generally are laterally extensive, show relatively abrupt changes in thickness, and are low in ash yield and sulfur contents (Cecil et al., 1985), attributes that are consistent with the laterally extensive ombrogenous peat deposits on the coastal plains of equatorial Indonesia and Malaysia. These attributes are not consistent, however, with autogenic controls on swamps in deltaic, fluvial, or back-barrier sedimentary environments. Thus, the ever-wet climatic controls on sedimentation in the central Sumatra basin appear to be a good analogue for coal beds in Lower through mid-Middle Pennsylvanian coal-bearing strata in the Appalachian basin. If so, these coal beds were primarily derived from extensive, low-ash, low-sulfur domed peat deposits that formed on poorly drained coastal plains in a foreland basin setting during relatively low stands of sea level, and ever-wet periods of climate cycles (Cecil, 1990). During these pluvial periods, vegetative cover and extensive leaching within catchment basins restricted fluvial siliciclastic and dissolved-sediment flux, conditions that are prerequisites for the formation of high-quality peat. Thus, the allocyclic effects of tectonics, climate, and eustasy are believed to have controlled sedimentary sequences with climate being the primary control on nonmarine sedimentary geochemistry including conditions necessary for extensive peat formation and variations in siliciclastic sediment supply.

ACKNOWLEDGMENTS

We thank James R. Beerbower, Edward H. Clifton, James M. Coleman, John C. Ferm, and Stephen F. Greb for their constructive reviews of the manuscript. We also thank BHP Minerals and PT Caltex Pacific Indonesia for their technical and logistical support while we were in the field.

REFERENCES CITED

Ahrens, D. A., 1991, Meteorology Today: St. Paul, Minnesota, West Publishing Co., 576 p.

Anderson, J.A.R., 1961, The ecology and forest types of peat swamp forests of Sarawak and Brunei in relation to silviculture [Ph.D. thesis]: Edinburgh, University of Edinburgh, p. 1–191.

Anderson, J.A.R., 1964, The structure and development of peat swamps of Sarawak and Brunei: Tropical Geography, v. 18, p. 7–16.

Anderson, J.A.R., 1983, The tropical peat swamps of western Malaysia, *in* Gore, A.J.P., ed., Mires: Swamp, bog, fen, and moor: Amsterdam, Elsevier, Ecosystems of the World, 4B, p. 181–200.

Beerbower, J. R., 1964, Cyclothems and cyclic depositional mechanisms in alluvial plain sedimentation: Kansas State Geological Survey Bulletin 169, v. 1, p. 32–42.

Belt, E. S., and Lyons, P. C., 1989, A thrust-ridge paleodepositional model for the Upper Freeport coal bed and associated clastic facies, Upper Potomac coal field, Appalachian basin, U.S.A.: International Journal of Coal Geology, v. 12, p. 293–328.

Berner, E. K., and Berner, R. A., 1987, The global water cycle: Geochemistry and environment: Englewood Cliffs, New Jersey, Prentice-Hall, 397 p.

Cameron, N. R., Ghazali, S. A., and Thompson, S. J., 1982, The geology of the Bengkalis and Siak Sri Indrapura-Tanjungpinang Quadrangles, Sumatra: Bandung, Indonesia, Ministry of Mines and Energy, Directorate General of Mines, The Geologic Research and Development Centre, p. 1–26.

Cecil, C. B., 1990, Paleoclimate controls on stratigraphic repetition of chemical and siliciclastic rocks: Geology, v. 18, p. 533–536.

Cecil, C. B., Stanton, R. W., Neuzil, S. W., Dulong, F. T., Ruppert, L. F., and Pierce, B. F., 1985, Paleoclimate controls on late Paleozoic peat formation and sedimentation in the central Appalachian basin (USA): International Journal of Coal Geology, v. 5, p. 195–230.

Coleman, J. M., Gagliano, S. M., and Smith, W. G., 1970, Sedimentation in a Malaysian high tide tropical delta, *in* Morgan, J. P., ed., Deltaic sedimentation modern and ancient: Tulsa, Oklahoma, Society of Economic Paleontologists and Mineralogists Special Publication 15, p. 185–197.

Cross, A. T., 1975, The Dunkard in perspective: Geology, sedimentation, and life, *in* Barlow, J. A., ed., The age of the Dunkard: First I. C. White Memorial Symposium: Morgantown, West Virginia Geological Survey, p. 297–304.

Donaldson, A. C., Renton, J. J., and Presley, M. W., 1985, Pennsylvanian deposystems and paleoclimates of the Appalachians: International Journal of Coal Geology, v. 5, p. 167–193.

Dulong, F. T., and Cecil, C. B., 1989, Stratigraphic variation in the bulk sample mineralogy of Pennsylvanian underclays from the central Appalachian basin: Carboniferous geology of the eastern United States: Washington, D.C., American Geophysical Union, Field Trip Guidebook T143, p. 112–118.

Dyer, 1986, Coastal and estuarine sediment dynamics: John Wiley and Sons, p. 231–265.

Ferm, J. C., and Horne, 1979, Carboniferous depositional environments in the Appalachian region: Columbia, University of South Carolina, Department of Geology, 760 p.

Geyh, M. A., Kudrass, H. R., and Streif, H., 1979, Sea-level change during the late Pleistocene and Holocene in the Strait of Malacca: Nature, v. 278, p. 441–443.

Hamilton, Warren, 1979, Tectonics of the Indonesian Region: U.S. Geological Survey Professional Paper 1078, 345 p.

Haslam, D. W., 1983, One Fathom bank to Singapore Strait: Taunton, England, British Admiralty Chart 1358.

Hem, J. D., 1989, Study and interpretation of the chemical characteristics of natural waters: U.S. Geological Survey Water-Supply Paper 2254, 263 p.

Huntington, Ellsworth, 1907, Some characteristics of the glacial period in non-glaciated regions: Bulletin of the Geological Society of America, v. 18, p. 351–388.

Jansen, J. C., Diemont, W. H., and Koenders, N. J., 1985, Peat development for power generation in West Kalimantan: Rotterdam, Netherlands Economic Institute, p. 1–105.

Keller, G. H., and Richards, A. F., 1967, Sediments of the Malacca Strait, Southeast Asia: Journal of Sedimentary Petrology, v. 37, no. 1, p. 102–127.

Kennett, J. P., 1982, Marine Geology: Prentice-Hall, Englewood Cliffs, New Jersey, 812 p.

Klein, G. deV., 1990, Pennsylvanian time scales and cycle periods: Geology, v. 18, p. 455–457.

Klein, G. deV. and Willard, D. A., 1989, Origin of the Pennsylvanian coal-bearing cyclothems of North America: Geology, v. 17, p. 152–155.

McCabe, P. J., 1984, Depositional environments of coal and coal-bearing strata, *in* Rahmani, R. A., and Flores, R. M., eds., Sedimentology of coal and coal-bearing sequences: Oxford, Blackwell, Special Publication of International Association of Sedimentologists No. 7, p. 13–42.

McCabe, P. J., 1987, Facies studies of coal and coal-bearing strata: *in* Scott, A. C., ed., Coal and Coal-bearing strata: Recent advances: London, Blackwell, Geological Society of London Special Publication 32, p. 51–66.

Mörner, Nils-Axel, 1976, Eustasy and geoid changes: The Journal of Geology, v. 84, p. 123–151.

Neuzil, S. G., and Cecil, C. B., 1984, A modern analog of low-ash, low-sulfur Pennsylvanian age coal: Geological Society of America Abstracts with Programs, v. 16, no. 3, p. 184.

Perlmutter, M. A., and Matthews, M. D., 1989, Global cyclostratigraphy—A model, *in* Cross, A. T., ed., Quantitative dynamic stratigraphy: Englewood Cliffs, New Jersey, Prentice-Hall, p. 233–260.

Phillips, T. L., and Peppers, R. A., 1984, Changing patterns of Pennsylvanian coal-swamp vegetation and implications of climate control on occurrence: International Journal of Coal Geology, v. 3, p. 205–255.

Retallack, G. J., 1989, Soils of the past: Unwin Hyman, Winchester, Massachusetts, 520 p.

Schopf, J. M., 1975, Pennsylvanian climate in the United States, *in* McKee, E. D., et al., eds., Paleotectonic investigations of the Pennsylvanian System of the United States: U.S. Geological Survey Professional Paper 853, p. 23–31.

Smith, A.H.V., 1962, The paleoecology of Carboniferous peats based on the miospores and petrography of bituminous coals, *in* Proceedings, Yorkshire Geological Society: Yorkshire Geological Society, v. 33, p. 423–474.

Soil Survey Staff, 1975, Soil taxonomy: U.S. Department of Agriculture Handbook Number 436, 754 p.

Tankard, A. J., 1986, Depositional response to foreland deformation in the Carboniferous of eastern Kentucky: The American Association of Petroleum Geologists Bulletin, v. 70, p. 853–868.

Umbgrove, J.H.F., 1949, Structural history of the East Indies: Cambridge, Cambridge University Press, p. 1–12.

Van Wagoner, J. C., Posamentier, H. W., Mitchum, R. M., Vail, P. R., Sarg, J. F., Loutit, T. S., and Hardenbol, J., 1988, An overview of the fundamentals of sequence stratigraphy and key definitions, *in* Wilgus, C. K., Hastings, B. K., Kendall, C.G.St.C., Posamentier, H. W., Ross, C. A., and Van Wagoner, J. C., eds., Sea-level changes: An integrated approach: Tulsa, Oklahoma, Society of Economic Paleontologists and Mineralogists Special Publication 42, p. 40–45.

Verstappen, H. Th., 1975, On palaeo climates and landform development in Malesia: Modern Quaternary Research in Southeast Asia, v. 1, p. 3–35.

Wanless, H. R., and Shepard, F. P., 1936, Sea level and climate changes related to late Paleozoic cycles: Bulletin of the Geological Society of America, v. 47, p. 1177–1206.

White, D., 1913, Climates of coal forming periods, *in* White, C. D. and Theissen, R., eds., The origin of coal: U.S. Bureau of Mines Bulletin 38, p. 68–79.

White, D., 1925, Environmental conditions of deposition of coal: American Institute of Mining and Metallurgical Engineers transactions, v. 71, p. 3–34.

Wyrtki, Klaus, 1971, Oceanographic atlas of the International Indian Ocean Expedition: Washington, D.C., National Science Foundation, 531 p.

MANUSCRIPT ACCEPTED BY THE SOCIETY JANUARY 14, 1993

Printed in U.S.A.

Geological Society of America
Special Paper 286
1993

Inorganic geochemistry of domed peat in Indonesia and its implication for the origin of mineral matter in coal

Sandra G. Neuzil
U.S. Geological Survey, 956 National Center, Reston, Virginia 22092
Supardi
Directorate of Mineral Resources, Jl. Soekarno-Hatta 444, Bandung, West Java 40254, Indonesia
C. Blaine Cecil and Jean S. Kane*
U.S. Geological Survey, 956 National Center, Reston, Virginia 22092
Kadar Soedjono
Directorate of Mineral Resources, Jl. Soekarno-Hatta 444, Bandung, West Java 40254, Indonesia

ABSTRACT

The inorganic geochemistry of three domed ombrogenous peat deposits in Riau and West Kalimantan provinces, Indonesia, was investigated as a possible modern analogue for certain types of low-ash, low-sulfur coal. Mineral matter entering the deposits is apparently limited to small amounts from the allogenic sources of dryfall, rainfall, and diffusion from substrate pore water. In the low-ash peat in the interior of the deposits, a large portion of the mineral matter is authigenic and has been mobilized and stabilized by hydrological, chemical, and biological processes and conditions.

Ash yield and sulfur content are low through most of the peat deposits and average 1.1% and 0.14%, respectively, on a moisture-free basis. Ash and sulfur contents only exceed 5% and 0.3%, respectively, near the base of the deposits, with maximum concentrations of 19.9% ash and 0.56% sulfur. Peat water in all three deposits has a low pH, about 4 units, and low dissolved cation concentration, averaging 14 ppm. Near the base, in the geographic interior of each peat deposit, pH is about two units higher and dissolved cation concentration averages 110 ppm. Relative concentrations of the inorganic constituents vary, resulting in chemical facies in the peat. In general, Si, Al, and Fe are the abundant inorganic constituents, although Mg, Ca, and Na dominate in the middle horizon in the geographic interior of coastal peat deposits.

The composition of the three deposits reported in this paper indicates that domed ombrogenous peat deposits will result in low ash and sulfur coal, probably less than 10% ash and 1% sulfur, even if marine rocks are laterally and vertically adjacent to the coal.

INTRODUCTION

Mineral matter in precursor peat is often interpreted as the major source of the mineral matter in coal (Cecil et al., 1982). Yet little is known about mineral matter sources and sinks in the largest tropical peatlands found on Earth in the lowlands of Indonesia and Malaysia. These Holocene peatlands may be modern analogues to low-ash coal deposits that are interpreted to have originated in a tropical ever-wet climate such as the coal deposits of the Lower and lower Middle Pennsylvanian in the Appalachian basin (Neuzil and Cecil, 1984; Cecil et al., 1985). Information on the geochemistry of the Holocene tropical peatlands may therefore provide critical information explaining areal and vertical distribution of major and trace elements in much of

---

*Present address: Office of Standard Reference Materials, Building 202, Room 215-A, Gaithersburg, Maryland 20899.

Neuzil, S. G., Supardi, Cecil, C. B., Kane, J. S., and Soedjono, K., 1993, Inorganic geochemistry of domed peat in Indonesia and its implication for the origin of mineral matter in coal, *in* Cobb, J. C., and Cecil, C. B., eds., Modern and Ancient Coal-Forming Environments: Boulder, Colorado, Geological Society of America Special Paper 286.

the world's coal resources that formed under tropical ever-wet climatic conditions.

Several pathways allow allogenic (externally derived) inorganic constituents to enter domed ombrogenous peat deposits as solid particles. Minerals can be incorporated in the bottom of the peat, in the initial accumulation of topogenous peat, by detrital influx and by plant root bioturbation. Throughout peat formation, solids can be incorporated at the upper surface by infall of airborne terrestrial dust, volcanic ash, and cosmic dust, either as dryfall or in rain. The relative proportion of the elements entering the peat as minerals by these pathways should be similar to their distinction in crustal rocks (Damman, 1987); in decreasing abundance, the elements found in igneous rocks, sandstone, and shale are Si, Al, Fe, Ca, K, Mg, Na, Ti, P, and Mn (Hem, 1989).

In addition to allogenic minerals, inorganic constituents can enter domed ombrogenous peat deposits as dissolved mineral matter. Dissolved solids can enter the peat deposit in marine aerosols. Aerosol concentration and hence dissolved solids decrease with increasing distance from the coast (Damman, 1987). The composition of dissolved solids in rain is similar to that in seawater; in decreasing concentration, these elements are Cl, Na, Mg, S, Ca, and K. Discharge of ground water can move dissolved cations up into peat by advective mass transport, with cation concentrations nearly equal to those in the ground water. Ground-water discharge may vary within and among ombrogenous peat deposits (Siegel and Glaser, 1987; Siegel, 1988a). Diffusion along a concentration gradient may move cations from higher concentration substrate pore water up into lower concentration peat pore water. Finally, dissolved cations can be removed from the peat by advective mass transport in ground-water recharge to the underlying substrate.

Autogenic processes within a peat deposit result in mobilization and stabilization of mineral matter. The quantity and location of authigenic mineral matter within the peat are determined by (1) chemical dissolution and precipitation of mineral matter in peat (Bennett and Siegel, 1987; Bennett and others, 1988), (2) desorption and adsorption of cations on peat by cation exchange and chelation (Andriesse, 1988), and (3) decay and growth of plants and microbes (Sillampää, 1972). Further discussion and references cited can be found in Damman (1978), Shotyk (1988), and Hill and Siegel (1991). Although analyses discussed in this study are of total mineral matter in peat, inferences are made for the autogenic processes controlling mineral matter composition.

The objective of this study was to examine the chemical composition in three domed peat deposits in Indonesia—Siak Kanan, Bengkalis Island, and Teluk Keramat, hereinafter referred to as Siak, Bengkalis, and Keramat (Figs. 1A, 1B, and 1C, respectively). These peat deposits occur in the intertropical convergence zone near the equator and are in close proximity to marine conditions. This study considered (1) content of 10 inorganic elements in the peat; (2) concentration of the same 10 inorganic elements in the peat pore water; and (3) vertical and lateral variation of these elements through the three extensive,

thick rainfall-dominated low-ash peat deposits. This paper reports on the low-level input of mineral matter to the peat from both terrestrial and marine sources, the apparent leaching and loss of some of these inorganic constituents, and the distribution of mineral matter in the peat resulting from the rainfall dominance and domed geometry of the peat deposits rather than the predecessor and coeval marine environments associated with the peat.

Terminology

In this study, *peat* refers to an organic soil with ash yield <25% dry weight. *Topogenous peat* forms when there is a standing water table above the peat surface with the peat receiving significant water inflow from surface water and/or ground water resulting in a peat deposit with a planar upper surface (Bates and Jackson, 1987). *Ombrogenous peat* forms when precipitation exceeds evapotranspiration throughout the year and rain is the dominant or only source of water in the peat resulting in a raised upper surface and domed geometry (Bates and Jackson, 1987). The peat overlies a mineral *substrate* that has an ash yield >25% dry weight and may be organic rich. In vertical profile, the peat can be considered as three horizons, *top, middle,* and *bottom,* distinguished by chemical composition, similar to soil horizons. Peat and water samples are designated *top* (<1 m below the upper peat surface), *middle* (>1 m below the upper surface and for water samples >1 m above the substrate or for peat samples <5% ash), and *bottom* (<1 m above the substrate and, for peat, >5 percent ash). The *edge* of a peat deposit is where the peat thins to <1 m thick. The *margin* of a peat deposit refers to geographic locations <2 km from the edge. The *interior* of a peat deposit refers to geographic locations >2 km from the nearest edge. *Inland* peat refers to locations >15 km from the open sea (this does not include estuaries and narrow straits) and includes all sites sampled in Siak. *Coastal* peat refers to locations <15 km from the open sea and includes all of the Bengkalis and Keramat peat deposits. *Ash yield* is the residue of ashing at 550 °C for 16 hours, expressed as wt% of dry peat. *Mineral matter* in peat is the sum of all the inorganic constituents in peat that are isolated from the peat and constitute the residue of the high-temperature ashing procedure (regardless of whether the components of the mineral matter were originally present in peat as dissolved cations, inorganic-organic complexes, or inorganic compounds of biologic or geologic origin). For each inorganic element the concentration in peat is calculated by multiplying the percent ash in dry peat times the percent oxide in ash times the percent element in oxide and reported as ppm on a whole dry-peat basis (Table 1).

STUDY AREA

Characteristics of domed peat deposits in Indonesia and Malaysia

Equatorial peat deposits greater than 1 m thick cover about 190,000 km² in Indonesia and Malaysia (Bord na Mona, 1984). As early in 1913, White and Theissen (1913) suggested that the peat swamps in Sumatra are a modern analogue for Pennsylvanian Appalachian Basin coal beds. More recently, Neuzil and Cecil (1984) and Cecil et al. (1985) suggested that the domed peat deposits on the coastal lowlands of Indonesia and Malaysia are excellent modern analogues for the low-ash, low-sulfur coal beds of the Lower to mid-Middle Pennsylvanian of the Appalachian Basin. This comparison was based on the deposit geometry, the low-ash and low-sulfur contents of the peat, acid swamp conditions, and the woody and well-preserved nature of the peat (Anderson, 1961, 1964, 1983; Driessen et al., 1975; Suhardjo and Widjaja-Adhi, 1976; Driessen, 1977). Furthermore, the chemical stratigraphy of Lower to mid-Middle Pennsylvanian strata (Cecil et al., 1985) is consistent with the tropical ever-wet climate of Indonesia and Malaysia (Schmidt and Ferguson, 1951).

Carbon-14 ages show that accumulation of Holocene peat on the coastal lowlands of Indonesia and Malaysia started about 5,000 years ago (Wilford, 1959; Supardi et al., this volume) when rising sea level stabilized, or dropped slightly (Biswas, 1973; Tjia, 1975; Geyh et al., 1979; Tjia et al., 1984), resulting in exposure of large, relatively flat areas of marine sediments (Cecil et al., this volume).

Domed peat deposits in Indonesia and Malaysia lie within the intertropical convergence zone where rising air masses and maritime influences result in heavy precipitation. Rainfall generally exceeds 2.5 m annually in these peat areas. There is a "rainy season" and a "dry season" according to local Indonesian terminology, but even the dry season is wet, with a minimum rainfall of about 100 mm per month, which exceeds evapotranspiration in the peat-swamp forests (Schmidt and Ferguson, 1951; Nieuwolt, 1965, 1968; Brunig, 1971; Whitmore, 1975; Morley, 1981). Areas that receive less than 60 mm of rain per month for two or more consecutive months do not have domed ombrogenous peat deposits in Indonesia and Malaysia (Schmidt and Ferguson, 1951; Morley, 1981).

Studies of floral associations and ecology in Indonesian and Malaysian domed peat deposits show a change in floral communities as peat accumulates (Anderson and Muller, 1975; Morley, 1981). These studies interpreted the initial flora as either brackish-water mangroves and nipa palms or fresh-water herbaceous aquatic plants, each representative of a topogenous peat. The topogenous peat deposit may go through a transition to a domed ombrogenous peat deposit if the climate is ever-wet (Anderson, 1964).

Floral ecology changes with the transition to ombrogenous peat. In Indonesia and Malaysia, numerous species of large angiosperm trees populate the ombrogenous peat and are collectively referred to as mixed peat-swamp forest (Anderson, 1961, 1964, 1976, 1983; Morley, 1981). Further changes in floral ecology, decrease in species diversity, and decrease in plant stature may occur during ombrogenous peat accumulation, resulting in a pole or padang forest (Anderson, 1961, 1964, 1976, 1983; Anderson and Muller, 1975; Morley, 1981). Trees in the mixed peat-swamp forest and pole forest (Anderson, 1961, 1964) have spreading,

buttressed, and prop roots, which are generally confined to a root mat 50–80 cm thick at the top of the peat and do not penetrate to the deeper peat or mineral sediments below thick peat.

Before the present study, little was known about water flow and the source and composition of mineral matter at depth in tropical domed peat deposits. Although Anderson (1961, 1964, 1976) asserted that many of the peat deposits in Indonesia and Malaysia are ombrogenous based on the shape, surface drainage, low ash yield of the deposits, and the prevailing ever-wet climate, there was little data from the interior of the deposits to determine whether the water source was exclusively rain or included ground-water discharge.

Previous analyses of mineral matter in tropical domed peat deposits have focused on (1) nutrient status of the margins and top of the peat for agriculture and forestry (Anderson, 1961, 1983; Driessen and Suhardjo, 1976; Suhardjo and Widjaja-Adhi, 1976; Andriesse, 1988) or (2) minor- and trace-element content at depth in a peat deposit (Cameron et al., 1989). Energy-resource studies have considered the entire peat deposit, but the analytical data are generally restricted to economically important properties such as heating value, proximate and ultimate analyses, and ash yield (Bord na Móna, 1984; Supardi and Priatna, 1985; Andriesse, 1988; Supardi et al., this volume). A petrographic study of a domed peat deposit in East Malaysia as a modern analogue of coal discussed ash yield but not mineral-matter composition (Esterle et al., 1989).

Site location

Field study areas were chosen in Siak and Bengkalis, Riau province, Sumatra, Indonesia, and Keramat, West Kalimantan province, Kalimantan, Indonesia (Fig. 1). This selection was based on the tropical setting, areal extent, thickness, and domed geometry of the peat deposits (Supardi and Priatna, 1985; Soedjono et al., 1988; Supardi et al., this volume).

The Siak peat covers approximately 1,100 km^2 and is a portion of a large, continuous peat deposit on the coastal lowlands of Sumatra between the Siak River and Kampar River estuaries (Supardi et al., this volume). The Bengkalis peat extends for 144 km^2 and is one of five partially coalesced peat domes that cover a total area of 665 km^2 on Bengkalis Island, which lies 10 km off the coast of Sumatra in the Strait of Malacca (Supardi et al., this volume). The Keramat peat extends for 175 km^2 and is a single peat dome on a peninsula between the Sambas River estuary and the South China Sea on the west coast of Kalimantan (Soedjono et al., 1988).

The domed geometry and elevation relative to mean sea level of the peat deposits in the study areas is based on level surveys of these deposits (Supardi et al., this volume; Soedjono et al., 1988) (Figs. 2A, 2B, and 2C). The doming causes radial drainage and precludes flooding by nearby river/estuary and marine systems. The water table within the peat deposit fluctuates throughout the year. During the dry season, the water table may drop below the peat surface (Anderson, 1964), and oxygenated

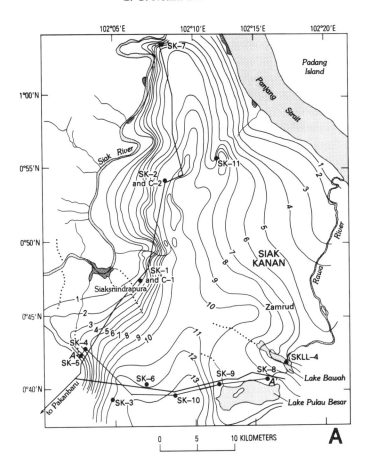

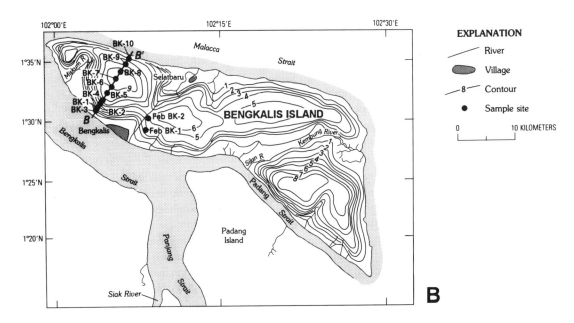

Figure 1. Location of study areas and sample sites (on this and facing page). A, Siak Kanan, Riau Province, Sumatra, Indonesia (modified from Supardi et al., this volume); B, Bengkalis Island, Riau Province, Sumatra, Indonesia (modified from Supardi et al., this volume). C, Teluk Keramat, West Kalimantan Province, Kalimantan, Indonesia (modified from Soedjono et al., 1988); and D, index maps of locations shown in A, B, and C (modified from Supardi, 1988). Contours (in meters) show isopach thickness of peat.

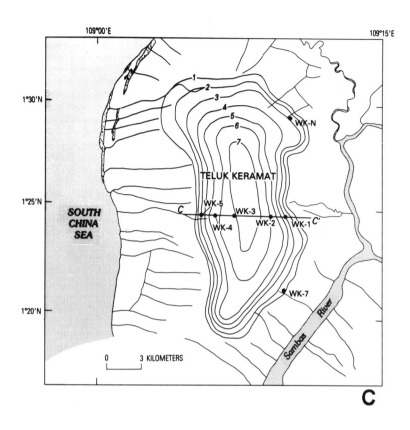

C

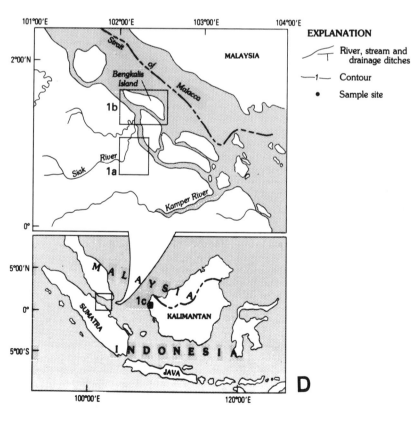

EXPLANATION

River, stream and drainage ditches

—1— Contour

• Sample site

D

S. G. Neuzil and others

TABLE 1. PEAT DATA FOR SIAK KANAN, BENGKALIS ISLAND, AND TELUK KERAMAT PEAT DEPOSITS

	HTA (%)	Sulfur (%)	Si (ppm)	Ti (ppm)	Al (ppm)	Fe (ppm)	Mn (ppm)	Mg (ppm)	Ca (ppm)	Na (ppm)	K (ppm)	P (ppm)
All peat except bottom												
Average	1.13	0.14	1,369	26	619	629	17	1,130	756	275	108	93
Minimum	0.13	0.08	55	5	51	91	2	57	39	1	5	20
Maximum	4.66	0.30	13,806	243	3,333	2,779	133	8,210	11,351	1,787	667	653
Count	*95*	*94*	*83*	*83*	*80*	*83*	*83*	*83*	*83*	*83*	*72*	*83*
Top peat												
Average	1.52	0.14	3,314	38	766	663	18	779	1,001	125	173	228
Minimum	0.40	0.10	252	6	104	95	2	140	57	8	49	90
Maximum	4.66	0.19	13,806	164	3,333	2,779	133	4,547	11,351	327	667	653
Count	*32*	*31*	*32*	*32*	*32*	*32*	*32*	*32*	*32*	*32*	*29*	*32*
Middle peat												
Average	1.07	0.14	935	23	584	622	17	1,209	702	309	91	63
Minimum	0.13	0.08	55	5	51	91	2	57	39	1	5	20
Maximum	3.66	0.30	6,143	243	2,362	2,408	105	8,210	7,986	1,787	307	199
Count	*63*	*63*	*51*	*51*	*48*	*51*	*51*	*51*	*51*	*51*	*43*	*51*
Bottom peat												
Average	11.79	0.30	35,956	859	16,920	3,335	98	3,204	2,633	1,344	1,522	117
Minimum	5.21	0.17	2,938	72	832	930	20	176	133	175	147	63
Maximum	19.87	0.56	64,398	1,797	30,949	4,714	274	9.777	23,949	3,142	3,133	165
Count	*13*	*11*	*10*	*10*	*10*	*10*	*10*	*10*	*10*	*8*	*7*	*10*
Substrate												
Average	48.16	1.60	142,198	2,614	44,201	14,736	163	6,666	5,985	2,102	8,123	207
Minimum	41.24	0.24	101,495	1,569	32,892	8,082	128	4,022	4,133	797	3,992	155
Maximum	73.89	2.30	258,410	3,463	63,361	20,092	218	8,455	12,833	2,785	11,073	277
Count	*5*	*3*	*5*	*5*	*5*	*5*	*5*	*5*	*4*	*5*	*5*	*5*
Top margin inland												
Average	2.33	0.15	6,020	70	2,178	1,560	19	207	304	112	241	142
Minimum	1.19	0.14	1,556	34	1,428	881	5	140	147	8	182	90
Maximum	4.66	0.17	13,806	123	3,333	2,779	133	365	613	327	526	226
Count	*6*	*6*	*6*	*6*	*6*	*6*	*6*	*6*	*6*	*6*	*3*	*6*
Top interior inland												
Average	1.41	0.13	4,203	27	459	483	14	455	495	39	125	226
Minimum	0.70	0.10	1,585	16	196	351	4	210	168	10	81	131
Maximum	2.40	0.16	7.849	46	940	699	41	638	1,287	52	290	419
Count	*8*	*8*	*8*	*8*	*8*	*8*	*8*	*8*	*8*	*8*	*8*	*8*
Top margin coast												
Average	1.85	0.15	2,653	47	482	592	30	1,421	2,855	145	150	298
Minimum	0.96	0.10	318	14	132	289	8	443	297	48	49	117
Maximum	4.41	0.19	12,142	164	1,060	925	75	4,547	11,351	214	303	653
Count	*9*	*8*	*9*	*9*	*9*	*9*	*9*	*9*	*9*	*9*	*9*	*9*
Top interior coast												
Average	0.63	0.13	831	14	178	166	8	938	235	202	211	232
Minimum	0.40	0.10	252	6	104	95	2	355	57	136	60	103
Maximum	1.41	0.16	1,705	22	261	365	26	1,990	776	317	667	590
Count	*9*	*9*	*9*	*9*	*9*	*9*	*9*	*9*	*9*	*9*	*9*	*9*
Middle margin inland												
Average	1.24	0.17	2,389	64	1,542	985	4	90	98	44	119	52
Minimum	0.73	0.13	753	24	1,174	786	3	57	78	1	71	41
Maximum	2.03	0.27	6,143	243	2,112	1,220	6	115	134	104	215	64
Count	*7*	*7*	*6*	*6*	*6*	*6*	*6*	*6*	*6*	*6*	*5*	*6*
Middle interior inland												
Average	0.62	0.13	921	13	407	684	13	460	205	102	63	67
Minimum	0.13	0.08	187	5	70	150	3	103	39	19	17	20
Maximum	1.39	0.24	3,843	43	883	2,408	71	1,685	632	357	126	199
Count	*22*	*22*	*17*	*17*	*17*	*17*	*17*	*17*	*17*	*17*	*12*	*17*

TABLE 1. PEAT DATA FOR SIAK KANAN, BENGKALIS ISLAND, AND TELUK KERAMAT PEAT DEPOSITS (continued)

	HTA (%)	Sulfur (%)	Si (ppm)	Ti (ppm)	Al (ppm)	Fe (ppm)	Mn (ppm)	Mg (ppm)	Ca (ppm)	Na (ppm)	K (ppm)	P (ppm)
Middle margin coast												
Average	1.33	0.19	1,716	47	993	765	36	1,245	924	289	105	78
Minimum	0.42	0.10	195	10	261	209	7	353	276	154	48	39
Maximum	3.31	0.30	5,514	146	2,362	1,636	105	3,545	2,658	705	169	107
Count	*8*	*8*	*8*	*8*	*8*	*8*	*8*	*8*	*8*	*8*	*6*	*8*
Middle interior coast												
Average	1.37	0.14	219	11	271	398	18	2,253	1,277	596	97	57
Minimum	0.29	0.10	55	6	51	91	2	514	50	166	5	39
Maximum	3.66	0.19	1,271	28	951	1,441	68	8,210	7,986	1,787	307	82
Count	*26*	*26*	*20*	*20*	*17*	*20*	*20*	*20*	*20*	*20*	*20*	*20*

HTA = high-temperature ash; Average = average weighted by sample thickness; Minimum = minimum value; Maximum = maximum value; Count = actual number of samples (not expressed as a percentage or in ppm).

rainwater entering the top of the peat results in aerobic, oxidizing conditions in the root mat. For example, the water table was observed 0.5–1 m below the surface in Siak, 0.3–0.9 m below the surface in Bengkalis, and 0.0–0.1 m below the surface in Keramat near the end of the dry season in July 1987, August 1987, and August 1988, respectively. The water table is generally at or slightly above the peat surface during the rainy season. Excess water drains out of the peat in blackwater streams.

Field data and samples were collected at 11 sites in Siak (SK- samples), 12 sites in Bengkalis (BK- samples), and 7 sites in Keramat (WK- samples) (Figs. 1A, 1B, and 1C, respectively). In each study area, most of the sample sites were located along a transect crossing the thick geographic interior of the peat deposit, A–A′, B–B′, and C–C′ in Siak, Bengkalis, and Keramat, respectively.

METHODS

Peat samples

Peat samples were collected with a Macaulay-type auger. (Any use of trade, product, or firm names in this publication is for descriptive purposes only and does not constitute endorsement by the U.S. Government or the Geological Society of America.) At the upper surface of the peat, the interval from 0.0 to 0.5 m was divided at 0.25 m to study the chemistry of the surface peat and living root mat in more detail. Starting at 1.0 m depth, two consecutive samples were combined to provide samples from 1-m intervals. Near the base of the peat, samples were collected on the basis of changes in megascopic visual appearance; this strategy often resulted in thinner sample intervals in the transition from peat to mineral sediments below the peat. Detailed methods for peat sample collection and preparation are discussed in USGS Open-File Report 92-205 (Neuzil et al., 1993).

Splits of peat samples were dried to constant weight (60 °C) and ground to pass a 40-mesh sieve (opening size 420 microns). The ash yield of 93 peat and 5 mineral substrate samples was determined by ashing in a muffle furnace at 550 °C for 16 hours (Day et al., 1979). Using a Leco sulfur analyzer, 90 dry peat and 3 substrate samples were analyzed for total sulfur. The ash yield and sulfur content determined by Dickinson Laboratories, Inc., El Paso, Texas, will be used in this report for an additional 15 samples.

Ten major elements (major with respect to concentration in ash) were determined by inductively coupled plasma–atomic-emission spectrometry on ash of 93 peat and 5 mineral substrate samples, after being fused and dissolved by mixed lithium metaborate/lithium tetraborate fusion methods (Shapiro and Brannock, 1962; Kane and Neuzil, this volume; Neuzil et al., 1993). Analyses of major elements are reported as percent oxides in ash and calculated to express mineral matter in peat (ppm in dry peat). In this paper the averages for ash, sulfur, and major-element concentrations in peat have been weighted by peat sample thickness (Table 1).

Water samples

Water samples were collected by pumping water out of shallow cased wells made from one-half inch inside diameter polyvinyl chloride pipes (Chason and Siegel, 1986). Water samples were analyzed for pH and specific conductance (Table 2). Concentrations of 10 major cations were determined by inductively coupled plasma–atomic-emission spectrometry after filtering through 0.45 micron nuclepore filters (Neuzil et al., 1993).

RESULTS OF ANALYSES

Peat ash yield

A cross section of each peat deposit has been prepared with ash-yield data and isopleths (Figs. 2A, 2B, and 2C). The top and middle peat contain low ash (<5%) and very low ash (<2%), with very low ash constituting the major portion of the top and middle peat. Average ash yield for the 32 top peat samples is 1.5% and

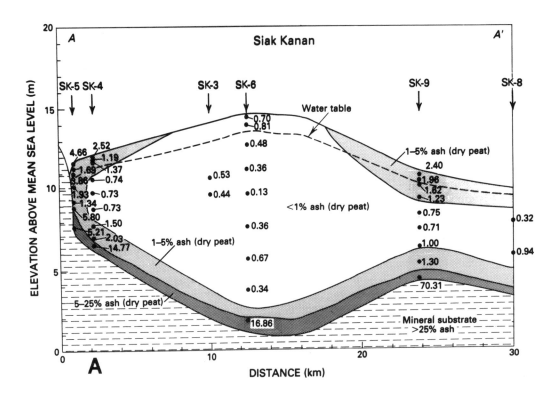

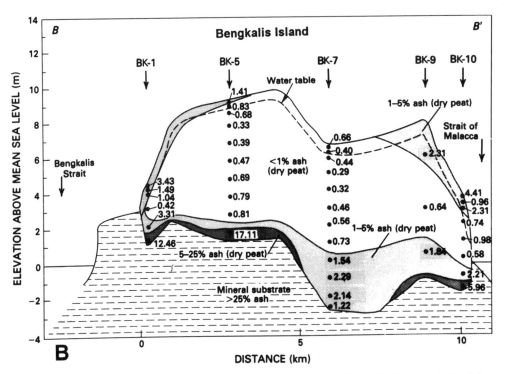

Figure 2. Generalized cross section of ash yield in peat deposit (on this and facing page). A, Siak Kanan; B, Bengkalis Island; C, Teluk Keramat. Dots indicate center of sample interval. Data is percent ash in dry peat. Shading indicates increasing ash yield.

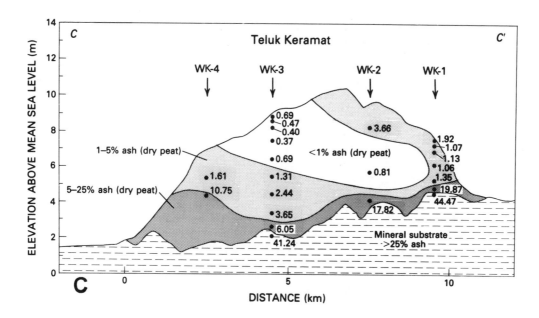

1.1% for the 63 middle peat samples (Table 1). The bottom peat is moderate ash (5%–15%) to high ash (15%–25%) in thin increments, generally <1 m thick, in the transition from low-ash peat to mineral substrate.

Average ash yield tends to be slightly higher at the margin than in the interior for both top and middle horizons in both inland and coastal deposits (Table 1; Fig. 2). In these peat deposits, the lowest ash yield is usually in the middle interior of the deposit (Fig. 2).

Composition of mineral matter in peat

Ash yield and concentrations of 10 elements in peat increase toward the bottom and increase slightly toward the top of each deposit (Fig. 3). Ash yield is positively correlated with percentages of SiO_2 and TiO_2 in ash (all three peat deposits) and with percent Al_2O_3 in ash (Bengkalis and Keramat). Ash yield is negatively correlated with percent MgO in ash (all three peat deposits), percent Na_2O in ash (Bengkalis and Keramat), and percent P_2O_5 in ash (Keramat) (Kane and Neuzil, this volume). Thus, Si, Ti, and Al concentration profiles are similar to ash-yield profiles, whereas Mg, Na, and P concentration profiles are dissimilar to ash-yield profiles (Fig. 3).

Mineral-matter composition changes among the horizons in the peat deposits (Fig. 4). The low-ash top and middle peat are relatively depleted in Si, Al, Ti, and K and enriched in Fe, Mg, Ca, Na, P, and Mn compared to the bottom peat and substrate (Fig. 4). This trend is generally more pronounced in middle than top peat. The composition of mineral matter in bottom peat is the most similar to substrate composition (Fig. 4). The relative concentration of elements in the substrate is similar to average crustal abundances.

For the top and middle horizons, mineral-matter composition varies depending on location in the deposit (margin or interior) and location of the deposit (coastal or inland) (Figs. 5A and 5B). The average mineral-matter content of the peat from the interior of the deposit is generally enriched in Mg, Ca, and Na (and P?) and depleted in Si, Al, and Fe (and Ti, Mn, and K?), compared to the margins for both top and middle peat horizons (Figs. 5A and 5B). Also, peat from coastal deposits is enriched in Mg, Ca, and Na and depleted in Si, Al, and Fe when compared to the inland deposit (Figs. 5A and 5B).

Peat sulfur content and composition

Total sulfur content is very low, generally <0.3%. Cross sections for each of the peat deposits with sulfur data and isopleths illustrate that the relative sulfur content is low to moderate at the top and at the margins of the deposits, lowest in the middle interior of the deposits, and highest at the bottom (Figs. 6A, 6B, and 6C). Ninety-four samples from the top and middle of the deposits have an average sulfur content of 0.14% (Table 1). Eleven samples from the bottom of the peat deposits do not exceed 0.6% sulfur and average 0.30%. The highest sulfur content measured, 2.3%, was in a substrate sample yielding 41% ash from the interior of Keramat. Organic sulfur, expressed as percent of total sulfur, averages 83% and ranges from 67% to 95% for 34 peat samples (Neuzil, 1990; Supardi et al., this volume). Thus, organic sulfur was the dominant sulfur form, which is typical for both low-sulfur peat and low-sulfur coal (Casagrande, 1987).

TABLE 2. WATER DATA FOR SIAK KANAN, BENGKALIS ISLAND, AND TELUK KERAMAT PEAT WATER AND SUBSTRATE WATER

	Field (pH)	Lab (pH)	Lab S.C.	Si (ppm/s)	Ti (ppb/s)	Al (ppm/s)	Fe (ppm/s)	Mn (ppb/s)	Mg (ppm/s)	Ca (ppm/s)	Na (ppm/s)	K (ppm/s)	P (ppb/s)	TDS (ppm/s)
All peat water except bottom interior														
Average	3.80	3.51	90	4.21	17	0.37	0.93	30	1.31	0.67	9.65	4.34	179	13.67
Minimum	3.05	2.85	36	2.50	10	0.10	0.11	10	0.05	0.02	0.73	0.80	12	1.00
Maximum	5.87	6.28	650	8.00	110	2.30	3.80	150	17.00	6.50	110.00	13.00	680	145.36
Count	*90*	*95*	*95*	*8*	*30*	*21*	*9*	*24*	*44*	*43*	*45*	*9*	*23*	*45*
BDL				*37*	*15*	*24*	*36*	*21*	*1*	*2*	*0*	*36*	*22*	*0*
All top peat water														
Average	3.43	3.11	77		13	0.49	0.16	19	0.55	0.62	3.48	2.00	214	5.02
Minimum	3.10	2.85	48		10			10	0.05	0.05	2.00		31	2.32
Maximum	3.92	3.52	101		16			30	0.95	2.00	4.80		580	7.07
Count	*13*	*15*	*15*	*0*	*5*	*1*	*1*	*4*	*8*	*7*	*8*	*1*	*4*	*8*
BDL				*8*	*3*	*7*	*7*	*4*	*0*	*1*	*0*	*7*	*4*	*0*
Middle margin														
Average	3.48	3.26	88		11	0.23	0.11	21	0.91	0.33	6.25	4.38	348	10.87
Minimum	3.05	2.93	62		10	0.13		15	0.37	0.12	3.60	0.80	200	5.37
Maximum	3.84	3.69	129		13	0.39		35	1.50	0.55	12.00	13.00	590	25.62
Count	*12*	*12*	*12*	*0*	*3*	*5*	*1*	*5*	*6*	*6*	*6*	*4*	*4*	*6*
BDL				*6*	*3*	*1*	*5*	*1*	*0*	*0*	*0*	*2*	*2*	*0*
Middle interior														
Average	3.92	3.58	94	4.92	18	0.23	0.15	12	1.42	0.63	12.19	4.93	87	16.05
Minimum	3.08	2.90	36	2.50	10	0.10		10	0.07	0.02	0.88	2.00	12	1.00
Maximum	5.87	6.28	650	8.00	110	0.47		16	17.00	6.50	110.00	7.00	280	145.36
Count	*54*	*54*	*54*	*5*	*15*	*6*	*1*	*7*	*21*	*21*	*22*	*3*	*10*	*22*
BDL				*17*	*7*	*16*	*21*	*15*	*1*	*1*	*0*	*19*	*12*	*0*
Bottom margin														
Average	4.01	3.91	91	3.03	19	0.53	1.32	57	2.02	1.03	11.23	4.80	199	17.41
Minimum	3.07	3.08	51	2.60	10	0.13	0.16	19	0.08	0.16	0.73		37	5.94
Maximum	5.25	5.12	174	3.40	50	2.30	3.80	150	3.80	2.20	23.00		680	43.49
Count	*11*	*14*	*14*	*3*	*7*	*9*	*6*	*8*	*9*	*9*	*9*	*1*	*5*	*9*
BDL				*6*	*2*	*0*	*3*	*1*	*0*	*0*	*0*	*8*	*4*	*0*
Bottom interior														
Average	5.44	5.36	475	18.10	16	0.41		59	12.93	5.71	76.45	15.50	523	112.57
Minimum	4.40	3.69	45	8.20	15			22	0.76	0.74	7.80	14.00	51	9.42
Maximum	6.33	6.69	1,170	28.00	17			130	33.00	15.00	190.00	17.00	1,400	260.76
Count	*10*	*10*	*9*	*2*	*3*	*1*	*0*	*3*	*4*	*4*	*4*	*2*	*4*	*4*
BDL				*2*	*1*	*3*	*4*	*1*	*0*	*0*	*0*	*2*	*0*	*0*
Substrate margin														
Average	5.66	5.54	504	13.37	19	0.16		120	6.46	3.51	55.53	9.67	291	83.10
Minimum	4.15	4.04	24	3.20	16	0.13		48	0.05	0.15	1.10	4.00	52	4.70
Maximum	6.67	6.43	828	32.00	23	0.18		250	14.00	6.20	100.00	17.00	530	169.30
Count	*5*	*6*	*6*	*3*	*3*	*2*	*0*	*3*	*4*	*4*	*4*	*3*	*2*	*4*
BDL				*1*	*1*	*2*	*4*	*1*	*0*	*0*	*0*	*1*	*2*	*0*
Substrate interior														
Average	6.10	6.04	761	13.98	28	0.26		147	34.58	13.58	183.65	30.50	469	262.16
Minimum	4.81	4.40	73	7.40	11	0.18		13	1.00	1.70	7.90	14.00	18	11.23
Maximum	6.70	6.77	1,530	28.00	100	0.37		480	143.00	56.00	860.00	72.00	1,700	1,140.13
Count	*9*	*9*	*9*	*4*	*6*	*3*	*0*	*6*	*6*	*6*	*6*	*4*	*5*	*6*
BDL				*2*	*0*	*3*	*6*	*0*	*0*	*0*	*0*	*2*	*1*	*0*
Detection limit				2.0	10	0.10	0.10	10	0.05	0.05	0.05	2.0	10	4.4

S.C. = specific conductance (microsiemens); ppm/s = parts per million in solution; ppb/s = parts per billion in solution; TDS = total dissolved solids (sum of cations measured); Average = average value for samples above detection limit; Minimum = minimum value; Maximum = maximum value; Count = actual number of samples analyzed (not ppm/s or ppb/s); BDL = number of samples below detection limit (not ppm/s or ppb/s).

Water chemistry

The pH of pore water tends to increase near the base of the peat and in the mineral substrate in each deposit (Figs. 7A, 7B, and 7C). The increase in pH with depth in the peat column is more pronounced in the interior of each peat deposit than at the margins (Table 2; Figs. 7A, 7B, and 7C). Profiles from representative sample sites from the interior of each of the three peat deposits show the trend of increasing pH near the base of the peat (Figs. 8A, 8B, and 8C). Unfortunately, our water sampling device was not able to penetrate below 10 m, and so we can only speculate on the pH near the base of the peat in the interior of the Siak peat deposit where peat exceeded 10 m in thickness (Fig. 7A).

Specific conductance measurements were made on all the water samples as an approximation of the concentration of total dissolved ions in the water (Hem, 1989). The pH increase near the base of the peat and in the substrate is probably caused by buffering of the organic acids by the increased concentration of dissolved solids (Figs. 8A, 8B, and 8C).

Concentrations of major cations were measured in 49 peat-water samples and 10 substrate–pore-water samples; averages are reported in parts per million in solution (ppm/s) or parts per billion in solution (ppb/s) with respective detection limits (Table 2). More than half of the peat-water samples analyzed had concentrations of Si, Al, Fe, and K below the detection limit (Table 2). The highest concentrations of total dissolved cations in peat water are in the bottom of the peat in the interior of each peat deposit (Table 2; Figs. 8A, 8B, and 8C). Near the margins of the peat deposits, the increase of total cation concentration near the base of the peat is less pronounced (Table 2). Concentrations of dissolved Si, Ti, Mn, Mg, Ca, Na, K, and P are generally higher in substrate pore water than in peat water, whereas concentrations of dissolved Al and Fe may be higher in the peat water than in the substrate pore water (Table 2).

Na^+ has the highest concentration and, in general, Mg^{++} is greater than Ca^{++} in the peat-water samples (Table 2). In 10 out of 49 peat-water samples analyzed for dissolved cations, the general trend is reversed so that $Ca^{++} > Mg^{++}$ (Neuzil et al., 1993). These 10 water samples are from sites that are inland, near cultivation or roads, or near the margin of the peat deposit. Pore water in the sediments below the peat deposits shows $Mg^{++} > Ca^{++}$ in 8 out of 10 samples (Neuzil et al., 1993).

DISCUSSION OF PEAT GEOCHEMISTRY

An allogenic source and autogenic emplacement of mineral matter in the peat deposits are consistent with the observed mineral-matter content and composition and the hydrology of the peat deposits. Analogous mineral-matter facies in coal may be derived from peat deposits such as these.

Allogenic sources of mineral matter in peat

Mineral-matter composition in bottom peat appears to be similar to that in the substrate (Fig. 4). The average abundances of major elements in five substrate samples show a relative concentration of Si > Al > Fe > K, Mg, Ca > Ti, Na > P, Mn (Table 1), which is similar to average crustal abundance. The substrate is generally silt- to clay-size sediments, and its mineralogy is dominated by quartz with lesser amounts of kaolinite and other clays, determined by X-ray diffraction (Cecil et al., this volume). We interpret that the mineral matter in bottom peat is dominated by allogenic aluminosilicate minerals from the substrate or perhaps from laterally adjacent sediments. This allogenic mineral matter was incorporated during the initial planar topogenous stage of peat formation by bioturbation of roots penetrating to the substrate and possibly by flooding of surface water from laterally adjacent sedimentary environments.

A mixed allogenic source for the mineral matter in middle and top peat is suggested by the presence of mineral grains in peat and the Mg:Ca ratio in peat water. Amorphous and crystalline mineral matter such as glass shards, quartz, mica, and potassium feldspar, has been observed to constitute approximately 10% of the low-temperature ash of Siak peat and is interpreted to be infall from dust or volcanic ash (Ruppert et al., this volume). The marked decrease from bottom to middle peat of ash, Si, Ti, Al, and K (Table 1 and Figs. 3 and 4) suggests a decrease of aluminosilicate minerals from allogenic sources. As the peat thickness increased, root bioturbation no longer penetrated to the substrate and as the upper surface of the peat became raised or domed, surface waters no longer flowed into the peat deposit carrying detrital sediments. Airborne dust and volcanic ash infall may be the major or only source of allogenic aluminosilicate minerals in the middle and top peat.

The source of water, and thus of dissolved cations entering the peat, appears to be dominated by rainfall. The low pH and low concentration of total dissolved cations through the peat column, almost to the mineral substrate, suggest that these peat deposits are rainfall dominated and the direction of ground-water flow is downward as recharge. Similar observations and interpretations have been made in temperate-climate peat deposits by Shotyk (1989) and Siegel (1988a, 1988b). Ground-water discharge from the substrate up into the peat seems unlikely, given the cross-sectional geometry of the peat domes, and appears to be a relatively minor mechanism of mass transport of cations into the peat. The water table in a domed peat deposit on an island (Bengkalis) and a peninsula (Keramat) should preclude upward ground-water discharge except at the margins, because the island or peninsula is above the surrounding base level, and there is no regional ground-water flow. The increase in pH and total dissolved cations toward the bottom of the peat in the interior of Bengkalis and Keramat is probably the result of diffusion from the substrate pore water to the peat pore water along the concentration gradient and advective circulation cells within the peat as suggested for Bengkalis by Romanowicz et al. (1991).

Seawater appears to be the source of alkali metal and alkaline earth cations in rain and in the substrate pore water. Marine aerosols carry salts from seawater into the atmosphere, where they may become sites of nucleation for rain drops. The sedi-

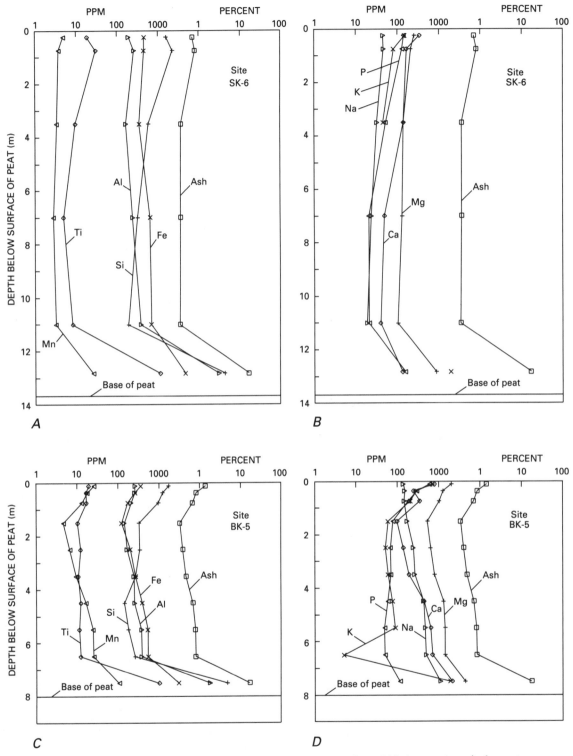

Figure 3. Ash yield (percent in dry peat, log scale) and concentrations of 10 elements (ppm in dry peat, log scale) versus depth for three representative interior sites (on this and facing page). A and B, Siak Kanan site SK-6 (inland); C and D, Bengkalis Island site BK-5 (coastal); E and F, Teluk Keramat site WK-3 (coastal). Figures 3A, 3C, and 3E illustrate Si, Ti, Al, Fe, Mn, and ash yield. Figures 3B, 3D, and 3F illustrate Mg, Ca, Na, K, P, and ash yield.

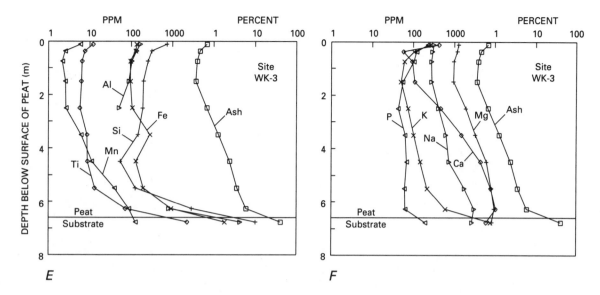

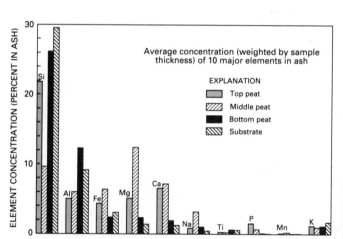

Figure 4. Bar graph of average concentration (weighted by sample thickness) of 10 major elements in ash from top, middle, and bottom peat and substrate samples (from Table 1) normalized to 100% ash.

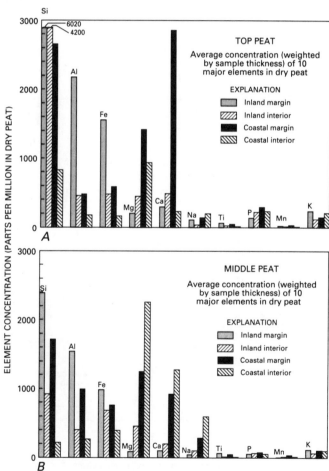

Figure 5. Bar graph of average concentration (weighted by sample thickness) of 10 major elements in peat from inland margin, inland interior, coastal margin, and coastal interior sites. A, top peat horizon; B, middle peat horizon.

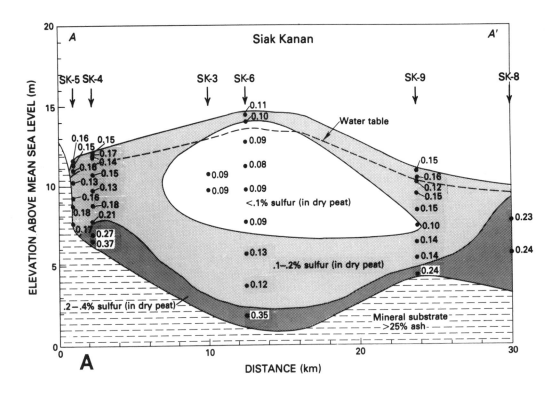

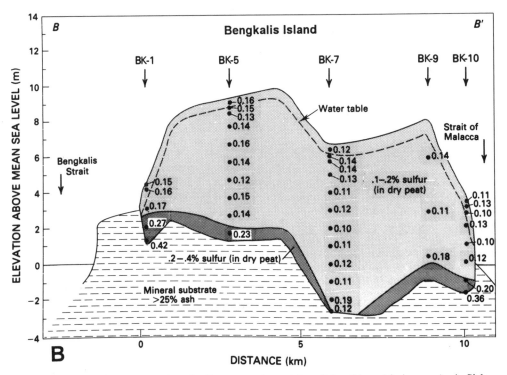

Figure 6. Generalized cross section of sulfur content in peat deposit (on this and facing page). A, Siak Kanan; B, Bengkalis Island; C, Teluk Keramat. Dots indicate center of sample interval. Data is percent sulfur in dry peat. Shading indicates increasing sulfur content.

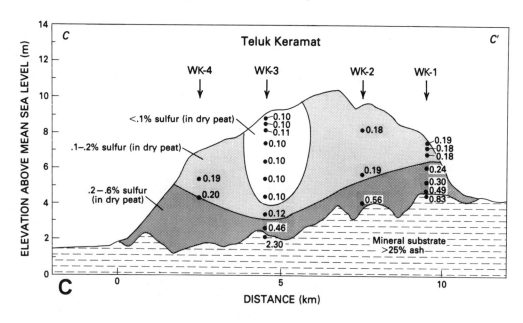

ments below the peat in this study are known to have been deposited in shallow marine conditions (Cecil et al., this volume), and marine or brackish water could have been trapped in the pore spaces. Since deposition of the substrate, sea level has dropped slightly and the original pore water has been diluted by rainwater infiltration, either directly on exposed subaerial surfaces or indirectly through the peat deposits. The dissolved alkali metal and alkaline earth elements tend to increase in concentration with depth in the peat (Figs. 8A, 8B, and 8C) and the increase is greater in the interior than at the margins of the deposits (Table 2). The concentration of alkali metal and alkaline earth cations in peat water is less than that in substrate pore water, which in turn is less than that in seawater. However, the ratio of the cations Mg:Ca:Na:K is similar in each of these types of water (Table 3). Thus, alkali metal and alkaline earth elements appear to enter the peat deposit as dissolved cations via rainfall or by diffusion from substrate pore water at ratios similar to those in seawater and at concentrations much lower than seawater.

Autogenic process/emplacement of mineral matter in peat

In the very-low-ash peat samples that contain minor amounts of crystalline allogenic minerals (Ruppert et al., this volume), the contribution of noncrystalline inorganic material to the mineral matter in peat needs to be evaluated (Kane and Neuzil, this volume). Mineral matter in peat is calculated from major oxide composition of ash. The ash may contain (1) S and N that were organic S and N in peat, (2) inorganic constituents that were dissolved ions in peat water, (3) inorganic constituents that were exchanged or complexed onto organic compounds of peat, (4) discrete inorganic grains and inorganic constituents from organo-metallic compounds of biogenic origin, and (5) discrete crystalline minerals of geologic origin that have all been concen-

trated from wet peat by the ashing process. We did not distinguish between (1) organically bound and readily soluble metals (sulfate and nitrate compounds that are artifacts of ashing, metals that were dissolved ions in peat water, metals that were cation exchanged and chelated on peat, and metals that were in organo-metallic compounds) and (2) acid-leach insoluble minerals (biogenic and geologic minerals) in the ash as suggested by Hill and Siegel (1991). However, dissolution in water of low-temperature ash from low-ash peat revealed that large portions of the ash, approximately 90%, were soluble in water (Ruppert et al., this volume).

Sulfur (and nitrogen?) in the peat may be partially retained in the high-temperature ash and constitute an average of 15% of the ash as sulfates (nitrates? and hydrated compounds?) (Kane and Neuzil, this volume).

Dissolved cations in peat water may contribute to ash yield because they are dried onto the peat during sample preparation for ashing. However, calculations indicate that Na was the only element with a high enough concentration in peat water to sometimes contribute >10% of that element's mineral-matter content in peat (Neuzil et al., 1993). Sodium tends to be a small component of the mineral matter in low-ash peat samples (Table 1), and thus Na and other cations dissolved in peat water are a minor component of the mineral matter in peat.

Metals that are held on the peat by cation exchange may make a significant contribution to the mineral matter of the low-ash middle horizon of ombrogenous peat in Indonesia. Inorganic constituents in peat held by organic compounds through processes of cation exchange or chelation (organo-metallic complexes) are isolated in the ashing process. In Table 4, average concentrations of elements in each peat horizon are converted to units of milliequivalents per 100 g dry peat (meq/100 g dry peat). In the middle horizon the average concentration of each of the 10

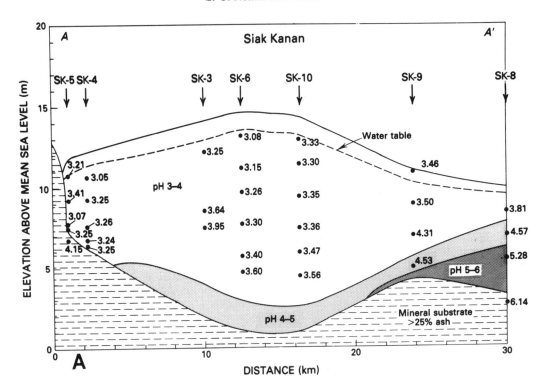

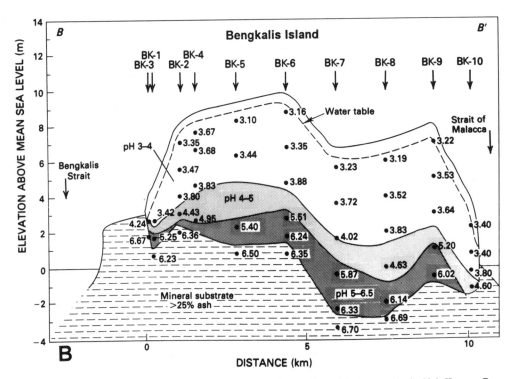

Figure 7. Generalized cross section of pH in peat deposit (on this and facing page). A, Siak Kanan; B, Bengkalis Island; C, Teluk Keramat. Shading indicates increasing pH.

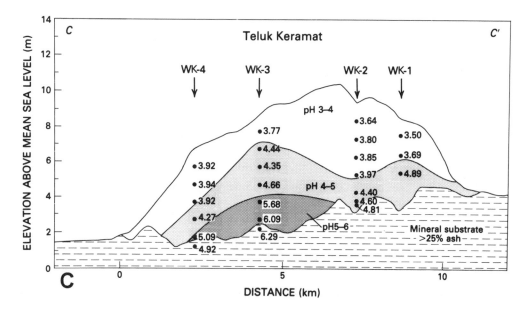

elements, except possibly Si, is generally <10 meq/100 g dry peat. These concentrations are within a range that can plausibly be attributed to cation exchange. Suhardjo and Widjaja-Adhi (1976) measured the cation exchange capacity of the surface 30 cm of peat in four peat deposits in Riau, Sumatra, and found a range of 114–137 meq/100 g dry peat. However, the base saturation was <10%. Thus, the peat contained about 10–15 meq/100 g dry peat of base cations held by cation exchange. Cations may be removed from the peat water and exchanged onto the peat until the peat becomes saturated (base saturation). The cation exchange capacity of peat is enhanced by three factors, which may increase with depth in the peat deposits in this study: (1) an increase in pH, (2) an increase in cation concentration, and (3) an increase in humification of peat (Supardi et al., this volume). Therefore, the amount of cations held by cation exchange in the middle peat is probably at least 10–15 meq/100 g dry peat and may account for more than a third of the mineral matter in the middle peat (Table 4).

Mineral matter in peat is also the result of biological processes. Amorphous or crystalline mineral matter of biologic origin, such as siliceous plant phytoliths and chrysophyte cysts, has been observed in low-temperature ash of Siak peat (Ruppert et al., this volume). The presence of siliceous plant phytoliths and chrysophyte cysts in Siak peat indicates that some plants and microorganisms are capable of precipitating silica from low concentration of silica in peat water. Furthermore, the high Ca, K, and P concentration at the surface of the peat in the root-mat zone (Figs. 3B, 3D, and 3F) may be the result of "element lifting" and recycling by plants (Sillampää, 1972) that concentrate inorganic constituents in leaf litter and live rootlets.

The low allogenic aluminosilicate mineral content in the middle peat may be caused in part by leaching of silicon and aluminum from quartz, glass shards, and aluminosilicate minerals. Etching, dissolution, and precipitation features have been observed on quartz and aluminosilicate minerals in the substrate by scanning electron microscopy (Ruppert et al., this volume). This is consistent with observations of quartz dissolution (Bennett and Siegel, 1987; Bennett et al., 1988) and enhanced aluminum mobility (Lind and Hem, 1975; Fein, 1991) in a near neutral pH range in the presence of organic acids. Silicon also may be leached from mineral matter where conditions are near neutral in the middle horizon. Silicon content in peat decreases with depth in the middle peat horizon. The decrease is most pronounced at sites in the interior of the deposits (Figs. 3A, 3C, and 3E). The lowest amount of Si is in the interior of the coastal deposits, where the increase of pH with depth is greatest (Neuzil, et al., 1993). Dissolved silicon concentrations are highest near the base of the peat where conditions are more neutral (pH 5–6; Table 1). The dissolved silicon may be from leaching and dissolution of aluminosilicate minerals in the peat. However, it is also possible that the source of the dissolved silicon is diffusion from substrate pore water.

Coal presursor mineral-matter facies

The inorganic constituents in a peat deposit are the primary source for mineral matter in coal (Spears, 1987). Therefore, mineral-matter facies in peat may result in mineral-matter facies in coal. A higher ash yield is observed in the top as compared to the middle, and at the margins as compared to the interior of each peat deposit. The higher ash yield observed in the top peat may be the reuslt of plant recycling that concentrated mineral matter in the root mat and leaf litter and/or dust infall that has not been leached. If the input of allogenic mineral matter from eolian dust and rainfall is the same at the margins and the interior of each peat deposit, then the higher ash yield at the margins may be attributed to a lower net preservation of organic matter at the thinner margins (Supardi et al., this volume) or a higher rate of

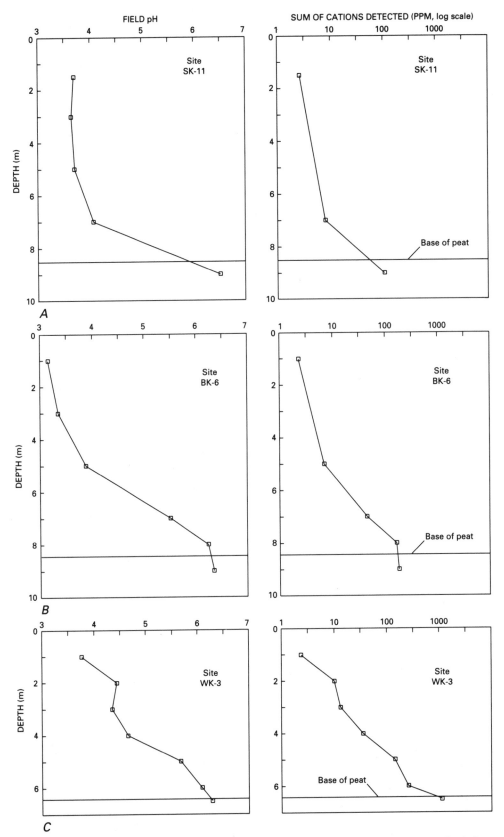

Figure 8. Profiles illustrating pH and total dissolved cation concentration (log scale) versus depth for three representative interior sites. A, Siak Kanan site SK-11 (inland); B, Bengkalis Island site BK-6 (coastal); C, Teluk Keramat site WK-3 (coastal).

TABLE 3. CONCENTRATION OF Mg, Ca, Na, AND K AND RATIO OF Mg:Ca:Na:K

	Refs.*	No. of Samples	Mg (ppm)	Ca (ppm)	Na (ppm)	K (ppm)	Ratio Mg:Ca:Na:K (relative concentration)			
							Mg :	Ca :	Na :	K
Rain (coastal)	A	n.a.[†]	0.4	0.3	3.0	0.2	0.13	0.10	1.00	0.07
Seawater	B	n.a.[†]	1,350	400	10,500	380	0.13	0.04	1.00	0.04
Igneous rocks	C	n.a.[†]	17,600	36,200	28,100	25,700	0.63	1.29	1.00	0.91
Sandstones	C	n.a.[†]	8,100	22,400	3,870	13,200	2.09	5.79	1.00	3.41
Top peat water	Table 2	8	0.55	0.62	3.48	2.0	0.16	0.18	1.00	0.57
Middle peat water	Table 2	28	1.30	0.56	10.91	4.6	0.12	0.05	1.00	0.42
Bottom peat water	Table 2	13	5.38	2.47	31.29	11.9	0.17	0.08	1.00	0.38
Substrate water	Table 2	10	23.34	9.56	132.40	21.6	0.18	0.07	1.00	0.16
Top peat	Table 1	32	779	1,001	125	173	6.23	8.01	1.00	1.38
Middle peat	Table 1	51	1,209	702	309	91	3.91	2.27	1.00	0.29
Bottom peat	Table 1	10	3,204	2,633	1,344	1,522	2.38	1.96	1.00	1.13
Substrate	Table 1	5	6,666	5,985	2,102	8,123	3.17	2.85	1.00	3.86

*A = Berner and Berner, 1987; B = Hem, 1989; C = Horn and Adams, 1966.
[†]n.a. = not available.

TABLE 4. AVERAGE ELEMENT CONCENTRATIONS IN PEAT HORIZONS WEIGHTED BY SAMPLE THICKNESS

	P	Si	Ti	Al	Fe	Mn	Mg	Ca	Na	K
Parts per million in dry peat										
Top peat	228	3,314	38	766	663	18	779	1,001	125	173
Middle peat	63	935	23	584	622	17	1,209	702	309	91
Bottom peat	117	35,956	859	16,920	3,335	98	3,204	2,633	1,344	1,522
Substrate	207	142,198	2,614	44,201	14,736	163	6,666	5,985	2,102	8,123
Milliequivalents per 100 grams dry peat										
Top peat	3.7	47	0.3	9	2.4	0.1	6	5	0.5	0.4
Middle peat	1.0	13	0.2	6	2.2	0.1	10	4	1.3	0.2
Bottom peat	1.9	512	7.2	188	11.9	0.4	26	13	5.8	3.9
Substrate	3.3	2,025	21.8	491	52.8	0.6	55	30	9.1	20.8

mineral dissolution and leaching in the interior, and perhaps redeposition near the margin, resulting in the lateral gradient in mineral-matter concentration.

The abundance of the six dominant inorganic elements in the top and middle peat horizons illustrates the change in mineral-matter composition among the peat deposits. The amounts of Si, Al, and Fe are greater in inland than in coastal deposits for each sample group (margin and interior locations) in both the top and middle horizons (Figs. 5A and 5B). The amounts of Mg, Ca, and Na are greater in coastal than in inland deposits for both margin and interior locations in the middle horizon (Fig. 5B), and generally also in the top horizon (Fig. 5A). We do not know why there is a difference in the distribution of Si, Al, Fe, Mg, Ca, and Na among the different peat deposits. One possible explanation is that deposits farther from the sea receive a *slightly* larger influx of dust, whereas deposits geographically closer to the sea receive a *slightly* larger influx from either marine

aerosols or from dilute fossil marine substrate pore water. It is also possible that the influx of dust is the same for inland and coastal peat deposits but that the higher influx of base cations from marine aerosols or diffusion from the substrate leads to slightly less acidic conditions favorable to slightly enhanced leaching of aluminosilicate mineral matter (Bennett and Siegel, 1987; Bennett et al., 1988; Lind and Hem, 1975; Fein, 1991) in coastal deposits.

The potential for a change in ash mineralogy within a coal bed is clearly demonstrated by the observed change in composition of mineral matter within each peat deposit. Although the average ash yield and Si content decrease from the top to the middle peat, the average content (ppm in dry peat), and thus also the proportion, of Al, Fe, Mg, Ca, and Na may increase from the top to the middle peat in coastal deposits or interior locations (Table 1; Figs. 5A and 5B). Similarly, the Si, Al, and Fe contents decrease from the margin to the interior in both top and middle horizons of both inland and coastal deposits (Table 1; Figs. 5A and 5B). However, the Mg, Ca, and Na contents increase from margin to interior locations in the middle horizon of both inland and coastal deposits (Table 1; Figs. 5A and 5B). Thus, low-temperature ash of a coal bed resulting from peat deposits similar to the domed ombrogenous peat in Indonesia might show a decrease in quartz and an increase in the proportion of aluminosilicate minerals such as kaolinite and illite in the middle horizon compared to the top of the coal bed. Ultimately, Ca, Mg, and Fe may reside in carbonate minerals and comprise a larger proportion of the minerals in the middle horizon, interior locations, or coastal settings of coal beds.

These peat deposits appear to receive inorganic constituents from marine sources, via marine aerosols or diffusion from fossil substrate seawater. However, seawater does not enter the peat directly because of the domed geometry and hydraulic head of the fresh-water lens in the peat. Sulfate introduced at low concentrations from marine sources would be assimilated easily by plants and incorporated into the peat. However, sulfur content in peat is low, <1%, and averages 0.13%–0.14% for top and middle peat in the geographic interior of all three of the deposits (Table 1). High sulfur content, >1%, was not found in any peat samples and only in one substrate sample. These peat deposits are adjacent to and coeval with marine environments and lie on marine sediments. However, no significant difference in sulfur content was found for the interior peat samples as a function of distance from the sea (Neuzil, 1990).

The quality of these peat deposits demonstrate that massive amounts of low-ash peat can develop close to marine conditions and above a marine substrate without high sulfur or pyrite contents. During burial and coalification, marine or high-sulfate ground water may influence the peat and overprint the original sulfur content and forms present in the peat stage. However, the greater porosity of peat, ~85%–97% (Päivänen, 1976), compared to unconsolidated sand, silt, or clay, ~25%–70% (Freeze and Cherry, 1979), will likely result in pore fluids being expressed from peat rather than entering peat during initial burial and compaction. The resulting coal may remain low in sulfur and pyrite, similar to the original peat. This type of domed ombrogenous peat may explain the origin of some low-sulfur coal deposits that are overlain by rocks of marine origin.

CONCLUSIONS

The low-ash yield and low-sulfur content of the peat and the low pH and low dissolved cation concentration of the peat water show that the net flux of inorganic constituents into the three tropical domed ombrogenous peat deposits is highly restricted. Mineral-matter facies observed in the peat deposits include a low-ash and low-sulfur horizon at the top; a very- low-ash and low-sulfur horizon through the middle; and a medium- to high-ash and moderate-sulfur horizon near the bottom of the peat profile. In the middle horizon in the geographic interior of these domed ombrogenous peat deposits, the ash yield and sulfur content are very low (average 0.6% ash and 0.13% sulfur in an inland

deposit) and increase only slightly with proximity to the sea (average 1.4% ash and 0.14% sulfur in two coastal deposits).

The relative proportions of the six dominant inorganic constituents show general trends within the peat. Peat deposits that are closer to the sea tend to have higher concentrations of dissolved Na, Mg, and Ca; higher amounts of Mg, Ca, and Na in mineral matter; and lower amounts of Si, Al, and Fe in mineral matter than deposits that are farther from the coast. In general, Si dominates near the top, the proportions of Mg, Ca, Al, and Fe increase as Si decreases down through the middle of the peat profile, and Si and Al are dominant at the bottom. The middle interior of each deposit has a lower content of Si, Al, and Fe and a higher content of Mg, Ca, and Na than the middle margin.

Prediction of coal quality trends within a coal bed may be possible based on quality trends within modern analogue peat deposits. In domed, rainfall-dominated peat deposits, the geochemical controls on mineral matter are dominantly autogenic, independent of surrounding depositional environments. Therefore, quality predictions for coal derived from domed peat deposits cannot be based on facies relations with enclosing sedimentary rocks. Prediction of coal quality should be based on autogenic geochemical processes and controls of peat formation, which can be recognized in the composition and spatial distribution of mineral matter in peat and coal. Future studies of mineral matter in coal to interpret the original peat type should focus on distinguishing between mineral matter from the peat stage and subsequent additions or losses through diagenesis.

ACKNOWLEDGMENTS

The U.S. Geological Survey (USGS) and Directorate of Mineral Resources (DMR) supported their respective personnel in the field during three field investigations (February 1987, July–August 1987, and July–August 1988) to collect the peat and water samples discussed in this paper. Gary Hill and Hal Gluskoter at the USGS and Salman Padmanagara and Hardjono at the DMR supported this international, cooperative research project, which studied nearly inaccessible Indonesian peat deposits as modern analogue settings for ancient coal deposits. Generous logistical support was received both from P. T. Caltex Pacific Indonesia, coordinated by Purnomo Prijosoesilo, for the Siak Kanan study area in 1987 and from P. T. Arutmin, coordinated by David Mathew, for all three investigations. Field work was far from easy and we owe much to the labors of those who participated, especially Limg Meng Sze Wu (P. T. Arutmin); A. D. Subekty, T. Widjaya, M. Romli, and M. Tjahyana (DMR); and C.F. Eble, T. A. Moore, W. F. Orem, and C. Wnuk (USGS). Laboratory sample preparation and analyses were carried out by J. Pontolillo, C. Skeen, M. Doughten, M. Rubin, and J. McGeehin (USGS). The text benefitted from many reviewers' insightful critiques; I would like to thank reviewers S. Altschuler, R. Finkelman, and especially D. I. Siegel. Many other people contributed to this project in Indonesia and in the United States, and although we can not thank them all individually, to them we are indebted.

REFERENCES CITED

Anderson, J.A.R., 1961, The ecology and forest types of the peat swamp forests of Sarawak and Brunei in relation to their silviculture [Ph.D. thesis]: Edinburgh, University of Edinburgh, 117 p.

Anderson, J.A.R., 1964, The structure and development of the peat swamps of Sarawak and Brunei: Journal of Tropical Geography, v. 18, p. 7–16.

Anderson, J.A.R., 1976, Observations on the ecology of five peat swamp forests in Sarawak and Kilamantan, *in* Peat and podzolic soils and their potential for agriculture in Indonesia: Bogor, Indonesia, Soil Research Institute Bulletin 3, p. 45–55.

Anderson, J.A.R., 1983, The tropical peat swamps of western Malesia, in Gore, A.J.P., ed., Mires: Swamp, bog, fen, and moor: Amsterdam, Elsevier, Ecosystems of the World, v. 4B, p. 181–200.

Anderson, J.A.R., and Muller, J., 1975, Palynological study of a Holocene peat and a Miocene coal deposit from NW Borneo: Review of Paleobotany and Palynology, v. 19, p. 291–351.

Andriesse, J. P., 1988, Nature and management of tropical peat soils: Rome, Food and Agriculture Organization of the United Nations FAO Soils Bulletin 59, 165 p.

Bates, R. L., and Jackson, J. A., eds., 1987, Glossary of geology, third edition: Alexandria, Virginia, American Geological Institute, 708 p.

Bennett, P., and Siegel, D. I., 1987, Increased solubility of quartz in water due to complexing by organic compounds, Nature, v. 326, p. 684–688.

Bennett, P. C., Melcer, M. E. Siegel, D. I., and Hassett, J. P., 1988, The dissolution of quartz in dilute aqueous solutions of organic acids at 25 °C: Geochmica et Cosmochimica Acta, v. 52, p. 1521–1530.

Berner, E. K., and Berner, R. A., 1987, The global water cycle: Geochemistry and environment: Englewood Cliffs, New Jersey, Prentice Hall, 397 p.

Biswas, B., 1973, Quaternary changes in sea level in the South China Sea: Geological Society of Malaysia Bulletin 6, p. 229–256.

Bord na Móna, 1984, Fuel peat in developing countries: The World Bank: Dublin, Ireland, Bord na Móna, 136 p.

Brunig, E. F., 1971, On the ecological significance of drought in the equatorial wet evergreen (rain) forest of Sarawak (Borneo), *in* Transactions, First Aberdeen-Hull Symposium on Malesian Ecology: Hull, England, University of Hull, Department of Geography Miscellaneous Series 11, p. 66–97.

Cameron, C. C., Esterle, J. S., and Palmer, C. A., 1989, The geology, botany, and chemistry of selected peat-forming environments from temperate and tropical latitudes: International Journal of Coal Geology, v. 12, p. 105–156.

Casagrande, D. J., 1987, Sulfur in peat and coal, *in* Scott, A. C., ed., Coal and coal-bearing strata: Recent advances: London, Blackwell, Geological Society of London Special Publication 32, p. 87–105.

Cecil, C. B., Stanton, R. W., Dulong, F. T., and Renton, J. J., 1982, Geologic factors that control mineral matter in coal, *in* Filby, R. H., Carpenter, B. S., and Ragaini, R. C., eds., Atomic and nuclear methods in fossil energy research: New York, Plenum Publishing Corporation, p. 323–335.

Cecil, C. B., Stanton, R. W., Neuzil, S. G., Dulong, F. T., Ruppert, L. F., and Pierce, B. S., 1985, Paleoclimate controls on Late Paleozoic sedimentation in the central Appalachian basin (U.S.A.): International Journal of Coal Geology, v. 5, p. 195–230.

Chason, D. B., and Siegel, D. I., 1986, Hydraulic conductivity and related physical properties of peat, Lost River Peatland, northern Minnesota: Soil Science, v. 142, p. 91–99.

Damman, A.W.H., 1978, Distribution and movement of elements in ombrotrophic peat bogs: Oikos, v. 30, p. 480–495.

Damman, A.W.H., 1987, Variation in ombrotrophy: Chemical differences among and within ombrotrophic bogs, *in* Rubec, C.D.A., and Overend, R. P., compilers, Proceedings, Symposium '87 Wetlands/Peatlands, August 23–27, 1987: Edmonton, Alberta, Canada, International Peat Society et al., p. 85–94.

Day, J. H., Rennie, P. J., Stanek, W., and Raymond, G. P., eds., 1979, Peat testing manual: Natural Research Council of Canada, Associate Committee on Geotechnical Research Technical Memorandum No. 125, 193 p.

Driessen, P. M., 1977, Peat soils, *in* Proceedings, Soils and Rice Symposium: Manilla, p. 763–779.

Driessen, P. M., and Suhardjo, H., 1976, On the defective grain formation of swamp rice on peat, *in* Peat and podzolic soils and their potential for agriculture in Indonesia: Bogor, Indonesia, Soil Research Institute Bulletin 3, p. 20–44.

Driessen, P. M., Soepraptohardjo, M., and Pons, L. J., 1979, Formation, properties, reclamation and agricultural potential of Indonesian ombrogenous lowland peats, *in* Schallinger, K. M., ed., Proceedings, International Symposium, Peat in Agriculture and Horticulture, Bet Dagan, Israel, 1975: Institute of Soils and Water, Special Publication 205, p. 67–84.

Esterle, J. S., Ferm, J. C., and Yiu-Liong, T., 1989, A test for the analogy of tropical domed peat deposits to "dullingup" sequences in coal beds—Preliminary results: Organic Geochemistry, v. 14, p. 333–342.

Fein, J. B., 1991, Experimental study of aluminum-oxalate complexing at 80 °C: Implications for the formation of secondary porosity within sedimentary reservoirs: Geology, v. 19, p. 1037–1040.

Freeze, R. A. and Cherry, J. A., 1979, Groundwater: Englewood Cliffs, New Jersey, Prentice-Hall, 604 p.

Geyh, M. A., Kudrass, H. R., and Streif, H., 1979, Sea-level change during the late Pleistocene and Holocene in the Strait of Malacca: Nature, v. 278, no. 5703, p. 441–443.

Hem, J. D., 1989, Study and interpretation of the chemical characteristics of natural water, third edition: U.S. Geological Survey Water-Supply Paper 2254, 264 p.

Hill, B. M., and Siegel, D. I., 1991, Groundwater flow and the metal content of peat: Journal of Hydrology, v. 123, p. 211–224.

Horn, M. K., and Adams, J.A.S., 1966, Computer-derived geochemical balances and element abundances: Geochimica et Cosmochimica Acta, v. 30, p. 279–297.

Lind, C. J., and Hem, J. D.,1975, Effects of organic solutes on chemical reactions of aluminum: U.S. Geological Survey Water-Supply Paper 1827-G, 83 p.

Morley, R. J., 1981, Development and vegetation dynamics of a lowland ombrogenous peat swamp in Kalimantan Tengah, Indonesia: Journal of Biogeography, v. 8, p. 383–404.

Neuzil, S. G., 1990, Domed, rainfall-dominated peat deposits in Indonesia as a model for low-sulfur coal [abs.]: U.S. Geological Survey Circular 1060, p. 57 and 59.

Neuzil, S. G., and Cecil, C. B., 1984, A modern analog of low-ash low-sulfur, Pennsylvanian Age coal [abs.]: Geological Society of America Abstracts with Programs, 1984, v. 16, no. 3, p. 184.

Neuzil, S. G., Supardi, Cecil, C. B., and Kane, J. S., 1993, Inorganic geochemistry of peat and water in the Siak Kanan, Bengkalis Island, and Teluk Keramat peat deposits, Riau and Kalimantan Barat, Indonesia: U.S. Geological Survey Open-File Report 92-205 (in press).

Nieuwolt, S., 1965, Evaporation and water balance in Malaya: Journal of Tropical Geography, v. 20, p. 34–53.

Nieuwolt, S., 1968, Uniformity and variation in equatorial climate: Journal of Tropical Geography,v. 27, p. 23–39.

Päivänen, J., 1976, Bulk density as a factor describing other physical properties of peat, *in* Transactions, Working Group for Classification of Peat: Helsinki, Finland, International Peat Society Commission I, p. 40–45.

Romanowicz, E. A., Neuzil, S. G., and Siegel, D. I., 1991, Hydrogeology and geochemistry of a modern coal swamp, Bengkalis Island, Sumatra [abs.], *in* Final Program, Twelfth Annual Meeting of the Society of Wetland Scientists, May 28–31, 1991, Ann Arbor, Michigan: Ann Arbor, Society of Wetland Scientists, p. 47.

Schmidt, F. H., and Ferguson, J.H.A., 1951, Rainfall types based on wet and dry period ratios for Indonesia with western New Guinea: Kementerian Perhubungan Djawatan Meteorologi dan Geofisik, Verhandelingen no. 42, 77 p.

Shapiro, L., and Brannock, W. W., 1962, Rapid analysis of silicate, carbonate, and phosphate rocks: U.S. Geological Survey Bulletin 1144-A, 56 p.

Shotyk, W., 1988, Review of the inorganic geochemistry of peats and peatland waters: Earth Science Reviews, v. 25, p. 95–176.

Shotyk, W., 1989, The chemistry of peatland waters: Water Quality Bulletin, v. 14, p. 47–58, and 103.

Siegel, D. I., 1988a, A review of the recharge-discharge function of wetlands, *in* Hook, D. D. and 12 others, eds., The ecology and management of wetlands: Portland, Oregon, Timber Press, v. 1, p. 59–67.

Siegel, D. I., 1988b, The recharge-discharge function of wetlands near Juneau, Alaska: Part II: Geochemical investigations: Journal of Ground Water, v. 26, p. 580–596.

Siegel, D. I., and Glaser, P. H., 1987, Groundwater flow in a bog-fen complex, Lost River Peatlands, northern Minnesota: Journal of Ecology, v. 75, p. 743–754.

Sillampää, M., 1972, Distribution of trace elements in peat profiles: Proceedings, Fourth International Peat Congress, v. 5, p. 185–191.

Soedjono, K., Supardi, Wijaya, T., Gluskoter, H. J., Cecil, C. B., and Neuzil, S. G., 1988, Peat deposit at Teluk Keramat, West Kalimantan Province: Bandung, Indonesia, Directorate of Mineral Resources, Subdirectorate of Coal and Peat Exploration, 38 p.

Spears, D. A., 1987, Mineral matter in coals, with special reference to the Pennine coalfields, *in* Scott, A. C., ed., Coal and coal-bearing strata: Recent advances: London, Blackwell, Geological Society of London Special Publication 32, p. 171–185.

Suhardjo, H., and Widjaja-Adhi, I.P.G., 1976, Chemical characteristics of the upper 30 cm of peat soils from Riau, *in* Peat and podzolic soils and their potential for agriculture in Indonesia: Bogor, Indonesia, Soil Research Institute Bulletin 3, p. 74–92.

Supardi, 1988, Endapan gambut di Pulau Bengkalis, Riau: Bandung, Indonesia, Direktorat Sumberdaya Mineral, map, scale 1:100,000 (in Indonesian).

Supardi and Priatna, 1985, Endapan gambut didaerah Siakkanan, Riau (The peat deposit in the Siakkanan area, Riau): Bandung, Indonesia, Directorate of Mineral Resources, Department of Mines and Energy Project for coal inventory, 20 p. (in Indonesian).

Tjia, H. D., 1975, Holocene eustatic sea levels and glacio-eustatic rebound: Zeitschrift für Geomorphologie N.F., supplement, v. 22, p. 57–71.

Tjia, H. D., Sujitno, S., Suklija, Y., Harsono, R.A.F., Rachmat, A., Hainim, J., and Djunaedi, 1984, Holocene shorelines in the Indonesian Tin Islands: Modern Quaternary Research in Southeast Asia, v. 8, p. 103–117.

White, D., and Theissen, R., 1913, The origin of coal: U.S. Bureau of Mines Bulletin 38, 390 p.

Whitmore, T. C., 1975, Tropical rain forests of the Far East: Oxford, England, Clarendon Press, 282 p.

Wilford, G. E., 1959, Radiocarbon age determination of Quaternary sediments in Brunei and northeast Sarawak: British Borneo Geological Survey Annual Report 1959, p. 16–20.

MANUSCRIPT ACCEPTED BY THE SOCIETY JANUARY 14, 1993

Geological Society of America
Special Paper 286
1993

General geology and peat resources of the Siak Kanan and Bengkalis Island peat deposits, Sumatra, Indonesia

Supardi and A. D. Subekty
Directorate of Mineral Resources, Jl. Soekarno-Hatta 444, Bandung, West Java 40254, Indonesia
Sandra G. Neuzil
U.S. Geological Survey, 956 National Center, Reston, Virginia 22092

ABSTRACT

Peat deposits cover 48,000 km^2 on the lowlands of Riau Province, Sumatra, Indonesia. Two areas containing typical dome-shaped peat deposits were selected for study. These peat deposits are topographically highest in the geographic interior of the deposit and are drained radially outward by blackwater streams. The source of the water in the peat is precipitation, which exceeds evapotranspiration throughout the year. In cross section, the peat deposits are biconvex; they rest on a nearly level surface, which is within a few meters of sea level. The peat accumulated in the past 5,000 years after stabilization of sea level following the rapid sea-level rise during glacial retreat. In the interior area of the peat deposits, the initial peat accumulation rate was rapid (4–5 mm/yr) for approximately 1,000 years; the rate decreased to less than 2 mm/yr for the past 3,500–4,000 years.

These peat deposits have a fibric to hemic texture with slight to moderate humification, a low ash yield, and a low sulfur content; and they contain acid water. A thin layer at the bottom of the deposits tends to be more sapric in texture, more humified, higher in ash yield, higher in sulfur content, and less acid than the overlying peat. Proximate and ultimate analyses of a suite of samples from the interior of each peat deposit show no significant differences in peat quality between the Siak Kanan and Bengkalis Island peat deposits.

A primary goal of this study was to evaluate peat resources. The 6.6×10^9 m^3 of peat in the Siak Kanan peat deposit and 3.0×10^9 m^3 of peat in the Bengkalis Island peat deposit constitute a significant fuel resource. This resource study has contributed a three-dimensional framework and peat quality data that can provide insight into the earliest stage of certain types of coal formation.

INTRODUCTION

Riau Province (95,000 km^2) in northeast Sumatra includes several islands and is almost entirely less than 100 m above sea level (Fig. 1). Peat deposits cover 51% of Riau; 93% of the deposits are ombrogenous peat swamp forests (Whitten et al., 1987). A northwesterly trending belt of lowland peat extends 450 km along the coast in Riau Province and is only briefly interrupted by seven major rivers crossing the coastal plain. Peat deposits cover most of the coastal lowlands and the islands of Bengkalis, Padang, Tebingtinggi, Rangsang, Mendol, and Rupat (Fig. 1).

The general area of investigation within Riau Province is bounded on the southwest by low hills, on the west by the Siak River, on the northeast by the Strait of Malacca, and on the south by the Kampar River. Detailed investigations discussed in this paper are limited to two peat deposits—Siak Kanan, on the

Supardi, Subekty, A. D., and Neuzil, S. G., 1993, General geology and peat resources of the Siak Kanan and Bengkalis Island peat deposits, Sumatra, Indonesia, *in* Cobb, J. C., and Cecil, C. B., eds., Modern and Ancient Coal-Forming Environments: Boulder, Colorado, Geological Society of America Special Paper 286.

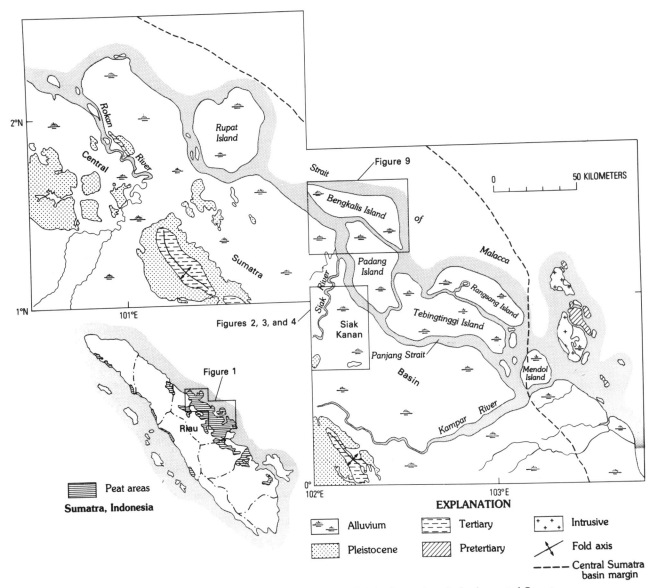

Figure 1. Location map of the Siak Kanan and Bengkalis Island peat deposits in the central Sumatra basin, Riau Province, Sumatra, Indonesia. Modified from Cameron et al. (1982a, 1982b) and Jansen et al. (1985).

coastal lowlands, and Bengkalis Island, both in Riau Province, Sumatra (Fig. 1). These study areas were chosen as representative of peat on the coastal lowlands and islands, respectively.

GENERAL GEOLOGY

Stratigraphic sketch

The peat is the youngest deposit in the central Sumatra basin. Pre-Tertiary rocks in the central Sumatra basin are dominantly metaquartzite, granite, and tuff. Tertiary rocks are mostly sedimentary (mudstones, siltstones, and sandstones, locally car-

bonaceous or tuffaceous, with wood, amber, and thin lignites). They are divided, oldest to youngest, into the Pematang, Sihapas, Telisa, and Petani Formations. These Tertiary units are economically important because they contain both oil and coal.

Quaternary deposits include Pleistocene and Holocene sediments. The Pleistocene consists of the Minas and Keramutan Formations, soft mudstones, siltstones, sands, and clayey to pebbly gravel (Cameron et al., 1982a, 1982b), which are exposed in the southwest part of the study area. Upper Pleistocene sediments called Older Alluvium are stabilized, partially lithified clays, silts, and clayey gravels having maximum elevations of approximately 5 to 7 m above present river levels. The Holocene Younger

Alluvium includes clay and silt on the coast; mud, sand, and gravel in the flood plains; and extensive freshwater peat deposits. Cameron et al. (1982a, 1982b) interpreted both the Older Alluvium and Younger Alluvium to have been deposited in paralic to fluviatile environments. New micropaleontology data, however, indicate that the sediments immediately below the peat (Younger Alluvium?) were deposited in marine to brackish conditions (Cecil et al., this volume). The term "alluvium" may, therefore, be a misnomer.

The timing is uncertain for the deposition of sediments forming the peat substrate on Bengkalis Island. One interpretation is that the sediments were deposited in a marine environment at a higher sea level during an interglacial episode. Then, during a glacial episode when sea level was lower, the coastal plain of Sumatra was dissected by downcutting processes and Bengkalis Island was cut off from the mainland as an erosional remannt (Cecil et al., this volume). Alternatively, these sediments may have been deposited from longshore currents in the Malacca and Panjang Straits as sea level rose after the last glacial maximum.

Holocene peat

Holocene peat deposits cover most of the coastal lowlands in the study area, possibly 90%. Peat accumulation started about 5,000 to 6,000 years ago on the coastal lowlands of Indonesia and Malaysia. After the last glacial cycle, sea level rose rapidly from approximately 12,000 to 6,000 years ago and has been fairly stable or has dropped slightly from 5,000 years ago to the present (Biswas, 1973; Geyh et al., 1979). Mangrove trees colonized coastal environments, trapping sediments in their root systems and extending the tidal flats (Chambers and Abdul Sobur, 1975; Whitten et al., 1987). Nipa palms colonized brackish water areas. Accumulating sediments (Andriesse, 1974; Hehuwat, 1982; Anderson, 1983) and plant debris, in combination with high rainfall and possibly a slight drop in sea level, transformed the nearly flat coastal plain of Sumatra into a large freshwater swamp. Plant debris that accumulated under these swampy conditions, in slight topographic depressions containing shallow surface water through most or all of the year, formed topogenous peat deposits. In the earliest phase of topogenous peat accumulation, plants were rooted in mineral sediments and received nutrients from both surface water and rain.

The second stage of peat formation resulted from ombrogenous conditions. Ombrogenous or domed peat deposits develop only where precipitation exceeds evapotranspiration throughout the year. Both Siak Kanan and Bengkalis Island are situated within the intertropical convergence zone throughout the year and receive rain from both the northeast and southwest monsoons so that precipitation exceeds evapotranspiration throughout the year. Plant debris accumulated rapidly, the peat increased in thickness, and the surface of the peat eventually rose above the influence of local surface water and became a raised or domed peat. Base level may have been lowered by a slight drop in sea level. In either case, rivers no longer drained into or flooded onto the peat deposits. Blackwater streams drained excess water out of the domed peat deposits in a radial drainage pattern, and the only water entering the peat deposits was rainfall. Plant roots could no longer penetrate through the thick peat to the mineral sediments below the peat. Thus, peat-forming plants were isolated from the nutrients and dissolved cations in river water and in the sediments below the peat. Nutrients available to plants were restricted to rainfall, airborne particles (dust and volcanic ash), and nutrients recycled from decayed vegetal matter. Trees continued to grow in the peat swamp forest and their plant litter accumulated as ombrogenous peat (Anderson, 1964, 1976, 1983).

Few bacteria and fungi are able to degrade organic matter in ombrogenous peat. Organic acids formed during partial degradation of plant debris remain unbuffered because of low concentrations of dissolved cations in rain and create an acid environment in ombrogenous peat. Thus, plant debris accumulates in the wet, acid environment of an ombrogenous, rainfall-dominated peat.

The degree of humification, or degradation, of peat can be categorized by three types, fibric, hemic, and sapric, based on von Post's scale of humification H1 through H10 (von Post, 1922; Farnham and Finney, 1965). Peat with low humification (H1–H3 is classified as fibric; plant remains are little humified, clearly recognizable, and not mushy. Peat with moderate humification (H4–H6) is classified as hemic; plant remains are recognizable, mushy, and not distinct. Peat with high humification (H7–H10) is classified as sapric; plant remains grade from present but not recognizable to completely humified. Transitions within types are gradual and somewhat subjective, e.g., a sample may be described as fibric H2-H3. Also, transitions between fibric, hemic, and sapric types are gradual and somewhat subjective. As a result, H3-H4 can be called fibric-hemic and H6-H7 can be called hemic-sapric. In this paper, the fibric, fibric-hemic, hemic, hemic-sapric, sapric classification scheme will be used in order to describe trends in the texture and decomposition of the peat.

THE SIAK KANAN PEAT DEPOSIT

Size and shape of the peat deposit

The 1200-km^2 peat deposit called Siak Kanan was investigated in 1984/1985 by the Directorate of Mineral Resources (DMR) (Figs. 2, 3, and 4). Approximately 230 holes were hand augured, the peat thickness was measured, and peat samples were collected and analyzed (Supardi and Priatna, 1985). Elevations and locations of the bore holes were measured by using oil-well sites for reference benchmarks to construct the topographic map (Fig. 2). An isopach map of the Siak Kanan peat deposit was compiled from the bore hole data (Fig. 3). A topographic map of the peat basal contact was constructed by subtracting the peat thickness from the peat surface topography (Fig. 4). In 1987, the study was continued at a more detailed level by the DMR in cooperation with the U.S. Geological Survey (USGS). Peat and water samples were collected at 11 sites (Fig. 3) and analyzed by

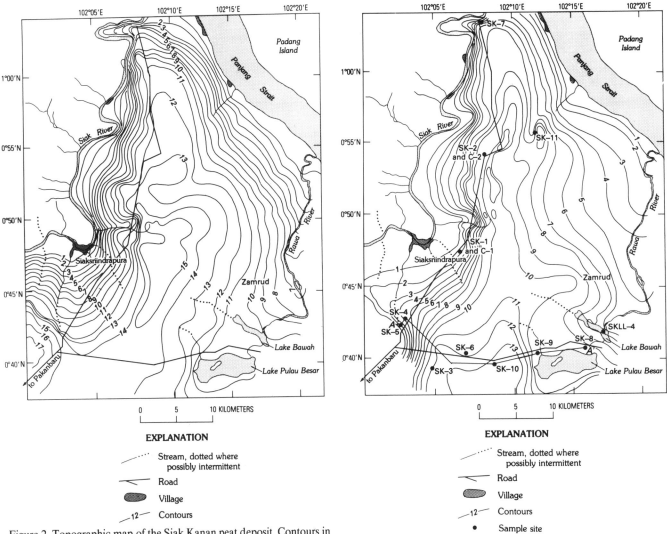

Figure 2. Topographic map of the Siak Kanan peat deposit. Contours in meters above mean sea level. Modified from Supardi and Priatna (1985).

Figure 3. Isopach map of the Siak Kanan peat deposit. Isopachs in meters. Section A–A′ is shown in Figure 8. Modified from Supardi and Priatna (1985).

DMR in Bandung, West Java, Indonesia; USGS in Reston, Virginia; and Dickinson Laboratories, El Paso, Texas. (Any use of trade names in this report is for descriptive purposes only and does not constitute endorsement by the U.S. Geological Survey or the Geological Society of America.)

The topographic map of the Siak Kanan peat deposit shows the domed or raised morphology of the peat, which has a maximum elevation of 15 m above sea level (Fig. 2). The surface of the peat slopes more steeply in the direction of the Siak River than on the seaward side toward Panjang Strait. Streams drain out of the peat dome on the west, north, and east sides in a radial drainage pattern characteristic of domed peat deposits. The peat dome continues to the south beyond the surveyed area. The radial drainage pattern on photogeologic maps (Cameron et al., 1982a) suggests that the peat deposit between the Siak and Kampar River estuaries (Fig. 1) consists of four coalesced peat domes.

The isopach map of the Siak Kanan peat deposit (Fig. 3) shows a maximum peat thickness of more than 13 m. Peat thickness, roughly corresponding to the surface elevation of the peat, thins rapidly at the margin near the Siak River and near the low hills in the southwest corner of the Siak Kanan study area. Its thickness decreases more gradually toward the seaward edge along Panjang Strait.

The elevation of the bottom of the peat (Fig. 4) ranges from 1 to 9 m above mean sea level. The highest elevations of the peat/substrate contact are along the west side of the peat deposit, on the east side of the Siak River. This ridge of sediments may have been deposited by flood waters of the Siak River as overbank deposits and crevasse splays. The sediments below the peat are mostly clay size but in some places near the Siak River are silt and fine-sand size. Alternatively, the ridge of sediments near the

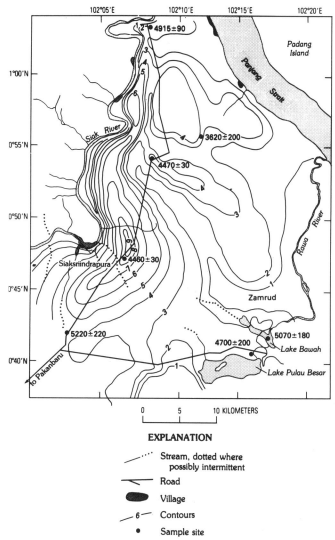

Figure 4. Map of the elevation of the base of the Siak Kanan peat deposit. Contours in meters above mean sea level. Carbon-14 dates (Table 1) are shown for samples from the base of the peat. Modified from Supardi and Priatna (1985).

Siak River may be an erosional remnant of Older Alluvium. The peat/substrate surface dips gently toward Panjang Strait, where a regression or eustatic lowering of sea level, exposing the sediments in this area to nonmarine conditions, may be the limiting factor for initiation of peat formation at lower elevations. If the coastal plain were progradational, then the age at the bottom of the peat would be older near the ridge and younger toward Panjang Strait, which does not appear to be the case.

Age and rate of peat accumulation

Carbon-14 dates were obtained for peat samples from the bottom of the peat at five sites in the Siak Kanan peat deposit (Meyer Rubin, personal communication; Table 1). The beginning of peat accumulation appears to have been almost simultaneous at four sites near the margins of the Siak Kanan peat deposit: SK-5, SK-7, SK-8, and SKLL-4 (Fig. 4). At site SK-11, the age at the bottom of the peat is somewhat younger. Previously, the gradual thinning of the peat toward the coast has been interpreted as evidence of a prograding shoreline contemporaneous with peat accumulation (Anderson, 1964, 1983). The available data for the Siak Kanan peat deposit do not support interpretations for either (1) a small peat deposit spreading laterally with time or (2) a peat deposit spreading toward the coast on a prograding coastal plain. Rather, the data indicate nearly simultaneous initiation of peat accumulation over a large area.

Diemont and Supardi (1987) studied the rate of peat accumulation at two sites in Siak Kanan (Table 2). Eleven peat samples from site C-2 with thick (10 m) peat in the interior of the peat deposit, approximately 4 km from the nearest margin, were dated by carbon-14 methods. Dates range from 4470 ± 30 yr B.P. near the basal contact to 690 ± 20 yr B.P. near the upper surface. Site C-1 may have been disturbed by humans through drainage and agriculture; it is near the west margin of the deposit where peat is thin (1.3 m). Two samples from site C-1 also indicate that peat formation started ca. 4500 yr B.P. Initial peat accumulation was rapid at both sites. At the thick interior sample site, the average rate of peat accumulation was approximately 4.3 mm/yr for the first 4 m of peat and decreased to an average of 1.7 mm/yr for the upper 6 m of peat (Fig. 5). At the thin margin of the peat deposit (C-1) if peat is still accumulating, the rate of peat accumulation is very slow, averaging 0.2 mm/yr for the upper 0.95 m (Fig. 6).

Forests are growing at the interior site C-2 and near the margin site C-1. However, in the interior of the peat deposit where the peat is thickest, the forest appears to be a padang or pole forest (Anderson, 1976) with a low canopy and small plant stature. Annual biomass production by the padang forest may be less than that in the mixed peat swamp forest at the margin. Therefore, the vegetation does not seem to explain the apparently faster rate of peat accumulation in the interior than at the margin of the pat deposit. There are several possible reasons for the difference in peat thickness and apparent rate of accumulation. In the interior of the peat dome, there may be better preservation of vegetal matter, resulting in a higher rate of peat accumulation. At the margin of the peat dome, there may be more oxidation and degradation of plants and peat or loss by flushing out fine material, resulting in a lower rate of peat accumulation. Another possible explanation for the apparently lower rate of peat accumulation at the margin of the peat deposit is that the peat was thicker at this site, but recently the uppermost peat layer has been compacted, degraded, or oxidized due to drainage and agriculture. For example, a postulated original peat surface 1 m above the present surface and a higher rate of peat accumulation, 0.4 mm/yr, are represented by dashed lines in Figure 6.

Lakes in the Siak Kanan peat

There are two lakes in the Zamrud area in the southeastern part of the Siak Kanan peat deposit (Fig. 2). Lake Pulau

**TABLE 1. CARBON-14 AGES OF PEAT SAMPLES FROM
THE SIAK KANAN AND BANGKALIS ISLAND PEAT DEPOSITS, RIAU, INDONESIA**

Lab Number*	Sample Site	Depth (m)	Average Height Above Base (m)	Sample Age (yr B.P.)	Average Rate of Peat Accretion† (mm/yr)
W-6183	SK-5	0.70 to 0.80	3.67	4,070 ± 190	
W-6189	SK-5	1.90 to 2.00	2.47	3,930 ± 200	3.2
W-6186	SK-5	3.00 to 3.10	1.37	4,900 ± 210	
W-6033	SK-5	4.33 to 4.43	0.04	5,220 ± 220	———
AA-4372	SK-7	0.695 to 0.72	0.64	4,340 ± 75	1.1
AA-4373	SK-7	1.300 to 1.31	0.04	4,915 ± 90	———
W-6034	SK-8	6.30 to 6.40	0.30	4,700 ± 200	
W-6035	SK-11	8.10 to 8.20	0.40	3,620 ± 200	
AA-4366	SKLL-4	0.060 to 0.070	0.57	1,105 ± 55	
AA-4367	SKLL-4	0.275 to 0.285	0.35	2,415 ± 60	0.14
AA-4370	SKLL-4	0.505 to 0.515	0.12	3,545 ± 70	- - - - -
AA-4371	SKLL-4	0.605 to 0.620	0.02	5,070 ± 180	0.07
W-6182	Siak River Levee S2 (buried wood, not in peat)		not applicable	5,560 ± 200	
W-5920	FebBK-1	3.40 to 3.50	0.05	5,730 ± 180	
W-5918	FebBK-1	4.74 to 5.00	-1.37	6,040 ± 180	
W-6026	BK-6	1.80 to 2.00	6.45	1,420 ± 200	
W-6192	BK-6	3.90 to 4.00	4.40	3,200 ± 180	1.2
W-6028	BK-6	4.80 to 5.00	3.45	4,200 ± 200	- - - - -
W-6193	BK-6	5.90 to 6.00	2.40	4,610 ± 200	5.1
W-6029	BK-6	7.74 to 7.89	0.67	4,740 ± 200	———
W-6080	Bengkalis Sta 28 (peat cliff) ~3		1.00	5,500 ± 200	

Age determinations by Meyer Rubin, U.S. Geological Survey, Reston, Virginia.
*W denotes dates from Radiocarbon Lab, U.S. Geological Survey, Reston, Virginia. AA denotes accelerator mass spectrometry (AMS) dates from University of Arizona, Tucson, Arizona.
†Rules bound the range of samples for which an average rate of peat accumulation is calculated. Heavier rules mark the top boundary, and lighter rules mark the lower boundary. Dashed rules indicate a transition to a different rate of accumulation.

Besar (Big Island) is approximately 8 km long and 2–3 km wide with several islands. Lake Bawah (lowest, bottom) is approximately 6 km long and 0.5–1 km wide. The origin of these lakes is not known. The surface elevation of the lakes is approximately 8–10 m above sea level, and their water depth ranges annually from approximately 4 m in the dry season to approximately 6 m in the wet season. Both lakes receive drainage from the surrounding peat in several blackwater streams. Blackwater is clear tea colored in transmitted light and dark brown in reflected light because of its load of dissolved organic matter. Water flows in a blackwater stream from Lake Pulau Besar into Lake Bawah. Lake Bawah is drained by the blackwater Rawa River, which flows out of the peat deposit into Panjang Strait throughout the year.

Cores were taken in the bottoms of both lakes: four drop cores in Lake Pulau Besar, five drop cores in Lake Bawah, and two piston cores in Lake Bawah. Five out of 11 cores penetrated organic-matter-rich clay-sized sediments below the peat. A fibrous, woody peat layer approximately 8–10 cm thick forms the base of the peat in four of five cores that penetrated to mineral sediments and in three of the six remaining cores. The woody peat was difficult to penetrate, but it appears to be fairly continuous over the mineral substrate within the lakes. This woody peat probably formed in place from plants growing in a peat swamp with very shallow water, initial topogenous peat. The fibric peat is overlain by approximately 30–50 cm of homogeneous, gelatinous, sapric peat (Fig. 7). The sapric peat appears to be the result of settling of fine suspended organic matter, brought into the lakes by streams draining the surrounding peat deposit. Peat petrography and organic geochemistry studies are in progress to

TABLE 2. CARBON-14 AGES OF PEAT SAMPLES FROM TWO PROFILES IN THE SIAK KANAN PEAT DEPOSIT, RIAU, SUMATRA, INDONESIA

Site and Sample Number	Average Sample Depth* (cm)	Bulk Density (kg dry matter per m³)	Ash (%)	Sample Age (yr B.P.)	Average Rate of Accretion† (mm/yr)
Site C-1 is the same location as site SK-1					
C-1-1	95	80	1.2	4,640 ± 30	
C-1-2	125	100	1.5	4,460 ± 30	
Site C-2 is the same location as site SK-2					
C-2-1	95	70	0.7	690 ± 20	———
C-2-2	195	100	0.5	1,260 ± 30	
C-2-3	295	60	0.4	1,980 ± 50	
C-2-3b	295	60	0.4	1,810 ± 50	1.7
C-2-4	395	60	0.5	2,170 ± 50	
C-2-5	495	80	0.6	2,880 ± 30	
C-2-6	595	50	0.4	3,550 ± 130	------
C-2-7	715	50	0.6	3,640 ± 50	
C-2-8	795	60	0.8	3,750 ± 50	4.3
C-2-9	895	40	0.5	3,500 ± 50	
C-2-10	995	70	7.0	4,470 ± 30	———

Data from Diemont and Supardi, 1987. Site C-1, thin peat (1.3 m), and Site C-2, thick peat (10 m).
*Each sample interval is 10 cm thick.
†Rules bound the range of samples for which an average rate of peat accumulation is calculated. Heavier rule marks the top boundary, and lighter rule marks the lower boundary. Dashed rule indicates a transition to a different rate of accumulation.

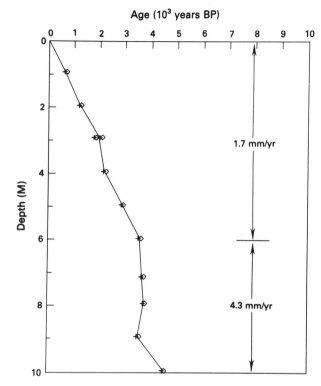

Figure 5. Rate of peat accumulation in thick peat, site C-2 (same location as site SK-2), Siak Kanan peat deposit. Symbols (diamond and plus) indicate plus and minus age range. From Diemont and Supardi (1987).

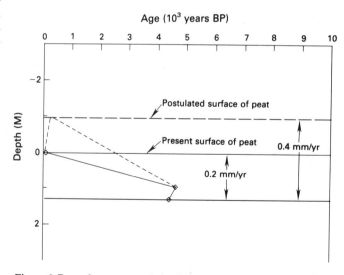

Figure 6. Rate of peat accumulation in thin peat, site C-1 (same location as site SK-1), Siak Kanan peat deposit. See Table 2 for plus and minus age range. From Diemont and Supardi (1987).

determine the source of the sapric peat, which may be derived from organic acids and colloids, transported fine organic matter, algae, or microorganisms. Preliminary nuclear magnetic resonance analyses confirm that the fine sapric peat is predominantly aliphatic compounds. The origin of these compounds could be either highly degraded plant material transported into the lake or algae (William Orem, 1990, personal communication). However, algal mats were not observed in either lake.

The low-temperature ash yield of the sapric peat in one drop core from Lake Bawah ranged from 1% to 6% (Ruppert et al., this volume). Siliceous sponge spicules, chrysophyte cysts, and phytoliths were observed in the low-temperature ash and indicate that aquatic fauna live in or near the lake and that remains of plants are incorporated in the lake peat.

Carbon-14 dating of a peat sample from a continuous piston core collected from the bottom of Lake Bawah at site SKLL-4 shows that the fibrous woody peat at the bottom of the lake started to accumulate at about the same time as peat at the bottom of the peat dome adjacent to the lake at site SK-8 (Table 1). The ages of three additional samples from the piston core show that the rate of peat accumulation was fairly constant during the deposition of the sapric peat in the lake, averaging 0.14 mm/yr (Table 1), a rate much lower than that in the peat dome (Table 1; Fig. 5).

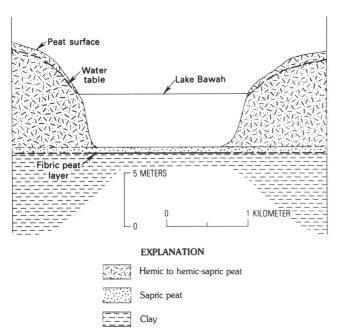

EXPLANATION

Hemic to hemic-sapric peat

Sapric peat

Clay

Figure 7. Generalized sketch of the peat at the bottom of Lake Bawah, a shallow lake in the Siak Kanan peat deposit.

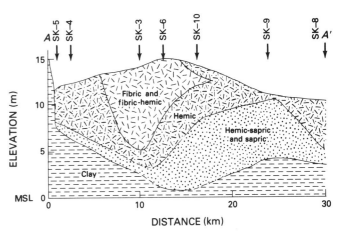

Figure 8. Cross section of Siak Kanan peat deposit showing decomposition of the peat. Arrows mark locations of peat sample profiles. Line of section is shown in Figure 3.

Peat quality

On the basis of the von Post classification, the peat in the Siak Kanan peat deposit can be divided into least degraded (fibric and fibric-hemic, H1 to H3-H4), moderately degraded (hemic, H4 to H6), and most degraded (hemic-sapric and sapric, H6-H7 to H10). A generalized cross section A–A' (Fig. 8) was constructed through the thick peat connecting seven sample sites along the south of the surveyed area (Fig. 3). Field descriptions of the peat humification show that the upper, interior portion of the peat dome is fibric and fibric-hemic and the bottom of the peat is mostly sapric and hemic-sapric.

In general, the top and middle layers of the peat deposit, which are fibric, fibric-hemic, and hemic, have a low ash yield (<2%), a low sulfur content (<0.2%), and water with low pH (pH 3–4; Neuzil et al., this volume). The sapric and hemic-sapric peat at the bottom of the peat deposit has higher ash yield (2–25%), higher sulfur content (0.2–0.4%), and higher pH in peat water (pH 4–6; Neuzil et al., this volume). The chemical composition of the mineral matter in the peat is discussed by Neuzil et al. (this volume) and Kane and Neuzil (this volume). Microscopic observations of discrete grains isolated from the peat by low-temperature ashing are discussed by Ruppert et al. (this volume). Peat petrography is discussed by Grady et al. (this volume).

Proximate and ultimate analyses and forms-of-sulfur data

for Siak Kanan peat are shown in Table 3 (as received and air-dry basis) and Tables 4a, 4b, and 4c (as received and dry basis). The 21 samples analyzed did not include any basal peat with higher ash yield. Data from the different laboratories show similar results calculated on a dry basis. Proximate analyses of these 21 peat samples show the following average values: moisture content, ~90% (as received); volatile matter, ~65% (dry basis); fixed-carbon content, ~33% (dry basis); ash yield, ~1% (dry basis); calorific value, ~5,800 cal/g (dry basis) (Tables 3 and 4a).

Ultimate analyses of 10 samples from the Siak Kanan peat deposit give the following average composition on a dry, ash-free basis: 61.5% C, 5.4% H, 0.9% N, 0.1% Cl, 0.2% S, and 31.9% O (Table 4b). Sulfur content in these samples is low, and organic sulfur is the dominant form, averaging 84% of total sulfur (Table 4c).

Peat resources

The peat resources of Siak Kanan were calculated by multiplying the average peat thickness by the area of the peat. The area of peat was determined for three thickness ranges: 1–5 m, 5–9 m, and >9 m (Fig. 3). Total peat resources in Siak Kanan are 6.6×10^9 m³ of in-situ peat (Table 5). Average bulk density of the peat is approximately 65 kg/m³ dry weight (Jansen et al., 1985; Diemont and Supardi, 1987), each cubic meter of in-situ peat contains 65 kg of dry peat or 100 kg of sod peat at 35% moisture. The Siak Kanan peat deposit is thus estimated to contain the equivalent of 6.6×10^8 tons of sod peat. To produce 1 megawatt (MW) of electricity, 10,000 tons of sod peat at 35% moisture are needed (Shell International, 1982). If half of the

**TABLE 3. PROXIMATE ANALYSES OF PEAT SAMPLES
FROM THE SIAK KANAN PEAT DEPOSIT, RIAU, SUMATRA, INDONESIA**

Sample	von Post Humification	Free Moisture (as received) (%)	Air-dry Basis				
			Moisture (%)	Volatile Matter (%)	Fixed Carbon (%)	Ash Yield (%)	Calorific Value (cal/g)
G. 12	H7–H9	86.6	11.8	53.9	32.4	1.9	4,975
G. 14	H5–H7	87.8	12.4	54.5	30.8	2.3	4,925
G. 18	H6–H8	86.2	11.0	53.8	34.5	0.7	5,070
G. 20	H5–H7	88.5	10.5	57.9	30.7	0.9	5,115
G. 28	H5-H7	88.8	12.6	55.4	32.4	1.5	5,020
G. 42	H5–H7	85.5	12.8	52.5	34.2	0.5	5,070
G. 43	H5–H7	85.4	11.2	53.0	35.3	0.5	4,975
G. 45	H6–H8	84.5	12.9	50.6	35.2	1.3	4,835
G. 46	H6–H8	82.4	14.1	50.5	34.7	0.6	4,835
G. 47	H5–H7	84.5	14.9	48.3	34.2	2.6	4,690
G. 50	H5–H7	85.2	12.9	52.0	34.4	0.7	4,880
Minimum		82.4	10.5	48.3	30.7	0.5	4,690
Maximum		88.8	14.9	57.9	35.3	2.6	5,115
Average		85.9	11.4	52.9	33.5	1.2	4,945

Analyses by DMR Laboratories, Bandung, Indonesia.

in-situ Siak Kanan peat deposit were converted to sod peat, it would be sufficient to produce 3.3×10^4 MW.

There is a significant component of hard wood in the Siak Kanan peat, perhaps as much as 20% by volume, which would be unsuitable for sod peat production. Therefore, resource estimates should probably be reduced by 20% with adjustments based on harvesting techniques.

THE BENGKALIS ISLAND PEAT DEPOSIT

Size and shape of the peat deposit

Bengkalis Island lies 10 km off the coast of Sumatra along the west side of the Strait of Malacca (Fig. 1). The island is almost flat and has a maximum surface elevation of approximately 10–15 m above sea level; it is 900 km^2, of which 665 km^2 is covered by peat more than 1 m thick.

The initial reconnaissance peat survey was conducted by the DMR in 1985/1986. A detailed survey of the surface elevations of the entire peat deposit on Bengkalis Island was not possible in the scope of this study. Maps of Bengkalis Island show streams draining out of the peat deposits in a radial drainage pattern characteristic of domed peat deposits (Fig. 9). The drainage patterns alone indicate that the peat deposit on the island is a group of peat domes each with a convex surface topography.

Approximately 60 holes were hand augered in a random distribution during the initial reconnaissance peat survey. The main goal of this reconnaissance work was to determine peat thickness and construct an isopach map of the peat deposit showing 1-m contour intervals (Fig. 9). The peat deposit is five domes of peat having maximum thicknesses of 6 to 10 m. The Kembung and Silan Rivers drain mineral soils that separate peat Area I in the southeast part of the island from the four coalesced peat domes in the north and west part of the island. This peat is designated Areas II, III, and IV for peat resource calculations. Area II covers the middle of the island and is two peat domes connected by a saddle. Area III lies west of the road connecting the town of Bengkalis on the south coast with the town of Selatbaru near the north coast. Area IV is west of the Miskum River on the west end of Bengkalis Island.

In 1987, the study was continued by a joint team of DMR and USGS geologists. A transect (B–B', Fig. 9) was surveyed across peat Area III in the western part of the island in order to accurately measure the surface elevation of the peat, measure the thickness of the peat, and calculate the elevation of the bottom of the peat (Fig. 10). Ten sample sites were established along the transect where field data and samples of peat, water, substrate sediment, and wood were collected. At the southern margin of the peat dome, the natural forest has been cleared in a zone approximately 3 km wide, and plantation crops are cultivated where the peat is less than 6 m thick (sites BK-1 through BK-4). The remainder of the transect was through a forest preserve that is relatively undisturbed (sites BK-5 through BK-10). Various analyses were conducted on selected samples (Neuzil et al., this volume).

Supardi and others

TABLE 4a. PROXIMATE ANALYSES OF PEAT SAMPLES FROM THE SIAK KANAN PEAT DEPOSIT, RIAU, SUMATRA, INDONESIA

Site-Sample	Peat Thickness (m)	Sample Interval Top (m)	Sample Interval Bottom (m)	Moisture (as received) (%)	Volatile Matter (%)	Fixed Carbon (%)	Ash Yield (%)	Calorific Value (cal/g)
SK-5-P4	4.42	1.00	2.00	91.89	63.97	34.10	1.93	5,785
SK-5-P5 and 6	4.42	2.00	4.00	90.70	61.80	32.40	5.80	5,500
SK-6-P4 and 5	13.65	1.00	3.00	91.07	67.53	31.00	0.48	5,633
SK-6-P7 and 8	13.65	4.00	6.00	92.09	64.86	35.01	0.13	5,785
SK-6-P11 and 12	13.65	8.00	10.00	93.41	68.12	31.21	0.67	5,842
SK-8-P4 and 5	6.65	1.00	3.00	90.15	65.28	34.40	0.32	6,004
SK-8-P6 and 7	6.65	3.00	5.00	88.89	63.88	35.18	0.94	5,858
SK-11-P4 and 5	8.55	1.00	3.00	91.36	66.99	32.22	0.79	5,828
SK-11-P7	8.55	4.00	5.00	93.25	65.14	34.09	0.77	5,807
SK-11-P10	8.55	7.00	8.00	91.72	64.62	34.70	0.68	5,714
Minimum	4.42			88.89	61.80	31.00	0.13	5,500
Maximum	13.65			93.41	68.12	35.18	5.80	6,004
Average	8.87			91.45	65.22	33.43	1.25	5,776

Analyses by Dickinson Laboratories Inc., El Paso, Texas.

TABLE 4b. ULTIMATE ANALYSES OF PEAT SAMPLES FROM THE SIAK KANAN PEAT DEPOSIT, RIAU, SUMATRA, INDONESIA

Site-Sample	Carbon (%)	Hydrogen (%)	Nitrogen (%)	Chlorine (%)	Sulfur (%)	Oxygen by Difference (%)
SK-5-P4	62.18	5.37	0.97	0.05	0.13	31.29
SK-5-P5 and 6	61.32	5.35	0.81	0.05	0.19	32.28
SK-6-P4 and 5	59.64	5.51	1.01	0.07	0.09	33.65
SK-6-P7 and 8	60.56	5.44	0.89	0.05	0.08	32.98
SK-6-P11 and 12	62.01	5.48	0.79	0.07	0.13	31.52
SK-8-P4 and 5	63.51	5.43	0.96	0.07	0.23	29.80
SK-8-P6 and 7	62.73	5.29	0.90	0.07	0.24	30.77
SK-11-P4 and 5	61.19	5.27	1.08	0.07	0.12	32.26
SK-11-P7	60.85	5.36	0.96	0.09	0.11	32.63
SK-11-P10	61.41	5.08	0.97	0.11	0.22	32.21
Minimum	59.64	5.08	0.79	0.05	0.08	29.80
Maximum	63.51	5.51	1.08	0.11	0.24	33.65
Average	61.54	5.36	0.93	0.07	0.16	31.94

Analyses by Dickinson Laboratories, Inc., El Paso, Texas.

TABLE 4c. FORMS OF SULFUR IN PEAT SAMPLES FROM THE SIAK KANAN PEAT DEPOSIT, RIAU, SUMATRA, INDONESIA

Site-Sample	Percent of Dry Peat				Percent of total Sulfur		
	Pyritic Sulfur (%)	Sulfate Sulfur (%)	Organic Sulfur (%)	Total Sulfur (%)	Pyritic Sulfur (%)	Sulfate Sulfur (%)	Organic Sulfur (%)
SK-5-P4	0.01	0.01	0.11	0.13	7.7	7.7	84.6
SK-5-P5 and 6	0.01	0.01	0.16	0.18	5.6	5.6	88.9
SK-6-P4 and 5	0.01	0.01	0.07	0.09	11.1	11.1	77.8
SK-6-P7 and 8	0.01	0.01	0.06	0.08	12.5	12.5	75.0
SK-6-P11 and 12	0.01	0.01	0.11	0.13	7.7	7.7	84.6
SK-8-P4 and 5	0.01	0.01	0.21	0.23	4.3	4.3	91.3
SK-8-P6 and 7	0.01	0.02	0.21	0.24	4.2	8.3	87.5
SK-11-P4 and 5	0.01	0.01	0.10	0.12	8.3	8.3	83.3
SK-11-P7	0.01	0.01	0.09	0.11	9.1	9.1	81.8
SK-11-P10	0.01	0.02	0.19	0.22	4.5	9.1	86.4
Minimum	0.01	0.01	0.06	0.08	4.2	4.3	75.0
Maximum	0.01	0.02	0.21	0.24	12.5	12.5	91.3
Average	0.01	0.01	0.13	0.15	7.5	8.4	84.1

Analyses by Dickinson Laboratories Inc., El Paso, Texas.

TABLE 5. PEAT RESOURCES IN THE SIAK KANAN PEAT DEPOSIT, RIAU, SUMATRA, INDONESIA

Area	km²	Thickness			Resource Volume x 10⁶ m³
		Minimum (m)	Maximum (m)	Average (m)	
I	386.5	1	5	2.5	965.5
II	463.5	5	9	6.5	3,012.3
III	263.5	9	13	10.0	2,635.0
Total	1,113.5				6,613.3

Along the transect, the base of the peat is above mean sea level in the southwest and below mean sea level in the northeast (Fig. 10). Although the base of the peat is below sea level, it is not infiltrated by sea water. The pH and dissolved cation concentrations in the peat water are higher in the bottom of the peat, especially where it is below mean sea level, than in the upper and middle layers of peat (Neuzil et al., this volume).

Today, sea level seems to be slightly higher than when peat started to accumulate. Erosion of peat is now occurring on the north coast of the Bengkalis Island, wher the base of the peat is lower than the base on the south coast. If the slope of the upper surface of the peat at the margin of the peat dome was originally the same on the north and south sides of the peat deposit, then a strip of peat approximately 0.5 km wide has been eroded away on the north side of the island at the site of the surveyed transect. The Strait of Malacca has apparently transgressed approximately 1 km onto Bengkalis Island in recent times. In contrast, the south coast appears to be more stable. This difference in stability is probably the result of the hydrodynamics of the inter-island straits. Sediments are accumulating on the west end of Bengkalis Island, extending the tidal flats. Sedimentation may be occurring here because of deposition from slack water on the leeward end of the island as the dominant tidal currents move to the northwest. Suspended-sediment load, erosion, and sedimentation patterns in the general study area are discussed further by Cecil et al. (this volume).

Age and rate of peat accumulation

The basal peat has been carbon-14 dated at three locations on Bengkalis Island (Table 1; Fig. 9). The dates range from 4740 ± 200 to 5730 ± 180 yr B.P. and are similar to or slightly older than the dates at the bottom of the peat in Siak Kanan (Table 1). It appears that peat accumulation started at about the same time on Bengkalis Island and in the Siak Kanan peat deposit. This timing suggests Bengkalis Island was an erosional remnant and negates progradation of the coastal plain during peat accumulation as a key factor in expanding the peat deposits to include Bengkalis Island.

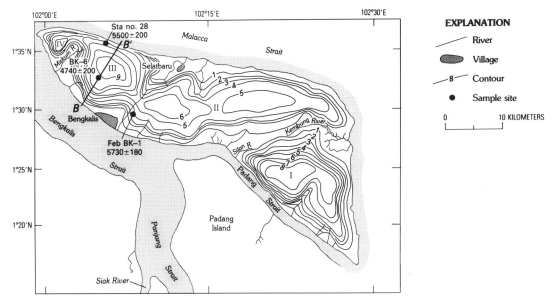

Figure 9. Isopach map of Bengkalis Island peat deposit (isopachs in meters) showing location of peat Areas I, II, III, and IV used to calculate peat resources in Table 8. Carbon-14 dates (Table 1) are shown for three samples from the base of the peat. Section B–B′ is shown in Figure 10. Map modified from Supardi (1988).

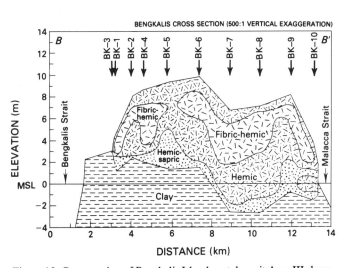

Figure 10. Cross section of Bengkalis Island peat deposit Area III showing decomposition of the peat. Arrows mark location of peat sample profiles. Line of section is shown in Figure 9.

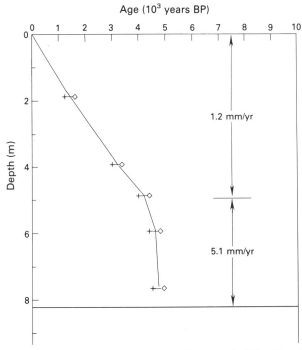

Figure 11. Rate of peat accumulation in thick peat, site BK-6, Bengkalis Island peat deposit. Symbols (diamond and plus) indicate plus and minus age range. Data from Table 1.

In the middle of the transect, at site BK-6, five carbon-14 dates in a profile show that the peat accumulation rate (Fig. 11) is similar to the peat accumulation rate in thick peat in Siak Kanan (Fig. 5). The initial 3 m of peat accumulated rapidly at an average rate of 5.1 mm/yr, and the upper 5 m of peat accumulated at a slower average rate of 1.2 mm/yr.

Peat quality

The peat from Bengkalis Island is brown (less humified) to dark brown (more humified). Most of the peat on Bengkalis is fibric-hemic to hemic-sapric. Generalized peat types are shown for a cross section along the transect (Fig. 10). The surface peat seems to be moderately humified over the entire deposit. Below the hemic surface peat, there is typically a layer of fibric-hemic peat, below which humification increases with depth through hemic to hemic-sapric peat near the base. This humification pattern is similar to one observed in Kalimantan by Driessen and Rochimah (1976), who suggested that the surface peat is more decomposed because of better aeration than in the subsurface peat. They also suggested that very fine and soluble decomposition products are gradually leached from the surface peat as the water table fluctuates; the leaching leaves a residual skeletal structure in peat by the time it is below the water table.

The ash yield and sulfur content are lowest in the middle of the interior of the peat deposits and highest at the bottom of the deposits. Most of the peat has an ash yield <2 % and a sulfur content <0.2%. The most humified hemic-sapric peat in general has the highest ash yield, highest sulfur content, and highest pH. Peat water becomes less acid with depth at each sample site. Approximately the upper half of the peat has a pH of 3–4. Below the middle of the peat profile, the pH is generally 4–5. The bottom of the peat, in the interior of the peat deposit although not at the margins, is pH 5–6.5 (Neuzil et al., this volume).

The trends in peat quality within the peat deposit are similar on Bengkalis Island and in the Siak Kanan deposit. The Bengkalis Island peat deposit is smaller and thinner than the Siak Kanan peat deposit, and a smaller portion of the Bengkalis Island peat deposit has a very low ash yield and sulfur content and is very acid. Sample spacing is closer on Bengkalis Island, and trends can be seen in more detail.

Proximate, ultimate, and sulfur analyses were conducted on a suite of 11 peat samples from Bengkalis Island (Tables 6a, 6b, and 6c). The sample suite represents only the middle layer of the peat deposit. No samples were within 1 m of the upper or lower boundaries of the peat; thus, they did not include any living root material at the upper surface of the peat or high-ash-yield samples at the bottom of the peat.

Proximate analyses of these 11 peat samples (Table 6a) show the following average values: moisture content, ~92% (as received); volatile matter, ~64% (dry basis); fixed carbon, ~34% (dry basis); ash yield, ~1.2% (dry basis); and calorific value, ~5,800 cal/g (dry basis). The peat on Bengkalis Island appears to have a higher calorific value (dry basis) than other peat deposits in Sumatra and Kalimantan, Indonesia (Table 7), where calorific value is reported on an air-dry basis with 10%–20% moisture in the peat samples. Also, the Bengkalis Island sample suite does not include any high-ash peat samples. Peat with a higher ash yield usually has a lower calorific value (Andriesse, 1988).

Ultimate analyses for the 11 peat samples from the cross section on Bengkalis Island (Table 6b) give the following average composition on a dry ash-free basis: ~62.0% C, ~5.3% H, ~1.0% N, ~0.1% Cl, ~0.1% S, and ~31.6% O. Ultimate analyses for Bengkalis Island and Siak Kanan peat are similar (Tables 4b and 6b). Total sulfur content is low (average ~0.1%) in the middle of the ombrogenous domed peat deposits. Forms of sulfur are dominated by organic sulfur. Average values are ~83% organic sulfur, ~8% pyritic sulfur and ~9% sulfate sulfur (Table 6c).

Peat resources

Peat resources on Bengkalis Island were calculated by the same method used for Siak Kanan. Nine thickness ranges were used based on 1-m contours (Table 8; Fig. 9). The area of peat for each contour interval in each peat area was measured and is shown as the upper numbers in Table 8. The volume of peat was calculated by multiplying the area of peat times the average peat thickness and is shown below the area of peat in Table 8.

The peat deposit on Bengkalis Island contains 3.0×10^9 cubic meters of peat. Most of the peat, 2.889×10^9 cubic meters, is greater than 2 m thick and is better suited for use as an energy resource than as an agricultural resource. Only 138×10^6 cubic meters of peat is less than 2 m thick and is recommended for other uses, especially agriculture and forest production.

Electric power production potential is calculated by the methods presented earlier. The average bulk density of peat on Bengkalis Island is about 80 kg/m^3 dry peat or 123 kg/m^3 sod peat at 30%–35% moisture. If half of the thicker peat (1.4×10^9 m^3 peat) can be exploited and converted into sod peat for energy production, this will yield 1.7×10^8 tons of sod peat, which could produce 1.7×10^4 MW of electricity.

SUMMARY

Drainage patterns and level-survey data show that the Siak Kanan and Bengkalis Island peat deposits are domed and have bases close to sea level. Carbon-14 dating indicates that peat accumulation started ca. 5000 yr B.P. In the thick interior portion

**TABLE 6a. PROXIMATE ANALYSES OF PEAT SAMPLES
FROM BENGKALIS ISLAND, RIAU, INDONESIA**

| Site-Sample | Peat Thickness | Sample Interval | | Moisture | Dry Basis | | | |
| | | Top | Bottom | (as received) | Volatile Matter | Fixed Carbon | Ash Yield | Calorific Value |
	(m)	(m)	(m)	(%)	(%)	(%)	(%)	(cal/g)
BK-1-P4	3.55	1.00	2.00	91.71	66.31	32.61	1.08	5,957
BK-1-P5	3.55	2.00	3.00	90.95	61.92	33.71	4.37	5,523
BK-5-P4 and 5	8.00	1.00	3.00	93.25	65.19	34.45	0.36	5,844
BK-5-P6 and 7	8.00	3.00	5.00	92.53	67.69	31.61	0.70	5,754
BK-5-P8 and 9	8.00	5.00	7.00	91.82	65.24	33.62	1.14	5,803
BK-7-P4 and 5	9.35	1.00	3.00	90.28	65.26	34.57	0.17	5,782
BK-7-P7	9.35	4.00	5.00	90.57	63.07	36.61	0.32	5,870
BK-7-P10	9.35	7.00	8.00	90.97	65.34	34.37	0.29	5,867
BK-9-P4 and 5	9.00	1.00	3.00	90.63	63.54	34.15	2.31	5,788
BK-9-P7 and 8	9.00	4.00	6.00	91.90	63.26	36.10	0.64	5,803
BK-9-P10	9.00	7.00	8.00	92.31	61.14	37.02	1.84	5,832
Minimum	3.55			90.28	61.14	31.61	0.17	5,523
Maximum	9.35			93.25	67.69	37.02	4.37	5,957
Average	7.83			91.54	64.36	34.44	1.20	5,802

Analyses by Dickinson Laboratories Inc., El Paso, Texas.

**TABLE 6b. ULTIMATE ANALYSES OF PEAT SAMPLES FROM
BENGKALIS ISLAND, RIAU, INDONESIA**

| Site-Sample | Dry Ash Free | | | | | |
| | Carbon | Hydrogen | Nitrogen | Chlorine | Sulfur | Oxygen by Difference |
	(%)	(%)	(%)	(%)	(%)	(%)
BK-1-P4	60.36	5.63	0.89	0.11	0.16	32.84
BK-1-P5	62.01	5.23	1.04	0.06	0.26	31.40
BK-5-P4 and 5	61.70	5.01	1.08	0.09	0.13	31.99
BK-5-P6 and 7	61.49	5.54	0.94	0.12	0.11	31.80
BK-5-P8 and 9	61.58	5.29	0.96	0.11	0.12	31.93
BK-7-P4 and 5	61.44	5.18	1.02	0.11	0.09	32.15
BK-7-P7	62.91	5.09	1.03	0.14	0.10	30.73
BK-7-P10	61.90	5.05	1.02	0.11	0.14	31.77
BK-9-P4 and 5	63.03	5.45	0.86	0.12	0.14	30.40
BK-9-P7 and 8	62.26	5.26	0.97	0.11	0.11	31.29
BK-9-P10	62.74	5.12	1.10	0.11	0.18	30.74
Minimum	60.36	5.01	0.86	0.06	0.09	30.40
Maximum	63.03	5.63	1.10	0.14	0.26	32.84
Average	61.95	5.26	0.99	0.11	0.14	31.55

Analyses by Dickinson Laboratories, Inc., El Paso, Texas.

**TABLE 6c. FORMS OF SULFUR IN PEAT SAMPLES
FROM BENGKALIS ISLAND, RIAU, INDONESIA**

Site-Sample	Percent of Dry Peat				Percent of total Sulfur		
	Pyritic Sulfur (%)	Sulfate Sulfur (%)	Organic Sulfur (%)	Total Sulfur (%)	Pyritic Sulfur (%)	Sulfate Sulfur (%)	Organic Sulfur (%)
BK-1-P4	0.01	0.01	0.14	0.16	6.3	6.3	87.5
BK-1-P5	0.01	0.02	0.22	0.25	4.0	8.0	88.0
BK-5-P4 and 5	0.01	0.01	0.11	0.13	7.7	7.7	84.6
BK-5-P6 and 7	0.01	0.01	0.09	0.11	9.1	9.1	81.8
BK-5-P8 and 9	0.01	0.01	0.10	0.12	8.3	8.3	83.3
BK-7-P4 and 5	0.01	0.01	0.07	0.09	11.1	11.1	77.8
BK-7-P7	0.01	0.01	0.08	0.10	10.0	10.0	80.0
BK-7-P10	0.01	0.03	0.10	0.14	7.1	21.4	71.4
BK-9-P4 and 5	0.01	0.01	0.12	0.14	7.1	7.1	85.7
BK-9-P7 and 8	0.01	0.01	0.09	0.11	9.1	9.1	81.8
BK-9-P10	0.01	0.01	0.16	0.18	5.6	5.6	88.9
Minimum	0.01	0.01	0.07	0.09	4.0	5.6	71.4
Maximum	0.01	0.03	0.22	0.25	11.1	21.4	88.9
Average	0.01	0.01	0.12	0.14	7.8	9.4	82.8

Analyses by Dickinson Laboratories Inc., El Paso, Texas.

of the deposits, peat accumulated at a rate of about 4–5 mm/yr for the first 600–1,000 years; the rate decreased to about 1–2 mm/yr for the past 3,500–4,000 years.

Both peat deposits are slightly to moderately decomposed and show fibric, fibric-hemic, and hemic textures except near the base, where they are moderately to highly decomposed and show hemic-sapric to sapric textures. The middle and upper portions of the peat deposits have extremely low ash yield and sulfur content, very acid water, and low to moderate humification. The bottom of each peat deposit has a higher ash yield, a higher sulfur content and less acid water and is more humified. In the smaller and thinner Bengkalis Island peat deposit, the basal layer of peat may represent a larger portion of the entire peat deposit. Proximate and ultimate analyses are similar for the interior portions of both the Siak Kanan and Bengkalis Island peat deposits.

Mapped area, measured thickness, and average heating value and bulk-density data have been used to calculate the volume and energy resource potential of each peat deposit. In addition to the peat resource information, a sense of the three-dimensional shape of the peat deposits has been defined in this study. The basic field descriptions, proximate and ultimate data, and further chemical and petrographical data can be placed in this framework to show facies relations within the peat. These spatial relations of peat quality can be used in models of the peat stage of coal formation.

ACKNOWLEDGMENTS

This paper was prepared following collaborative field studies in 1987 by personnel from the Directorate of Mineral Resources and the U.S. Geological Survey who were supported by their respective agencies. The researchers were particularly interested in the energy resource potential of the peat deposits as well as in examining one of the world's best modern analogues for some of the Carboniferous coal deposits of the eastern United States. We wish to acknowledge Ir. Salman Padmanagara, Director of the Directorate of Mineral Resources, and Drs. Hardjono, Head of the Subdirectorate of Coal and Peat Exploration, for supporting the DMR contribution to this paper. We thank Memed Romli and Maman Tjahyana of the DMR for their tireless field work conducting the level surveys. The field work and sample collection would not have been possible without the participation of Truman Wijaya (DMR) and Cortland Eble and Christopher Wnuk (USGS) and the willing labor of many local residents at each study area. We wish to acknowledge the assistance of Caltex Pacific Indonesia and Caltex personnel who provided logistical support while field work was conducted in the Siak Kanan peat deposit; Purnomo Prijosoesilo coordinated Caltex assistance. We thank P. T. Arutmin and their personnel for invaluable logistical assistance and, especially, Lim Meng Sze Wu for his participation in the 1987 field work.

Supardi and others

TABLE 7. PROXIMATE ANALYSES OF PEAT SAMPLES FROM OTHER PEAT DEPOSITS IN INDONESIA

Sample	Depth (m)	von Post Humification	Free Moisture (as received) (%)	Air-dry Basis				
				Moisure (%)	Volatile Matter (%)	Fixed Carbon (%)	Ash Yield (%)	Calorific Value (cal/g)
Siak Kiri, Riau, Sumatra*								
G. 27	1.71	H6–H8	88.5	10.7	55.4	32.4	1.5	5,020
G. 36	1.73	H6–H8	86.0	13.7	49.9	35.0	1.4	4,810
G. 38	1.82	H6–H8	82.8	14.3	49.1	34.5	2.1	4,690
Bukit Natu, Riau, Sumatra[†]								
G. 32	1.12	H7–H9	84.6	11.0	53.4	34.3	1.3	4,975
G. 34	0.72	H7–H9	82.1	11.7	50.6	29.3	8.4	4,690
Air Sugihan, South Sumatra[†]								
G. As. 33			79.1	14.7	48.0	22.9	14.4	4,005
G. As. 39			79.1	13.8	49.3	22.4	14.5	4,125
G. As. 41			80.5	11.4	39.6	16.3	32.7	3,200
G. As. 42			82.4	13.5	50.2	22.9	13.4	4,290
G. As. 44			84.5	13.2	49.1	22.7	15.0	4,055
G. As. 45			79.7	11.1	36.8	16.9	35.2	3,010
G. As. 48			83.0	14.7	50.1	22.6	12.6	4,075
Sambas, West Kalimantan[†]								
G. TK 1			77.7	13.6	54.3	26.6	5.5	4,995
G. TK 2			85.1	16.2	50.0	27.8	6.0	4,785
G. TK 3			91.5	15.1	52.8	29.0	3.1	4,690
G. TK 4			89.8	16.5	51.9	30.1	1.5	4,435
G. TK 6			85.1	16.4	51.4	30.8	1.4	4,575
G. TK 7			88.2	16.8	50.3	30.3	2.6	4,460
G. TK 8			90.5	15.2	52.8	28.9	3.1	4,575
G. TK 10			88.0	13.7	55.5	28.7	2.1	4,905
G. TK 11			91.2	14.3	50.8	25.9	9.0	4,365
G. TK 12			83.6	15.4	52.1	27.1	5.4	4,505
G. TK 14			75.7	13.5	50.2	25.3	11.0	4,340
G. TK 15			86.1	14.8	53.9	28.2	3.1	4,550
G. TK 16			85.8	14.9	55.0	26.3	3.8	4,765
Marambahan, South/Central Kalimantan[†]								
G. Bh 08			85.2	17.9	51.1	23.2	7.8	4,130
G. Bh 10			79.6	11.3	54.0	18.9	15.8	4,835
G. Bh 12			79.1	14.7	50.6	23.6	11.1	4,460
G. Bh 13			81.1	12.7	53.6	22.5	11.2	4,670
Kanamit, Central Kalimantan[†]								
G. Bh 27			82.9	14.6	51.9	31.4	2.1	4,785
G. Bh 28			84.8	14.9	51.2	32.0	1.9	4,715
G. Bh 29			89.4	14.3	50.9	30.3	4.5	4,690
G. Bh 30			87.7	18.7	49.3	29.7	2.3	4,410
G. Bh 31			87.4	15.4	52.0	30.3	2.3	4,855
G. Bh 33			86.0	13.9	52.2	30.1	3.8	4,925
G. Bh 37			76.2	13.8	39.3	22.8	24.1	3,950
G. Bh 38			86.6	14.9	51.3	31.5	2.3	4,645
G. Bh 41			85.3	15.1	50.6	30.8	3.5	4,670
G. Bh 42			84.9	18.7	47.6	30.5	3.2	4,270

*Data from Supardi and Priatna, 1985.
[†]Data from Priatna et al., 1984.

TABLE 8. PEAT RESOURCES, BENGKALIS ISLAND, RIAU, INDONESIA

Peat Area		Peat Thickness Range (m)									Total
		1–2	2–3	3–4	4–5	5–6	6–7	7–8	8–9	9-10	
I	Area*	25.4	27.3	27.1	27.2	24.6	24.1	29.8			185.5
	Volume†	38.1	68.3	94.9	122.4	135.3	156.7	223.5			839.1
II	Area*	50.1	38.4	56.5	79.0	67.6	23.5				315.1
	Volume†	75.2	96.0	197.8	355.5	371.8	152.8				1,249.0
III	Area*	12.8	10.6	14.5	10.2	14.7	21.8	25.0	18.6	15.5	143.7
	Volume†	19.1	26.5	50.8	45.9	80.9	141.7	187.5	158.1	147.3	857.7
IV	Area*	3.9	3.4	2.9	3.5	3.1	3.8				20.5
	Volume†	5.9	8.5	10.0	15.5	17.1	24.7				81.6
Total	Area*	92.2	79.7	101.0	119.9	110.0	73.2	54.8	18.6	15.5	664.8
	Volume†	138.2	199.3	353.3	539.3	605.0	475.8	411.0	158.1	147.3	3,027.3

The peat volume for each thickness interval was calculated by multiplying the average thickness by the area.
*Square kilometers.
†Million cubic meters.

REFERENCES CITED

Anderson, J.A.R., 1964, The structure and development of the peat swamps of Sarawak and Brunei: Journal of Tropical Geography, v. 18, p. 7–16.

Anderson, J.A.R., 1976, Observations on the ecology of five peat swamp forests in Sumatra and Kalimantan, *in* Peat and podzolic soils in Indonesia: Bogor, Indonesia, Soil Research Institute Bulletin 3, p. 45–55.

Anderson, J.A.R., 1983, The tropical peat swamps of western Malesia, *in* Gore, A.J.P., ed., Mires: Swamp, bog, fen, and moor: New York, Elsevier, Ecosystems of the World, v. 4B, p. 181–200.

Andriesse, J. P., 1974, Tropical lowland peats in south-east Asia: Amsterdam, Department of Agricultural Research of the Royal Tropical Institute Communication 63, 63 p.

Andriesse, J. P., 1988, Nature and management of tropical peat soils: Rome, Food and Agriculture Organization of the United Nations, FAO Soils Bulletin 59, 165 p.

Biswas, B., 1973, Quaternary changes in sea-level in the South China Sea: Geological Society of Malaysia Bulletin 6, p. 229–256.

Cameron, N. R., Ghazali, S. A., and Thompson, S. J., 1982a, The geology of the Siak Sri Indrapura–Tanjungpinang Quadrangles, Sumatra: Bandung, Indonesia, Ministry of Mines and Energy, Directorate General of Mines, The Geological Research and Development Centre, 26 p. and 2 maps, scale 1:250,000.

Cameron, N. R., Kartawa, W., and Thompson, S. J., 1982b, The geology of the Dumai Quadrangle, Sumatra: Bandung, Indonesia, Ministry of Mines and Energy, Directorate General of Mines, The Geological Research and Development Centre, 19 p., and 1 map, scale 1:250,000.

Chambers, M.J.G., and Abdul Sobur, A. S., 1975, The rates and processes of recent coastal accretion in the province of South Sumatra: A preliminary survey, *in* Proceedings, Regional Conference on Geology and Mineral Resources, SE Asia: Jakarta, Indonesia, Indonesian Association of Geologists, p. 165–174.

Diemont, W. H., and Supardi, 1987, Accumulation of organic matter and inorganic constituents in a peat dome in Sumatra, Indonesia [abs.]: Yogyakarta, Indonesia, International Peat Society Symposium on Tropical Peat and Peatlands for Development, February 9–14, 1987, Abstracts, p. 93.

Driessen, P. M., and Rochimah, L., 1976, The physical properties of lowland peats from Kalimantan, *in* Peat and podzolic soils in Indonesia: Bogor, Indonesia, Soil Research Institute Bulletin 3, p. 56–73.

Farnham, R. S., and Finney, H. R., 1965, Classification and properties of organic soils: Advances in Agronomy, v. 17, p. 115–162.

Geyh, M. A., Kudrass, H. R., and Streif, H., 1979, Sea-level changes during the late Pleistocene and Holocene in the Strait of Malacca: Nature, v. 278, no. 5703, p. 441–443.

Hehuwat, F., 1982, An overview of peat deposits in Indonesia: A potential new mineral energy resource, *in* Seminar on Peat for Energy Use, Bandung, Indonesia, June 29–30: Bandung, Indonesia, Directorate General of Mines, Mineral Technology Development Center.

Jansen, J. C., Diemont, W. H., and Koenders, N. J., 1985, Peat development for power generation in West Kalimantan: An ecological and economic appraisal: Rotterdam, The Netherlands, The Netherlands Economic Institute, 105 p.

Priatna, Supardi, and Subekty, A. D., 1984, Endapan gambut di Air Sugihan Sumatra Selatan: Bandung, Indonesia, Direktorat Sumberdaya Mineral (in Indonesian).

Shell International, 1982, Peat in Indonesia: London, Shell International Petroleum, Non-Traditional Business Division.

Supardi, 1988, Endapan gambut di Pulau Bengkalis, Riau: Bandung, Indonesia, Direktorat Sumberdaya Mineral, map, scale 1:100,000 (in Indonesian).

Supardi and Priatna, 1985, Endapan gambut didaerah Siakkanan Riau: Tahun anggaran 1984/1985: Bandung, Indonesia, Departemen Pertambangan dan Energi, Direktorat Jenderal Geologi dan Sumberdaya Mineral, Direktorat Sumberdaya Mineral, Proyek Inventarisasi Batubara, 9 p. (in Indonesian).

von Post, L., 1922, Sveriges geologiska undersoknings torvinventering och nogra av dess hittills vunna resultat: Bilaga Svenska Mosskulturforeningen Tidskrift, v. 1, p. 1–25.

Whitten, A. J., Sengli, J. D., Jazamul, A., and Nazaruddin, H., 1987, The ecology of Sumatra (second edition): Yogyakarta, Indonesia, Gadjah Mada University Press, 583 p.

Manuscript Accepted by the Society January 14, 1993

Geological Society of America
Special Paper 286
1993

Brown coal maceral distributions in a modern domed tropical Indonesian peat and a comparison with maceral distributions in Middle Pennsylvanian–age Appalachian bituminous coal beds

William C. Grady
West Virginia Geological and Economic Survey, P.O. Box 879, Morgantown, West Virginia 26507
Cortland F. Eble
Kentucky Geological Survey, 228 MMRB, University of Kentucky, Lexington, Kentucky 40506
Sandra G. Neuzil
U.S. Geological Survey, 956 National Center, Reston, Virginia 22092

ABSTRACT

Analyses of modern Indonesian peat samples reveal that the optical characteristics of peat constituents are consistent with the characteristics of macerals observed in brown coal and, as found by previous workers, brown-coal maceral terminology can be used in the analysis of modern peat. A core from the margin and one from near the center of a domed peat deposit in Riau Province, Sumatra, reveal that the volume of huminite macerals representing well-preserved cell structures (red, red-gray, and gray textinite; ulminite; and corpo/textinite) decreases upward. Huminite macerals representing severely degraded (<20 microns) cellular debris (degraded textinite, attrinite, and densinite) increase uniformly from the base to the surface. Greater degradation of the huminite macerals in the upper peat layers in the interior of the deposit is interpreted to be the result of fungal activity that increased in response to increasingly aerobic conditions associated with the doming of the peat deposit. Aerobic conditions concurrent with the activities of fungi may result in incipient oxidation of the severely degraded huminite macerals. This oxidation could lead to the formation of degradosemifusinite, micrinite, and macrinite maceral precursors in the peat, which may become evident only upon coalification. The core at the margin was petrographically more homogeneous than the core from the center and was dominated by well-preserved huminite macerals except in the upper 1 m, which showed signs of aerobic degradation and was similar to the upper 1 m of the peat in the interior of the deposit.

The Stockton and other Middle Pennsylvanian Appalachian coal beds show analogous vertical trends in vitrinite maceral composition. The succession from telocollinite-rich, bright coal lithotypes in the lower benches upward to thin-banded/matrix collinite and desmocollinite in higher splint coal benches is believed to reflect a progression similar to that from the well-preserved textinite macerals in the lower portions of the peat cores to severely fragmented and degraded cellular materials (degraded textinite, attrinite, and densinite) in the upper portions of the cores. This petrographic sequence from bright to splint coal in the Stockton and other Middle Pennsylvanian coal beds supports previous interpretations of an upward transition from planar to domed swamp accumulations.

Grady, W. C., Eble, C. F., and Neuzil, S. G., 1993, Brown coal maceral distributions in a modern domed tropical Indonesian peat and a comparison with maceral distributions in Middle Pennsylvanian–age Appalachian bituminous coal beds, *in* Cobb, J. C., and Cecil, C. B., eds., Modern and Ancient Coal-Forming Environments: Boulder, Colorado, Geological Society of America Special Paper 286.

INTRODUCTION

David White was the first to propose that "it is in our coastal swamps, particularly those in tropical regions of very heavy rainfall, that is to be found the closest similarity between the peat deposits of today and those ancient deposits of the great coal fields" (White and Thiessen, 1913, p. 66). White and Thiesen (1913) proposed that the fresh-water inland swamps of eastern Sumatra, described by Potonie and Koorders (1909), are models for coal-forming swamps because of their thick (as much as 9 m) low-ash peats, stagnant fresh-water cover, and incredibly luxuriant and dense plant cover contributing vegetal matter to the peat at a rate exceeding the rapid decay in the tropical environment.

Combining palynology and petrography, Smith (1968) developed an analogy between British Carboniferous coals and the domed swamps of Sumatra based upon the concept of "miospore phases." These phases depict a sequence of floristic changes as plant communities adapted to changes in edaphic conditions, in response to the gradual doming of the peat surface. Each miospore phase also corresponds to a specific range of petrographic composition, which Smith (1968) concluded is related to changing conditions in the swamp analogous to the modern domed swamps. The Stockton, Coalburg, and Fire Clay coal beds (in the Middle Pennsylvanian Kanawha Formation) have been inferred to have accumulated as domed peat complexes consisting

of laterally and vertically contiguous planar and domed swamp paleoenvironments (Grady, 1983; Grady et al., 1985, 1992, Eble et al., 1989; Eble and Grady, 1990, and this volume). Relations among macerals, palynomorphs, ash yield, sulfur content, and mineralogy that were used to infer the paleoenvironments are addressed in a companion paper (Eble and Grady, this volume).

The purpose of this investigation is to compare the distribution of organic petrographic components of a modern, domed peat deposit with the distribution of macerals in the Stockton coal bed (Middle Pennsylvanian, Kanawha Formation) in West Virginia that may have accumulated under similar geochemical, environmental, and climatic conditions (Fig. 1). Although the ancient flora of the Pennsylvanian peat swamps and the modern vegetation of Indonesian tropical swamps are dissimilar and would most likely produce different peat types, the agents of decomposition (primarily aerobic and anaerobic microbes and physicochemical oxidation) may function comparably in peats of diverse ages, especially in relation to preservation versus degradation of cellular materials. In this paper comparisons between the constituents of the peat and those of the coal concentrate on the origins and the relative degrees of degradation (versus preservation) exhibited by the components of each. This study applies specific coal maceral types (Grady et al., 1992; Eble and Grady, this volume), which, as defined in Stach (1982, p. 89), indicate differences in the origin of a maceral as deduced from its shape or structure.

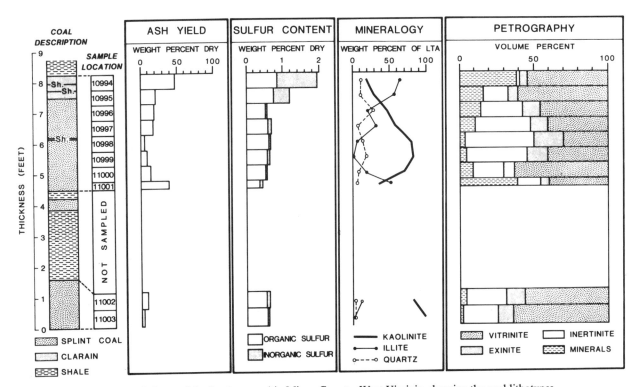

Figure 1. Column of the Stockton coal in Mingo County, West Virginia, showing the coal lithotypes, sample locations, and the distributions of ash yield, sulfur content, mineralogy, and petrographic components. The coal bed was split into upper and lower benches at this location by a mudstone parting, which was not sampled. From Grady et al. (1992).

Specific macerals of the Stockton coal bed reveal trends in abundance and structure that reflect the paleoenvironment and may be comparable to trends in modern peat components. Abundances of vitrinite maceral types reflect the degree of fragmentation and degradation of their precursors. Collinite as <50-micron bands or a matrix with inclusions separated by <50 microns is the dominant collinite maceral type in this coal bed (Fig. 2). This thin-banded/matrix collinite type and weaker reflecting desmocollinite reveal by the presence of inclusions that their precursors were cell and cell-wall fragments intimately mixed with spores, liptodetrinite, and inertodetrinite. Collectively, all these macerals are most abundant in the inertinite-rich intervals in the middle of the upper bench (Figs. 1 and 2). Collinite in >50-micron bands reveals cellular structures (cryptotelinite) upon etching with acidified potassium permanganate solution, indicating derivation from well-preserved "woody" fragments. This collinite type is most abundant in the mineral-rich, inertinite-poor intervals at the top and bottom of the upper bench and is a minor collinite type in the inertinite-rich intervals (Figs. 1 and 2).

Inertinite macerals are regarded as having formed principally from the same precursors as the vitrinite macerals but were subject to charring, oxidation, mouldering, or fungal attack prior to their deposition (Stach, 1982). Degradosemifusinite and minor degradofusinite are the predominant inertinite macerals in the Stockton coal bed (Fig. 3). Degradosemifusinite displays poorly defined cellular structure and, in the Stockton coal, commonly contains dispersed spore exines and liptodetrinite within an optically continuous semifusinite fragment (Grady et al., 1992). The diffuse structure and the inclusion of volatile exinite macerals support a slow, mouldering oxidative origin for the abundant degradosemifusinite (Teichmuller, 1982; Grady and others, 1992). Theorists on the origins of micrinite view this maceral as forming through oxidation of cell-wall tissues (Thiessen and Sprunk, 1936; Cohen, 1968). Maximum micrinite abundance occurs immediately below the interval of greatest total inertinite, where it is intimately mixed with thin-banded/matrix collinite containing exinite and inertodetrinite inclusions.

In the Stockton coal bed the lower increments of the upper bench (samples 11001–11000) have high ash yields, illite-dominant mineralogies, high vitrinite maceral contents, and low inertinite contents. Collinite in >50-micron bands is common, and thin-banded/matrix collinite, degradosemifusinite, and micrinite are low in abundance. This association of components has been interpreted to represent peats accumulated in planar, topogenous swamps (Eble and Grady, 1990, this volume). Higher increments in the upper bench (samples 10999–10996) display lower ash yields, kaolinite-dominant mineralogies, lower vitrinite maceral abundances, and higher exinite and inertinite abundances. These increments show an association of components that has been interpreted to represent peats accumulated in domed,

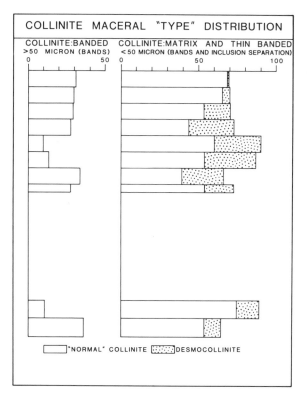

Figure 2. Collinite maceral type distribution in the Stockton coal column on a percent of total vitrinite basis. Collinite of "normal" reflectance is separated into collinite bands >50 microns in thickness (telocollinite) and collinite bands <50 microns in thickness or "matrix" with <50 microns separation between inclusions. Desmocollinite is a weaker reflecting (darker) matrix and thin-banded collinite submaceral. From Grady et al. (1992).

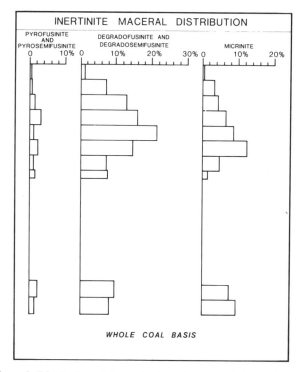

Figure 3. Distribution of the major inertinite macerals in the Stockton coal column. From Grady et al. (1992).

ombrogenous swamps (Eble and Grady, 1990, this volume). Very high inertinite maceral abundances, especially "degrado-" varieties of semifusinite and fusinite, and micrinite, are inferred to be important indicators of aerobic, ever-wet conditions (Smith, 1962, 1968; Grady et al., 1985, 1992).

Trends in the petrographic and mineralogic components depict the proposed vertical development of a domed, ombrogenous swamp on a topogenous peat substrate (Grady et al., 1992; Eble and Grady, this volume). Laterally a topogenous peat may persist without an overlying ombrogenous peat, or it may be supplanted by sediments (Grady et al., 1992). Similarly, a domed peat may accumulate without a substrate of topogenous peat, as shown in the lower bench (samples 11002-11003) of the Stockton coal bed (Fig. 1). In the top two increments of the coal bed, high ash yields and sulfur contents, illite-dominant mineralogies, and high vitrinite and low inertinite contents indicate a return to a planar, topogenous swamp and eventual burial of the peat by sediments (Fig. 1). This reversal is probably the result of an increased rate of subsidence (Smith, 1962; Grady et al., 1992).

Textural and size characteristics of the petrographic components of a modern domed peat deposit were compared to the distributions of maceral types observed in the Stockton coal bed in order to test the hypothesis that this, and other coal beds of the Middle Pennsylvanian (Eble and Grady, this volume), accumulated in domed, ombrogenous swamp complexes. Samples were obtained from a tropical, ombrogenous, domed peat swamp which is thick (as much as 16 m), has low ash (<5%), has low sulfur (<0.5%), is areally extensive (97,000 km^2), and is acidic (pH <4); the swamp is located on the island of Sumatra, Indonesia (Polak, 1950, 1975; Anderson, 1964; Andriesse, 1974; Whitmore, 1975; Driessen, 1977; Neuzil and Cecil, 1984). In sample collection, preparation, and microscopic analysis every attempt was made to parallel the methods of coal petrographic analysis so that the data of this study could be compared with data from Appalachian bituminous coal beds.

METHODS

Field methods

The peats examined in this study were collected from two sites on a domed peat deposit in Riau Province, Sumatra, Indonesia (Fig. 4). Sample site SK5 is located in 4.42 m of peat, less than 50 m from the fringe (1 m peat thickness) of the deposit. Sample site SK6 is 15 km east of SK5 on the crest of the domed deposit where the peat attains a thickness of 13.65 m. Core samples were obtained in consecutive half-meter or smaller lengths through the entire peat thickness with a hand-driven McCauley-type peat sampler. Half-meter segments of core were usually combined into 1 m thick sample increments (Table 1). The cores included all wood fragments that could be sampled. If logs were encountered that could not be sampled, another sample from the same depth was collected adjacent to the borehole. This unavoidable procedure was used once in SK5 (at a depth of ~4 m) and twice in

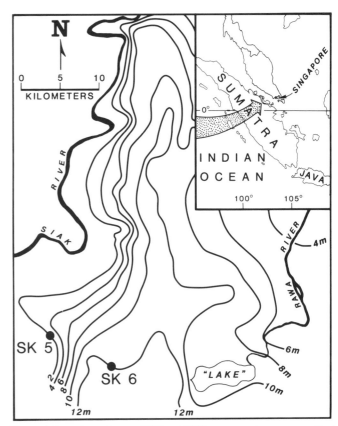

Figure 4. Locations of cores SK5 and SK6 superposed on a generalized isopach map of the Siak Kanan domed peat deposit, Sumatra, Indonesia. Modified from Supardi et al. (this volume).

SK6 (at depths of ~5 m and 8 m). Increment samples were described according to a modified U.S. Department of Agriculture classification system (Table 1) as fibric, fibric-hemic, hemic, hemic-sapric and sapric (Farnham and Finney, 1965; Supardi and others, this volume), and according to the von Post classification systems (von Post, 1924). The wet peat samples were sealed in plastic containers for shipment and were stored under refrigeration until preparation for petrographic analysis.

Sample preparation methods

Sample splits from the wet peat increments were obtained using a cone and quarter method of taking alternating small aliquots for petrographic and other analyses (Neuzil et al., this volume). Splits for petrographic analysis were freeze-dried to constant weight to reduce shrinkage and oxidation of the peat. The freeze-dried peat was screened on a 2-mm (10-mesh) sieve to pass the –2-mm fraction. The larger fraction was then repetitively crushed in stages in a mechanical blender and screened until all passed the 10-mesh sieve. This procedure minimized the fine fraction and produced a sample of predominantly 2-mm fragments. This top size was selected because of the feasibility of producing the minimum number of polished pellets (five) that

TABLE 1. SAMPLE SITE INFORMATION FROM LOCALITIES SK5 AND SK6

Sample Site SK5
Latitude 00°42'05"N; Longitude 102°01'57"E
Total peat depth = 4.42 m
Elevation (above mean sea level at top of peat) = 12 m
Water table = 0.90 m

Sample	Depth (m)	Peat Type*	Humification[†]	Wood Content[§]	Root-thread Content**	Fiber Content[‡]	Ash[§§] (%)	Sulfur[§§] (%)
SK5-P1	0.00–0.25	Hemic	4	3	3	2	4.66	0.16
SK5-P2	0.25–0.50	Hemic	4	3	3	2	1.69	0.15
SK5-P3	0.50–1.00	Hemic	4	3	3	2	3.86	0.16
SK5-P4	1.00–2.00	Hemic	4	2–3	2	2	1.93	0.13
SK5-P5	2.00–3.00	Hemic	5–6	1–2	2	2	1.34	0.18
SK5-P6	3.00–4.00	Hemic	6	1–2	1	2	5.80	0.18
SK5-P7	4.00–4.25	Hemic-sapric	6–7	1	0	1	5.21	0.17
SK5-P8	4.25–4.42	Hemic-sapric	6–7	1	0	1	n.a.	n.a.
N.S.	4.42–5.99	Clay, light gray, soft with organic matter						

Sample Site SK6
Latitude 00°39'30"N; Longitude 102°06'03"E
Total peat depth = 13.65 m
Elevation (above mean sea level at top of peat) = 14.5 m
Water table = 1.15 m

Sample	Depth (m)	Peat Type*	Humification[†]	Wood Content[§]	Root-thread Content**	Fiber Content[‡]	Ash[§§] (%)	Sulfur[§§] (%)
SK6-P1	0.00–0.25	Fibric	2–3	1	3	3	0.70	0.11
SK6-P2	0.25–0.50	Fibric	3	1	3	3	0.70	0.11
SK6-P3	0.50–1.00	Fibric	3	1	3	3	0.81	0.10
SK6-P4	1.00–2.00	Fibric	3	1	2	2	0.48	0.09
SK6-P5	2.00–3.00	Fibric	3	1	2	2	0.48	0.09
SK6-P6	3.00–4.00	Fibric	3	1	2	2	0.36	0.08
SK6-P7	4.00–5.00	Fibric-hemic	3–4	1	2	2	0.13	0.09
SK6-P8	5.00–6.00	Fibric-hemic	3–4	2	2	2	0.13	0.08
SK6-P9	6.00–7.00	Hemic	4–5	1	2	2	0.36	0.09
SK6-P10	7.00–8.00	Hemic	5–6	2	2	2	0.36	0.09
SK6-P11	8.00–9.00	Hemic	6	3	1	2	0.67	0.13
SK6-P12	9.00–10.00	Hemic	6	2	1	2	0.67	0.13
SK6-P13	10.00–11.00	Hemic	6	2	1	2	0.34	0.12
SK6-P14	11.00–12.00	Hemic-sapric	6–7	2	1	2	0.34	0.12
SK6-P15	12.00–13.65	Hemic-sapric	6–7	2	1	1	16.86	0.35
N.S.	13.65–14.25	Clay, light gray, soft with organic matter						

*U.S. Department of Agriculture method (Farnham and Finney, 1965)
[†]von Post (1924) Degree of Decomposition (1 = least; 9 = greatest).
[§]von Post wood content classes: 0 = no wood remainder; 1 = few wood remainders; 2 = many wood remainders; 3 = very many wood remainders.
**von Post root-thread content classes: 0 = no root threads; 1 = few root threads; 2 = many root threads; 3 = root threads forming main part of peat.
[‡]von Post fiber content classes: 0 = no fibers; 1 = few fibers; 2 = plenty of fibers; 3 = mainly fibers.
[§§]Data modified from Neuzil et al. (this volume).
N.S. = Not sampled.
n.a. = Not available.

were required for a quantitative analysis (Mackowsky, 1982, p. 296). The dried –2-mm peat was then coned and quartered to obtain approximately 5 g for petrographic analysis. This peat fraction was mixed with epoxy resin in molds 32 mm in diameter to construct five petrographic pellets, which were ground and polished according to standard coal petrographic procedures (American Society for Testing and Materials, 1985). This method was chosen to be representative of the entire increment for quantitative point-count analysis; the polished-block method used by Esterle et al. (1989, 1991) results in smaller subsamples that may not reflect the entire cored increment.

Petrographic methods

Previous studies have shown that brown-coal maceral terminology is equally applicable to constituents of both brown coal and peat (Teichmuller, 1982, p. 224; Esterle et al., 1989). The maceral terminology adopted for this study (Table 2) is a modification of the nomenclature of Teichmuller (1982, 1989) and Esterle et al. (1989). Inherent differences between brown coal and peat are responsible for some brown-coal maceral varieties (textinite A and B, for example) not being discriminated, whereas other types (red, red-gray, and gray textinite) of Esterle et al. (1989) are easily distinguished only in peat.

Huminite-group macerals (Table 2) are precursors to vitrinite in bituminous coals (Teichmuller, 1982, 1989). Textinite consists of ungelified cellular "woody" tissues that are microscopically similar to the cells of the living plant (Teichmuller, 1982). In the present study, textinite maceral types are defined on the basis of color, translucency, and other optical properties in reflected light (red, red-gray, and gray textinite), and the degree of fragmentation and degradation (e.g., degraded textinite) (Esterle et al., 1989). Red textinite, with its translucent deep red to yellow cell-wall color and strong internal (anisotropic) reflections, represents the best preserved woody cellular structures in the peat. In red-gray textinite, both translucent and homogeneous gray cell

TABLE 2. BROWN-COAL AND PEAT MACERALS APPLICABLE TO MODERN INDONESIAN PEATS

Maceral Group	Maceral Subgroup	Maceral Submaceral	Maceral Type
	Humotelinite	Textinite	Red textinite[*] Red-gray textinite[*] Gray textinite[*] Degraded textinite[*]
	Ulminite	Texto-ulminite	
			Eu-ulminite
Huminite	Humodetrinite	Attrinite Densinite	
	Humocollinite	Gelinite	
		Corpohuminite	Corpo/textinite[*] Corpo/detrinite[*]
Liptinite		Sporinite Cutinite Resinite Liptodetrinite	

Inertinite		Fusinite
		  Pyrofusinite
		  Degradofusinite
		  Primary fusinite
		Semifusinite
		  Pyrosemifusinite
		  Degradosemifusinite
		  Primary semifusinite	

		Sclerotinite Inertodetrinite	

From International Committee for Coal Petrology, 1971, 1975; Teichmuller, 1982, 1989; and Esterle et al., 1989.
[*]Esterle et al., 1989.

walls are present, whereas gray textinite consists entirely of homogeneous gray cell walls. In this study, degraded textinite is defined as woody tissue fragments consisting of degraded, but connected, cell walls with disrupted cell structure; the minimum size of textinite in the peat is defined as one intact cell. Smaller cell-wall fragments are assigned to the maceral attrinite. Ulminite represents an alteration of textinite through partial to complete gelification of the cell walls or cellular debris in the peat (Teichmuller, 1982) and is distinguished by its unstructured homogeneous gray appearance.

Humodetrinite macerals (attrinite and densinite) are very finely divided humic "detritus" and "gel" formed from the degradation of cell walls (Teichmuller, 1982). Attrinite detritus particles are formed from cell-wall fragments less than 20 microns in size that are loosely packed and well differentiated and have high porosity and reflectance equal to that of gray textinite. In densinite, "cementation" of colloidal-size components results in less visible porosity and poor differentiation of the dark gray gel particles (Teichmuller, 1982). Attrinite commonly contains small (1–20 microns), weakly reflecting (huminite reflectance) circular to elongate bodies probably derived from fungal hyphae and mycelia. Because of their size and intimate association with attrinite detritus, these fungal remains were included in the maceral attrinite.

Humocollinite macerals are cell infillings formed through biochemical and/or geochemical processes (Teichmuller, 1982). Gelinite displays a granular appearance, and corpohuminite appears homogeneous with a huminite reflectance. Following the terminology of Esterle et al. (1989), corpohuminite within textinite is called corpo/textinite, and bodies dispersed in the peat are corpo/detrinite.

Macerals of the liptinite group include sporinite (spore and pollen exines), cutinite, resinite, and liptodetrinite. Identifications of thin (<2 microns) exines, cuticles, or liptodetrinite particles are tentative because the particles are similar to cell-wall fragments in color and shape.

Macerals of the inertinite group are fusinite, semifusinite, sclerotinite, and inertodetrinite. Fusinite and semifusinite are represented by the maceral types pyrofusinite and pyrosemifusinite, degradofusinite and degradosemifusinite, and primary fusinite and semifusinite, depending upon their presumed origin (Teichmuller, 1982).). Pyrofusinite and pyrosemifusinite mainly result from the charring of wood and peat during fires and display high to very high homogeneous reflectances and distinct cellular structures (Teichmuller, 1982). Degradofusinite and degradosemifusinite form through slower dehydration and oxidation of the peat, or during mouldering caused by wood-decomposing fungi, and they usually display weak and variable reflectances and poorly defined cell structures (Teichmuller, 1982). Primary fusinites and semifusinites are highly reflecting clusters of cells formed within the living plant (Teichmuller, 1982). The maceral sclerotinite inclucdes all strongly reflecting fungal bodies (sclerotia, mycelia, plechtenchyme, etc.) (Teichmuller, 1982).

Optically identified minerals in the peat include quartz, zircon, rutile, and clay minerals. Pyrite and fragments of siliceous animal tests and sponge spicules are probably autochthonous.

Petrographic procedures require a point count of more than 1,000 counts per sample to obtain a representative volume-percent analysis (Mackowsky, 1982; American Society for Testing and Materials, 1985). Because of the large peat particle size (2 mm), it was necessary that five pellets of each sample be analyzed (Mackowsky, 1982, p. 296). The total number of counts required to cover the five pellets ranged from 779 to 2,509 per sample, and the average was 1,200. In conjunction with the point counts, factors relating to the degree of fragmentation, corrosion, and degradation of macerals were noted or measured for comparisons with maceral distributions. The percentages of the macerals greater than and less than 100 microns in size were recorded for each sample in order to determine the ratio of framework (plant organ and tissue fragments in the peat >100 microns in size) to matrix (tissue and cell wall fragments <100 microns in size) (Cohen and Spackman, 1977). The framework:matrix ratio was calculated from the volume-percent macerals (mainly textinite and ulminite fragments) greater than 100 microns in size and the volume-percent macerals less than 100 microns in size (mainly textinite, attrinite and densinite). Cuticles and spore/pollen exines, despite their reported toughness (Teichmuller, 1982), were observed to be susceptible to fragmentation, corrosion, and degradation. Cuticles were observed specifically for size, proportion of cuticles attached to textinite cellular structures, degree of corrosion (pitting and boring, possibly by microbes), and degradation (disintegration of the cuticle on a submicroscopic scale) exhibited by the outer surface and cuticular ledges.

Sclerotinite exhibited a wide range of morphologies, mainly attributable to anatomical features of fungi preserved in the peat, but were usually too rare to provide accurate point-count abundances. Because the relative abundance of sclerotinite may be an indicator of aerobic conditions, their frequency of occurrence relative to the volume of peat was estimated by recording the number of sclerotinite bodies present within the field of view (as opposed to directly under the cross hairs) at each point where a maceral was counted during the point count. Because the concentration of peat exposed on the polished surface varied between samples, the sclerotinite counts were prorated to number of sclerotinites encountered per 1,000 counts and plotted graphically as a survey of relative changes in sclerotinite abundances vertically through the peat of column SK6.

RESULTS

Huminite macerals are extremely abundant in both columns; liptinite and inertinite macerals and petrographic mineral matter are less abundant (Tables 3 and 4). The distribution of the huminite macerals and maceral types reveals a complex structure within the peat columns.

Petrography of the SK6 peat column

Contrasting trends are evident in the distribution profiles of the textinite maceral types (Fig. 5). Except in the surficial and

W. C. Grady and others

**TABLE 3. PETROGRAPHIC ANALYSES OF THE INCREMENT SAMPLES
FROM CORE SK6**

	P1	P2	P3	P4	P5	P6	P7	P8
Huminite	99%	98%	99%	99%	98%	99%	97%	99%
Humotelinite	48	66	70	76	76	79	77	72
Textinite	45	60	67	71	73	74	69	66
Red textinite	10	11	18	19	27	20	15	10
Red-gray textinite	4	10	8	7	12	16	21	19
Gray textinite	1	2	2	3	3	2	5	5
Degraded textinite	29	36	39	42	30	36	28	33
Ulminite	3	6	3	4	4	5	8	5
Texto-ulminite	3	6	3	4	4	5	8	5
Eu-ulminite	0	0	0	0	0	0	0	0
Humocollinite	1	1	2	2	2	4	6	2
Gelinite	0	0	1	1	1	1	3	1
Corpohuminite	1	1	1	1	1	2	4	1
Corpo-textinite	1	1	1	0	1	2	3	1
Corpo-detrinite	0	0	0	1	0	1	1	1
Humodetrinite	50	32	27	21	20	16	14	26
Attrinite	48	24	24	15	11	9	8	21
Densinite	2	8	4	6	9	7	7	5
Liptinite	1%	1%	1%	1%	1%	1%	3%	0%
Sporinite	0	0	0	0	0	0	0	0
Cutinite	0	0	0	0	0	1	1	0
Resinite	0	0	0	0	1	0	1	0
Liptodetrinite	0	0	0	1	0	0	0	0
Inertinite	1%	1%	0%	0%	1%	0%	0%	0%
Fusinite	0	0	0	0	0	0	0	0
Pyrofusinite	0	0	0	0	0	0	0	0
Degradofusinite	0	0	0	0	0	0	0	0
Semifusinite	0	0	0	0	0	0	1	0
Pyrosemifusinite	0	0	0	0	0	0	0	0
Degradosemifusinite	0	0	0	0	0	0	1	0
Primary semifusinite	0	0	0	0	0	0	0	0
Sclerotinite	1	0	0	0	0	0	1	0
Inertodetrinite	0	0	0	0	1	0	0	0
Mineral Matter	0%	0%	0%	0%	0%	0%	0%	0%
Clay minerals	0	0	0	0	0	0	0	0
Quartz	0	0	0	0	0	0	0	0
Siliceous spicules and "tests"	0	0	0	0	0	0	0	0
Pyrite	0	0	0	0	0	0	0	0
Framework:matrix ratio*	1:4	1:2	1:2	1:2	1:1	1:1	4:1	3:1

basal increments, the red and red-gray textinites are most abundant in the upper and lower parts of the SK6 peat column. Gray textinite shows a nearly steady decrease, and degraded textinite shows an almost uniform increase in abundance upward in the peat. The humodetrinite macerals attrinite and densinite are abundant in the basal peat increment (Fig. 6). Two upward increasing cycles in attrinite content are apparent and sharp reductions occur between increments P14 and P15 and P7 and P8. At three levels in the core attrinite abundance exceeds 20 volume percent, and, in the uppermost 0.25 m attrinite comprises nearly 50% of the peat. The distribution profile of densinite above the basal peat increment shows an irregular upward increasing trend with reduced abundance in the upper 2 m.

Ulminite and humocollinite are common in the middle of the column (P9–P11), coinciding with the decrease in red and red-gray textinite (Table 3).

The constitution of the peat in terms of relative size of the components is represented by the framework:matrix ratio (Fig. 7). The basal peat is predominantly matrix, but the overlying peat (12–4 m depth) is primarily framework (wood fragments from logs, stems, and roots). In the upper 4 m, especially the upper 2 m, the peat is primarily matrix components, and the greatest proportion of matrix is in the surface peat.

Liptinite macerals are minor constituents; cutinite, the most abundant component, is distributed throughout the SK6 peat column primarily as unattached, degraded fragments. Spore/

TABLE 3. PETROGRAPHIC ANALYSES OF THE INCREMENT SAMPLES FROM CORE SK6 (continued)

	P9	P10	P11	P12	P13	P14	P15
Huminite	**99%**	**99%**	**96%**	**99%**	**97%**	**92%**	**99%**
Humotelinite	63	84	83	91	91	74	55
Textinite	53	73	74	84	84	68	39
Red textinite	8	17	20	22	29	25	5
Red-gray textinite	12	16	19	20	19	12	8
Gray textinite	8	13	8	10	12	12	9
Degraded textinite	25	28	27	33	24	19	17
Ulminite	10	11	9	6	7	6	7
Texto-ulminite	10	11	9	6	7	6	7
Eu-ulminite	0	0	0	0	0	0	0
Humocollinite	4	4	2	2	2	2	4
Gelinite	1	2	1	1	1	0	0
Corpohuminite	3	2	1	1	1	1	4
Corpo-textinite	2	2	1	1	1	1	2
Corpo-detrinite	1	1	0	0	0	0	2
Humodetrinite	32	11	11	6	4	17	40
Attrinite	20	6	6	4	2	1	27
Densinite	12	5	5	2	2	16	13
Liptinite	**1%**	**1%**	**0%**	**1%**	**1%**	**1%**	**1%**
Sporinite	0	0	0	0	0	0	0
Cutinite	0	0	0	0	1	1	1
Resinite	0	0	0	0	0	0	0
Liptodetrinite	0	0	0	0	0	0	0
Inertinite	**1%**	**1%**	**3%**	**1%**	**2%**	**7%**	**0%**
Fusinite	0	0	0	0	0	0	0
Pyrofusinite	0	0	0	0	0	0	0
Degradofusinite	0	0	0	0	0	0	0
Semifusinite	0	0	3	0	1	6	0
Pyrosemifusinite	0	0	0	0	0	0	0
Degradosemifusinite	0	0	0	0	0	0	0
Primary semifusinite	0	0	3	0	1	6	0
Sclerotinite	0	0	0	0	0	0	0
Inertodetrinite	0	0	0	0	0	1	0
Mineral Matter	**0%**	**1%**	**0%**	**0%**	**0%**	**1%**	**2%**
Clay minerals	0	0	0	0	0	0	2
Quartz	0	0	0	0	0	0	0
Siliceous spicules and "tests"	0	0	0	0	0	0	0
Pyrite	0	0	0	0	0	1	0
Framework:matrix ratio[*]	2:1	6:1	5:1	1:1	2:1	3:1	1:2

Macerals are on a volume-percent mineral-matter-free basis, and minerals are on a whole-peat basis. Group maceral abundances may not equal the sums of individual macerals because of round-off differences.
[*]Ratio of "woody" peat particles >100 microns in size (framework) to <100 micron (matrix) constituents (after Cohen and Spackman, 1977).

pollen exines (sporinite), resinite, and liptodetrinite are extremely rare in the lower portions of the peat, and exines become increasingly common and more severely corroded and degraded to liptodetrinite in the upper increments (P1–P5).

Inertinite macerals are absent at the base of the peat, are low to moderate in abundance (one to seven volume percent) in the lower increments (P11–P14), and are minor constituents higher in the peat column (Table 3). In the lower increments, two types of possible "primary semifusinite" (Teichmuller, 1982) are pres-

ent and account for the modest inertinite abundances. Their origins are discussed later. Pyrofusinite and pyrosemifusinite fragments are extremely rare and were found only in the upper 0.5 m of the peat. Bright fungal remains (collectively termed sclerotinite) increase in abundance upward through the peat (Fig. 8).

Minerals are very rare and seldom encountered in the point count, except occasional clay mineral inclusions in the basal peat, which rests on a substrate rich in clay-sized quartz and organic matter (Table 1). Quartz grains of probable allochthonous origin

W. C. Grady and others

and pyrite are extremely rare and are confined to the uppermost and lowermost increments of the peat. Siliceous animal test fragments were observed in the upper 5 m (P1–P7) and near the base (P14) of the peat. Sponge spicules were observed only in the basal peat increment. The ash-yield and sulfur-content distributions reflect the observed trends in the quartz, clay minerals, and pyrite (Table 1).

Petrography of the SK5 peat column

The textinite macerals in the SK5 peat column (Fig. 9) show trends similar to those observed in the SK6 column; red and gray textinites generally decrease in abundance upward while red-gray and degraded textinites generally increase upward. The basal and surface peat increments (P8 and P1) are low in textinite macerals,

TABLE 4. PETROGRAPHIC ANALYSES OF THE INCREMENT SAMPLES FROM CORE SK5*

	P1	P2	P3	P4	P5	P6	P7	P8
Huminite	99%	97%	98%	96%	98%	98%	97%	99%
Humotelinite	37	58	51	68	76	71	82	53
Textinite	35	52	47	64	68	64	77	50
Red textinite	12	17	14	21	25	25	32	18
Red-gray textinite	8	10	13	12	12	9	17	11
Gray textinite	4	11	7	18	20	20	16	14
Degraded textinite	12	14	13	12	12	8	12	8
Ulminite	2	4	4	4	6	8	5	4
Texto-ulminite	2	4	4	3	6	8	5	4
Eu-ulminite	0	0	1	1	0	0	0	0
Humocollinite	2	7	3	3	4	3	3	7
Gelinite	1	2	1	1	1	1	1	5
Corpohuminite	2	5	2	2	3	2	2	2
Corpo-textinite	1	5	2	1	2	1	2	1
Corpo-detrinite	1	1	0	1	1	1	1	1
Humodetrinite	58	35	44	25	18	24	12	37
Attrinite	47	25	32	14	10	18	11	26
Densinite	11	10	12	11	8	6	2	11
Liptinite	2%	3%	2%	4%	2%	2%	2%	1%
Sporinite	0	0	0	0	0	0	0	0
Cutinite	1	2	1	3	1	0	2	1
Resinite	0	1	1	1	1	1	1	0
Liptodetrinite	1	0	0	0	1	0	0	0
Inertinite	0%	0%	0%	0%	0%	0%	1%	0%
Fusinite	0	0	0	0	0	0	0	0
Pyrofusinite	0	0	0	0	0	0	0	0
Degradofusinite	0	0	0	0	0	0	0	0
Semifusinite	0	0	0	0	0	0	1	0
Pyrosemifusinite	0	0	0	0	0	0	0	0
Degradosemifusinite	0	0	0	0	0	0	0	0
Primary semifusinite	0	0	0	0	0	0	1	0
Sclerotinite	1	0	0	0	0	0	1	0
Inertodetrinite	0	0	0	0	0	0	0	0
Mineral Matter	2%	1%	2%	0%	0%	1%	1%	8%
Clay minerals	0	0	0	0	0	0	0	4
Quartz	0	0	0	0	0	1	1	4
Siliceous spicules and "tests"	0	0	1	0	0	0	0	0
Pyrite	1	1	1	0	0	0	0	0
Framework:matrix ratio*	1:3	1:1	1:1	2:1	3:1	2:1	3:1	1:1

Macerals are on a volume percent mineral-matter-free basis and minerals are on a whole-peat basis. Group maceral abundances may not equal the sums of individual macerals because of round-off differences.
*Ratio of "woody" peat particles >100 microns in size (framework) to <100 micron (matrix) constituents (after Cohen and Spackman, 1977).

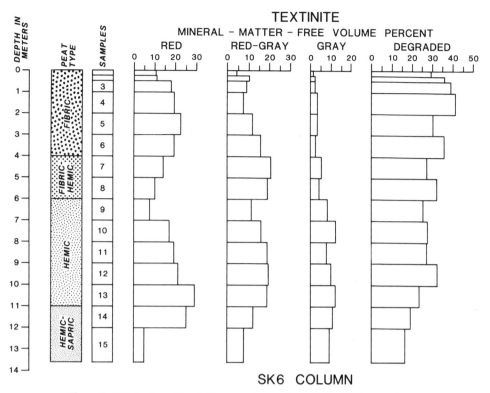

Figure 5. Distribution of textinite maceral types through the SK6 peat column.

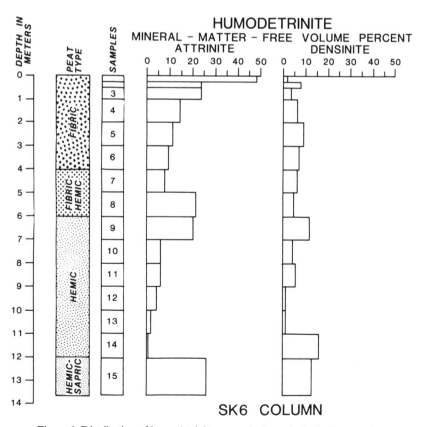

Figure 6. Distribution of humodetrinite macerals through the SK6 peat column.

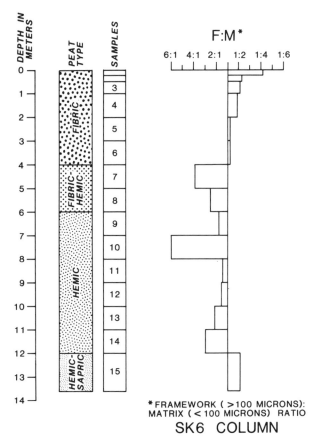

Figure 7. Framework:matrix ratio trends through the SK6 peat column. Framework particles (plant organs and tissue fragments >100 microns) and matrix (tissue and cell-wall fragments <100 microns) are calculated from the point-count data in combination with the petrographic maceral (size) descriptions.

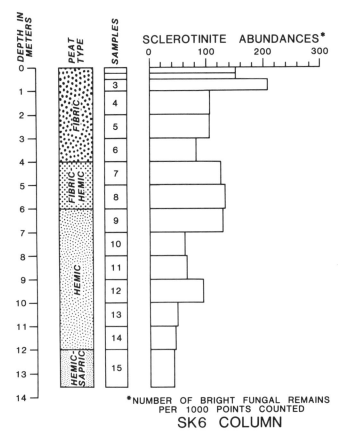

Figure 8. Distribution of semiquantitative sclerotinite maceral abundances through the SK6 peat column.

but the increment 0.25 m above the base (P7) contains abundant well-preserved red and red-gray textinites (Fig. 9). Humodetrinite macerals increase in abundance upward (Table 4); attrinite is abundant in the basal increment and in the upper 1 m (Fig. 10).

The framework:matrix ratio (Fig. 11) shows that the peat column is predominantly framework; only the surface layer is composed primarily of matrix components. Even the attrinite- and densinite-rich basal increment is constructed mainly of framework-size peat components.

Liptinite macerals are evenly distributed through the column (Table 4), predominantly as well-preserved cutinite, commonly attached to well-preserved red and red-gray textinites, and also as resinite fragments and cell fillings. Spore/pollen exines are present mainly in the upper 1 m of the peat.

Inertinite macerals are extremely rare in the SK5 column and occur almost exclusively as sclerotinite. Occurrences of "primary semifusinite" are rare; pyrofusinite is present only in the upper 0.5 m of the peat, occurring as inertodetrinite fragments less than 20 microns in size.

The ash yield and sulfur distributions (Table 1) parallel the trends in petrographic mineral matter observed throughout the SK5 peat column. Detrital clays and quartz are common only in the basal 0.25 m of the peat, but rare silt- and sand-size angular quartz grains occur dispersed throughout the column. Pyrite occurs infrequently in the upper 1 m of the peat and extends, in trace amounts, to a depth of 4 m. In these increments, most of the pyrite occurs as isolated 1–2 micron crystals, mainly in attrinite, but also as framboids 4–10 microns in diameter and as euhedral 10–20 micron crystals in textinite. In the upper 1 m of the peat, angular fragments of 10–200 micron siliceous animal tests and sponge spicules are dispersed in attrinite.

Comparison of the peat columns

A comparison of the means of the petrographic analyses of cores from the two peat sites reveals that the two columns contain similar abundances of huminite macerals (Table 5). However, the SK5 column contains more of the well-preserved (red + red-gray + gray) textinites (48.9% versus 38.4%) and less of the degraded textinite (11.1% versus 28.2%) than the SK6 column. Humodetrinite is unexpectedly more abundant in the better preserved peat of the SK5 column, primarily because the attrinite-rich surface and basal layers are proportionally of greater significance in the thinner peat. The SK5 peat also contains more liptinite (mainly

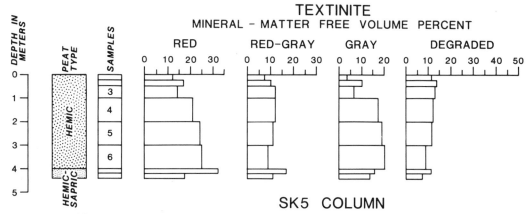

Figure 9. Distribution of textinite maceral types through the SK5 peat column.

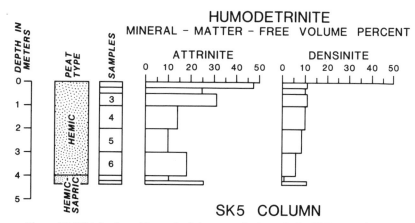

Figure 10. Distribution of humodetrinite macerals through the SK5 peat column.

resinite and well-preserved cutinite), minerals, and framework components relative to matrix than does the SK6 peat.

DISCUSSION

The precursors of vitrinite macerals are present in the modern Indonesian peats in overwhelming abundance. The formation of specific huminite/vitrinite macerals and maceral types appears to be a function of the composition of the original plant constituents and the peatification processes active within the peat. The initial stage of peatification, "humification," is a slow oxidation of the plant debris, accelerated by tropical heat and microbial activity (Teichmuller, 1982, 1989). Humification affects the upper "peatigenic layer" and may be accelerated by oxygen introduced through occasional desiccation or percolation of oxygenated water (Teichmuller, 1982). Previous petrographic studies have revealed "stages" of peatification characterized by changes in the physical appearance and chemistry of woody cell walls and tissues in peats and lignites (Barghoorn, 1952a, 1952b; Spackman and Barghoorn, 1966; Cohen, 1968; Cohen and Spackman,

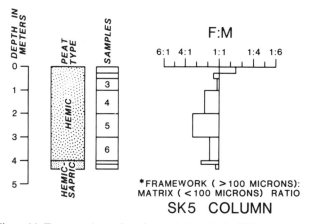

Figure 11. Framework:matrix ratio trends through the SK5 peat column. Ratio calculated by the method of Cohen and Spackman (1977).

**TABLE 5. MEAN PETROGRAPHIC COMPOSITION
OF TWO PEAT CORES**

	SK5 Column n = 8* (%)	SK6 Column n = 15* (%)
Huminite	95.8	97.4
Humotelinite	66.0	74.3
Textinite	60.0	66.0
Red textinite	21.7	17.1
Red-gray textinite	11.2	14.2
Gray textinite	16.0	7.1
R + RG + G textinite†	48.9	38.4
Degraded textinite	11.1	28.2
Ulminite	5.3	6.6
Humocollinite	3.3	2.9
Humodetrinite	26.5	20.1
Attrinite	18.0	12.9
Densinite	8.6	7.3
Liptinite	2.2	0.9
Inertinite	0.3	1.2
Mineral Matter	1.6	0.4
Framework:matrix ratio§	2:1	1:1
Precursors of >50 micron vitrinite	57.0	47.4
Precursors of <50 micron vitrinite	38.4	49.0

Calculations weighted by sample thickness.
*Number of increments.
†Combined red, red-gray, and gray textinite abundances.
§Ratio of "woody" peat particles >100 microns in size (framework) to constituents <100 micron (matrix). Ratio calculated by the method of Cohen and Spackman (1977).

1980). These recognized stages and the processes of alteration, degradation and gelification defined by Stout and Spackman (1987) affect the cell walls of woody peat components in the initial phases of peatification. The resultant products of these stages and processes are recognizable as tissue and by genus (Stout and Spackman, 1987), which in the context of the present study are constituted of structurally intact red, red-gray, and gray textinite macerals. Ensuing fragmentation and disintegration of the cell walls leading to the formation of degraded textinite, attrinite, and densinite, as shown in the present study, are equally important peatification processes, especially when considered from the viewpoint of the genesis of coal macerals.

Formation of vitrinite maceral precursors

Maceral and component size (framework:matrix ratio) distributions (Figs. 5–11) suggest that the peat in the lower two-thirds of the SK6 column (except P15) and the peat in most of the SK5 column are moderately degraded; they contain red, red-gray, and gray textinite maceral types formed from fragments of well-preserved woody tissues (Fig. 12, A and B). These textinite macerals are probably precursors to collinite in >50-micron bands (telocollinite). The peats in the upper one-third of the SK6 column and the upper 1 m of the SK5 column consist predominantly of severely degraded (i.e., degraded textinite) to

disintegrated (i.e., attrinite and densinite) cellular debris. The processes by which the peat components were degraded and disintegrated need further investigation to determine if this peat is analogous to the vitrinite-poor, exinite-rich, inertinite-rich durains and splint coals proposed to have accumulated in domed swamp complexes (Smith, 1962, 1968; Grady et al., 1985, 1992; Littke, 1987; Eble and Grady, this volume).

Degraded textinite increases in abundance upward in both columns (Figs. 5 and 9) and is the result of the fragmentation and degradation of woody cellular tissues. In the lower increments of SK6 (samples P11–P14) and SK5 (samples P3–P8), the degraded textinite appears to have formed from thin-walled tissues susceptible to mild degradation (Fig. 12C). In these increments, degraded textinite and the red, red-gray, and gray textinites contain minor amounts of both weakly and strongly reflecting (gray textinite and inertodetrinite reflectances) remains of fungal hyphae and mycelia (Fig. 12, B and D). In the upper peat layers of both columns (SK6 samples P1–P10; SK5 samples P1–P2), the degraded textinite, structurally and morphologically, appears to be derived from the same wood cells as the red and red-gray textinites. This degraded textinite contains significant amounts of red and red-gray cell fragments, liptodetrinite and inertodetrinite (Fig. 13A). Fungal remains, both gray hyphae and bright sclerotinites, are common; many fragments of degraded textinite are composed of >50% fungal remains (Fig. 13B).

Attrinite is an important constituent in the upper and basal increments of both columns and in the middle of the SK6 column. Attrinite in these peats contains numerous small (<50 microns) inclusions consisting of cell-wall fragments, shards of gelified cell walls, degraded textinite fragments, corpo/detrinite, liptodetrinite and spore/pollen exine fragments, inertodetrinite, and strongly reflecting fungal sclerotia and spores (Fig. 13C). Fused among the attrinite particles and inclusions are weakly reflecting gray fungal hyphae and mycelia, which constitute 10%–50% (rarely as much as 100%) of the attrinite volume (Fig. 13D). The concentration of fungal remains within attrinite in these peats suggests that the attrinite formed through severe fungal attack on cellular tissues. Aggregates of extremely fine angular particles, similar to the peat attrinite of this study, have been observed (by transmission electron microscopy) in proximity to possible fungal hyphae in subbituminous coals (Taylor and Liu, 1987). These aggregates also contained possible lipid-rich bacterial remains and rounded particles of humic matter, which have the appearance and properties of humic acids (Taylor and Liu, 1987). Collectively, the size of the attrinite detritus, the abundance of liptinite and inertinite maceral inclusions, and the possible inclusión of lipid-rich bacterial remains suggest that attrinite may be the precursor of matrix collinite or desmocollinite in bituminous coals.

Densinite is composed of aggregates of dark colloidal gel, which occur as small (<20 microns) blebs attached to degraded cell walls and also as a cement within attrinite. In the upper parts of the SK6 column, densinite forms homogeneous layers or pockets in the peat greater than 1,000 microns in size. These

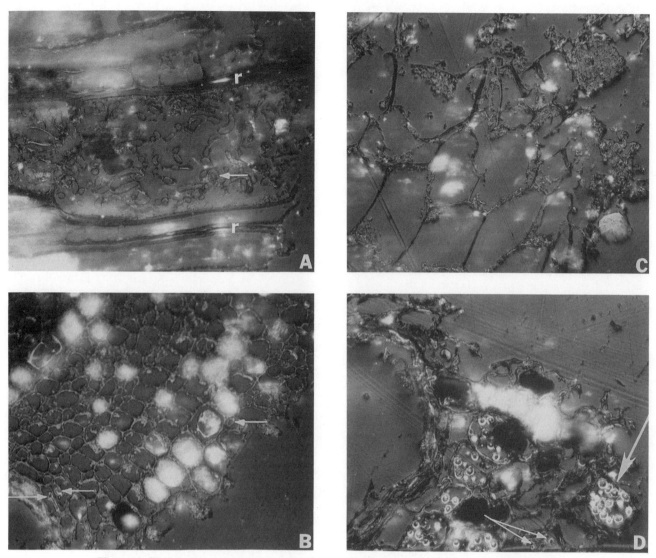

Figure 12. Textinite maceral types. A, Red textinite (r) cell walls containing fungal hyphae (arrow). Photo width = 200 microns. B, Gray textinite containing minor fungal hyphae (arrows). Photo width = 150 microns. C, Gray textinite grading into degraded textinite and attrinite in the upper right. Photo width = 150 microns. D, Gray textinite containing weakly and strongly reflecting fungal remains (arrows). Photo width = 150 microns. All photomicrographs with incident light under oil immersion.

homogeneous occurrences are weaker reflecting than attrinite or gray textinite and may also be the precursors to desmocollinite in bituminous coal.

The peats of both columns are predominantly a framework of well-preserved textinites in their lower increments and become increasingly matrix-dominant in the upper increments. These trends parallel the collinite distributions shown in the lower part (samples 11001–10998) of the main bench of the Stockton coal bed (Fig. 2). When the probable peat maceral precursors to the thick-banded collinite (red, red-gray, and gray textinite plus ulminite and corpo/textinite) are plotted opposite the probable precursors to thin-banded (<50 micron) and matrix collinite (degraded textinite plus humodetrinite, corpo/detrinite, and geli-

nite), (Fig. 14), the similarity is apparent. The upward trend, from well-preserved woody components to degraded and fragmented collinite maceral precursors in the peat columns, parallels the increase in thin-banded/matrix collinite in the Stockton coal bed (Fig. 2).

Formation of the exinite maceral precursors

Liptinite macerals are more abundant and better preserved in the SK5 column, where cutinite is commonly attached to well-preserved cellular tissues of leaves. In the degraded-textinite- and attrinite-rich upper increments of both peat columns, cutinite is fragmented, corroded, and devoid of attached textinite. Spore/pollen exines are also fragmented and corroded and occur

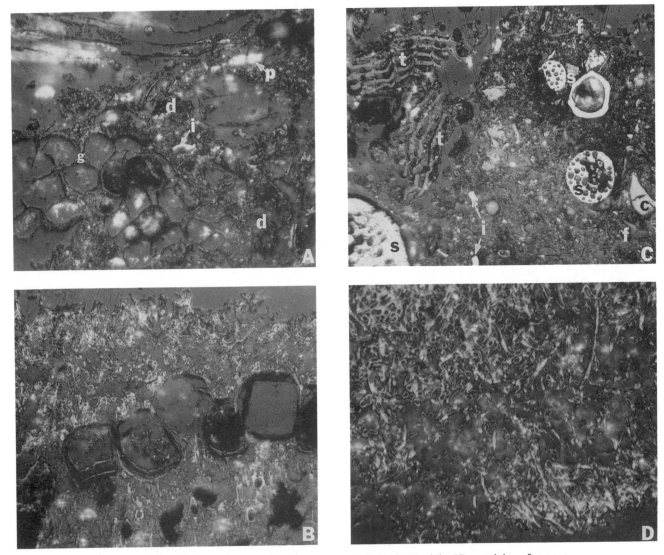

Figure 13. Textinite and humodetrinite maceral types. A, Degraded textinite (d) containing a fragment of gray textinite (g), inertodetrinite (i), and pyrite (p). Photo width = 250 microns. B, Fragment of degraded textinite with resistant gray textinite cells and extremely abundant (>50 volume percent) fungal remains. Photo width = 150 microns. C, Attrinite matrix containing fragments of red-gray textinite (t), cell wall fragments and shards (f), corpo/detrinite (c), sclerotinite (s), and inertodetrinite (i). Photo width = 250 microns. D, Attrinite containing extremely abundant fungal hyphae material. Photo width = 150 microns. All photomicrographs with incident light under oil immersion.

dispersed in attrinite as liptodetrinite. Liptinite maceral abundance in the peat (1%–4%) is similar to the 3.8% mean observed in Appalachian bituminous coals (Grady, 1979).

Formation of the inertinite maceral precursors

Inertinite macerals are rare in the peats at both sample sites and consist primarily of strongly reflecting sclerotinites (Fig. 13C). Only in the lower peat of the SK6 column (P11–P14) are inertinites minor in abundance (as much as 7%). Two types of semifusinite are common in these increments but are also observed throughout both cores. One type consists of thick, polyg-

onal cells with dark primary cell walls and secondary cell walls zoned in reflectance and hardness (Fig. 15A). The lumens are empty or contain fine granular (<1 micron) inertinite. The dark primary walls are commonly structurally attached to the cell walls of surrounding red textinite. The second type is very strongly reflecting with rectangular cellular structure (Fig. 15B). The dark, extremely thin cell walls surround a homogeneous inertinite that partially or completely fills the lumen. The cell walls are contiguous with the cell walls of surrounding red textinite (Fig. 15B). These structures form large (>2,000 microns) arcuate rings in the peat (Fig. 15B). The consistent morphology of

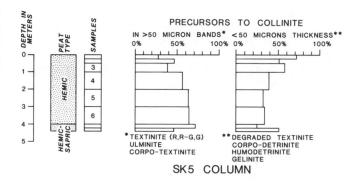

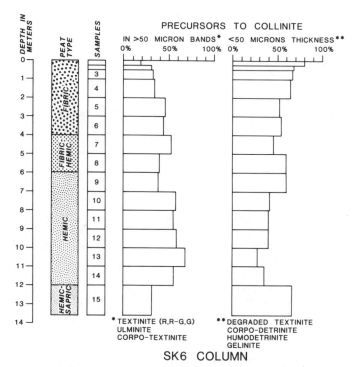

Figure 14. Probable distribution of the precursors to thick-banded (>50 microns) collinite, and matrix collinite in the SK5 and SK6 peat columns, if this peat is to be coalified as it presently exists.

these two semifusinite types, their occurrence in both peat columns, and their coupled association with textinite in roots or stems suggest that they are "primary semifusinites" and formed within the living tissues of modern plants (Teichmuller, 1982).

Strongly reflecting fungal sclerotinites are diverse and range from small multicelled and single-celled spheres to massive structures greater than 2,000 microns in size. Sclerotinites occur in textinites, but most are found within attrinite. Single-celled and multicelled types, and their fragments (inertodetrinite), are especially abundant (Fig. 13C) and are typical of types encountered rarely in Pennsylvanian coal beds (Teichmuller, 1982).

Of possible importance in the genesis of inertinite in peats and bituminous coals are sclerotinites that occur only within attrinite and are concentrated in the upper 2 m of core SK6.

These large (100–2,000 microns) fungal remains (mycelia?) consist of aggregates of irregular-sized, bright-rimmed "cells" (hyphae?) with lower reflecting granular inertinite in the "intercellular" spaces (Fig. 15C). The largest and brightest cells occur in the interior; the reflectance and size of the cells (hyphae?) decrease outward and blend into the surrounding attrinite (Fig. 15, C–E). In the intercellular spaces, the submicron-sized granular inertinite contains weakly reflecting gray cell walls and other inclusions that typically occur in attrinite. There is no evidence in the peat to indicate whether these bodies are the remains of the fungi or strongly oxidized metabolic products that formerly surrounded the hyphae. Other masses of fungal mycelia surround and incorporate the cell walls of gray and red textinite (Fig. 15F). Characteristics of these sclerotinite bodies include (1) size of 100 to greater than 2,000 microns; (2) variable cell size; (3) variable cell-wall thickness, shape, and reflectance; (4) indeterminate cell boundaries; (5) diffuse boundaries with surrounding macerals; and (6) inclusion of unoxidized macerals. These characteristics are remarkably similar to those of some degradosemifusinite forms in bituminous coals (Grady et al., 1992).

Micrinite and macrinite are unknown in peats, brown coals, and lignites. They are termed "rank inertinites" and are believed to achieve their inertinite reflectance only upon coalification to highly volatile bituminous rank (Teichmuller, 1982). Theories on the formation of micrinite and macrinite from degraded cellular materials formed within the peat (i.e., desmocollinite, degradinite) are summarized in Teichmuller (1982). In the upper 1 m of the SK5 column and the upper 7 m and the basal increment of the SK6 column, degraded textinite, attrinite, and densinite compose more than 50% of the peat volume. The great abundance (5–50 volume percent) of weakly reflecting (gray textinite reflectance) fungal hyphae and mycelia components contributing to the attrinite, and to a lesser degree within the degraded textinite, suggests that these macerals formed through severe degradation and disintegration of the cell-wall components via aerobic microbes, mainly fungi. The abundance of bright fungal remains in the upper levels of SK6 (Fig. 8) also supports an increasing presence of fungi upward in the peat. The varied reflectances of fungal hyphae (Figs. 12, B and D; 13, B and D, and 15, C–F) suggest that the fungal remains are oxidized to varying degrees. Therefore, it appears plausible that not only are the fungal remains oxidized, but that the cell-wall remnants represented by the degraded textinite, attrinite, and densinite may also be partially and variably oxidized. Opaque granular matter with similar characteristics in peats and brown coals has been postulated to be the precursor to micrinite, macrinite, and secretion sclerotinites in bituminous coals (Thiessen and Sprunk, 1936; Spackman and Barghoorn, 1966; Cohen, 1968; Mackowsky, 1975; Cohen and Spackman, 1980; Cohen et al., 1987).

The peat of the upper one-third of the SK6 column, which is compared in this study to the inertinite-rich layers of Middle Pennsylvanian bituminous coals, is very low in inertinite content. Possible explanations for this discrepancy include the following: (1) the upper layers of the modern peats may be too immature for

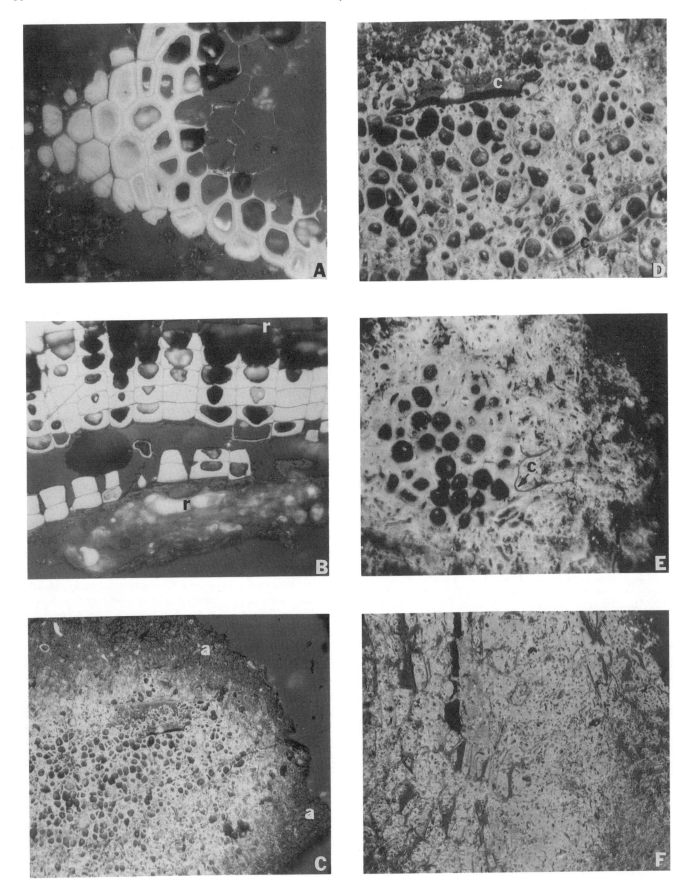

aerobic conditions to have produced inertinites similar to those in the coals; (2) the severity of the aerobic conditions may be less because of differences in climatic or hydrologic factors (e.g., less rainfall or less relief) on the modern domed deposit, resulting in less severe oxidation of the peat components; (3) precursors to "rank inertinites" (degradosemifusinite, micrinite, macrinite, and secretion sclerotinites) may be latent macerals in the peat and not directly evident.

CONCLUSIONS

The objective of this study was to determine whether sufficient petrologic evidence exists to affirm that peats accumulating in tropical, domed, ombrogenous peat swamps in Sumatra, Indonesia, are good analogues for low-ash, low-sulfur, Middle Pennsylvanian bituminous coal beds of the Appalachian region. Petrographic evidence from samples studied in reflected light under oil immersion reveals that peat components share sufficient characteristics with brown-coal macerals to warrant the application of brown-coal maceral terminology in the petrographic analysis of peat samples. Differences between the botanical composition of Pennsylvanian plants and modern vegetation preclude a direct quantitative comparison of tissue abundances. However, trends in maceral and maceral-type abundances in Middle Pennsylvanian coal beds are believed to reflect the degree of degradation of the plant debris. These trends result from changes in the geochemical, edaphic, and hydrological environment in response to the accumulation of domed ombrogenous peat (Eble and Grady, this volume). This study has revealed some comparable trends in the type and degree of degradation of modern and ancient peats, particularly pertaining to the vitrinite and exinite maceral precursors.

The abundance of framework-dominant peat, consisting of well-preserved textinite and liptinite macerals in the lower portions of the peat cores, is comparable to the abundance of vitrinite-rich, bright coal lithotypes in Middle Pennsylvanian coal beds. Framework-dominant peats and bright coal lithotypes are hypothesized to accumulate in planar, topogenous or slightly domed, ombrogenous swamps. These may represent the initial peat accumulations in a domed swamp as in the SK6 column, or the entire peat thickness of a planar or less-domed peat accumulation as in column SK5. Just as in this modern peat deposit, a coal bed consisting entirely of vitrinite-rich, bright coal lithotypes may correlate with the lower, bright benches of a domed peat accumulation (Grady et al., 1992; Eble and Grady, this volume).

The vitrinite maceral precursors in the matrix-dominant upper 4–7 m of peat in the SK6 column are primarily macerals formed from severely fragmented and degraded cell walls (degraded textinite, attrinite, and densinite). The macerals in this domed portion of the peat deposit are comparable in size and degree of degradation to the thin-banded/matrix collinite and desmocollinite in the splint coal (durain) benches of Middle Pennsylvanian coal beds. These benches contain very little telocollinite representing well-preserved woody tissues and are dominated by collinite types of a size consistent with fragments of cells and cell walls.

The geochemical conditions in the Pennsylvanian domed swamps that produced the durain and splint coal benches were probably aerobic and caused slow mouldering oxidation of the peat. Aerobic conditions are inferred from the abundance and morphology of the inertinite macerals (Eble and Grady, this volume). Although the modern peats contain few inertinite macerals, the upper 7 m of the SK6 column are rich in attrinite containing abundant (5–50 volume percent) weakly reflecting fungal hyphae and mycelia. These fungal remains are evidence of in situ aerobic conditions during peat accumulation and the subsequent degradation of the plant debris to attrinite. These conditions should be conducive to the formation of inertinite macerals. However, the exposure of the peat to aerobic conditions may not have been for sufficient duration or severity to produce inertinites similar to those in the coals. Alternatively, latent precursors to micrinite, macrinite and secretion sclerotinites, which are hypothesized to become evident upon coalification to bituminous coal rank (Thiessen and Sprunk, 1936; Spackman and Barghoorn, 1966; Cohen, 1968; Mackowsky, 1975; Cohen and Spackman, 1980; Cohen et al., 1987), may be present in the upper layers of this peat. Some of the bright fungal sclerotinites in the peat display characteristics that superficially resemble degradosemifusinite.

ACKNOWLEDGMENTS

We wish to acknowledge geologists of the Indonesian Directorate of Mineral Resources, Priatna, Supardi, A. D. Soebakti, and their coworkers for the initial study of the Siak Kanan peat deposit. The peat samples used in this study were collected during cooperative field investigations by the Directorate of Mineral Resources, Bandung, Indonesia, and the U.S. Geological Survey, Reston, Virginia, in July 1987. We especially thank Supardi for the field description of these samples, and Ronald W. Stanton, Timothy A. Moore, James C. Hower, and Richard D. Harvey for their capable manuscript review and suggestions. We acknowledge the support of the West Virginia Geological and Economic Survey in the petrographic analysis of the peat and coal samples and Ray Strawser for preparation of the diagrams.

Figure 15. Inertinite maceral types. A, Possible "primary" semifusinite similar to that observed in both peat columns. Cells on the left of this fiber cross section are filled with fine granular material. Photo width = 250 microns. B, "Primary" semifusinite displaying highly reflecting, homogeneous rectangular cells with dark cell walls merging into red textinite cell walls. Photo width = 250 microns. C, Part of a 700-micron sclerotinite (fungal mycelia?) within attrinite (a). Photo width = 250 microns. D, Higher magnification view of sclerotinite shown in C, showing granular structure and cell-wall inclusions (c). Photo width = 150 microns. E, A 200-micron fungal sclerotinite with largest and brightest cells at the center and containing inclusions of cell-wall fragments (c). Photo width = 150 microns. F, Cells of red and gray textinite enveloped in bright fungal mycelia displaying micrinite-like texture. Photo width = 250 microns. All photomicrographs with incident light under oil immersion.

REFERENCES CITED

American Society for Testing and Materials, 1985, Preparing coal samples for microscopical analysis by reflected light, *in* ASTM annual book of standards: Gaseous Fuels; Coal and Coke; Part 19: Philadelphia, American Society for Testing and Materials, p. 384–389.

Anderson, J.A.R., 1964, The structure and development of the peat swamps of Sarawak and Brunei: Journal of Tropical Geography, v. 18, p. 7–16.

Andriesse, J. P., 1974, Tropical lowland peats in Southeast Asia: Amsterdam, Department of Agricultural Research of the Royal Tropical Institute Communication 63, 63 p.

Barghoorn, E. S., 1952a, Degradation of plant materials and its relation to the origin of coal, Second Conference on the Origin and Constitution of Coal: Crystal Cliffs, Nova Scotia, Nova Scotia Department of Mines, p. 181–207.

Barghoorn, E. S., 1952b, Degradation of plant tissues in organic sediments: Journal of Sedimentary Petrology, v. 22, p. 34–41.

Cohen, A. D., 1968, Petrology of some peats of southern Florida: With special reference to the origin of coal [Ph.D. thesis]: State College, Pennsylvania State University, 352 p.

Cohen, A. D., and Spackman, W., 1977, Phytogenic organic sediments and sedimentary environments in the Everglades-Mangrove Complex: Part II, The origin, description and classification of the peats of southern Florida: Palaeontographica, v. 162 B, p. 71–114.

Cohen, A. D., and Spackman, W., 1980, Phytogenic organic sediments and sedimentary environments in the Everglades-Mangrove Complex of Florida: Part III, The alteration of plant material in peats and the origin of coal macerals: Palaeontographica, v. 172 B, p. 125–149.

Cohen, A. D., Spackman, W., and Raymond, R., 1987, Interpreting the characteristics of coal seams from chemical, physical and petrographic studies of peat deposits, *in* Scott, A. C., ed., Coal and coal-bearing strata: Recent advances: Geological Society of London Special Publication 32, p. 107–125.

Driessen, P. M., 1977, Peat soils, *in* Proceedings, Soils and Rice Symposium: Manila, p. 763–779.

Eble, C. F., and Grady, W. C., 1990, Paleoecological interpretation of a Middle Pennsylvanian coal bed in the central Appalachian basin, U.S.A.: International Journal of Coal Geology, v. 16, p. 255–286.

Eble, C. F., Grady, W. C., and Gillespie, W. H., 1989, Palynology, petrography and paleoecology of the Hernshaw—Fire Clay coal bed in the central Appalachian basin, *in* Cecil, C. B. and Eble, C. F., eds., Carboniferous geology of the Eastern United States: Washington, D.C., American Geophysical Union, Twenty-eighth International Geological Congress, Field Trip Guidebook T143, p. 133–142.

Esterle, J. S., Ferm, J. C., and Tie Y.-L., 1989, A test for the analogy of tropical domed peat deposits to "dulling-up" sequences in coal beds—Preliminary results: Organic Geochemistry, v. 14, p. 333–342.

Esterle, J. S., Moore, T. A., and Hower, J. C., 1991, A reflected-light petrographic technique for peats: Journal of Sedimentary Petrology, v. 61, p. 614–616.

Farnham, R. S., and Finney, H. R., 1965, Classification and properties of organic soils: Advances in Agronomy, v. 17, p. 115–162.

Grady, W. C., 1979, Petrography of West Virginia coals, *in* Donaldson, A., Presley, M. W., and Renton, J. J., eds., Carboniferous coal guidebook: West Virginia Geological and Economic Survey Bulletin B-37-1, v. 1, p. 240–277.

Grady, W. C., 1983, Petrography of West Virginia coals as an indicator of paleoclimate and coal quality [abs.]: Geological Society of America Abstracts with Programs, v. 15, p. 584.

Grady, W. C., Eble, C. F., and Gillespie, W. H., 1985, Relationships of palynology, petrography and coal quality in some Upper Kanawha Formation coalbeds of West Virginia [abs.]: Geological Society of America Abstracts with Programs, v. 17, p. 595.

Grady, W. C., Eble, C. F., and Ashton, K. C., 1992, Coal supplies for the 1990's: A reevaluation of Kanawha Formation splint coals in central and southern West Virginia, *in* Platt, J., Price, J., Miller, M., and Suboleski, S., eds., 1. 2: New perspectives on central Appalachian low-sulfur coal supplies: Fairfax, Virginia, TechBooks, p. 77–101.

International Committee for Coal Petrology, 1971, International handbook of coal petrology, 1st supplement to the second edition: Paris, Centre National de la Recherche Scientifique, not paginated.

International Committee for Coal Petrology, 1975, International handbook of coal petrology, 2nd supplement to the second edition: paris, Centre National de la Recherche Scientifique, not paginated.

Littke, R., 1987, Petrology and genesis of upper Carboniferous seams from the Ruhr region, West Germany: International Journal of Coal Geology, v. 7, p. 147–184.

Mackowsky, M.-Th., 1975, Comparative petrography of Gondwana and northern hemisphere coals related to their origin, *in* Proceedings, Third International Gondwana Symposium: Canberra, p. 195–220.

Mackowsky, M.-Th., 1982, Sampling and preparation of polished surface or thin sections, *in* Stach, E., Mackowsky, M.-Th., Teichmuller, M., Taylor, G. H., Chandra, D., and Teichmuller, R., eds., Stach's textbook of coal petrology: Berlin, Gebruder Borntraeger, p. 295–299.

Neuzil, S. G., and Cecil, C. B., 1984, A modern analog of low-ash, low-sulfur, Pennsylvanian-age coal [abs.]: Geological Society of America Abstracts with Programs, v. 16, p. 184.

Polak, B., 1950, Occurrence and fertility of tropical peat soils in Indonesia, *in* Proceedings, Fourth International Congress on Soil Sciences, Volume 2: Amsterdam, p. 183–185.

Polak, B., 1975, Character and occurrence of peat deposits in the Malaysian Tropics: Modern Quaternary Research in Southeast Asia, p. 71–81.

Potonie, H., and Koorders, S. H., 1909, Die tropen-sumpfflachmoor-natur der moore des produktiven carbons: Geolgisches Jahrbuch Landenstatt, v. 30, no. 1, p. 389–443.

Smith, A.H.V., 1962, The paleoecology of Carboniferous peats based on the miospores and petrography of bituminous coals, *in* Proceedings, Yorkshire Geological Society, Volume 33: Yorkshire Geological Society, p. 423–474.

Smith, A.H.V., 1968, Seam profiles and seam characters, *in* Murchison, D. G., and Westoll, T. S., eds., Coal and coal-bearing strata: Edinburgh, Oliver Boyd, p. 31–40.

Spackman, W., and Barghoorn, E. S., 1966, Coalification of woody tissue as deduced from a petrographic study of the Brandon Lignite, *in* Gould, R. F., ed., Coal science: Washington, D.C., American Chemical Society, p. 695–707.

Stach, E., 1982, The microscopically recognizable constituents of coal, *in* Stach, E., Mackowsky, M.-Th., Teichmuller, M., Taylor, G. H., Chandra, D., and Teichmuller, R., eds., Stach's textbook of coal petrology: Berlin, Gebruder Borntraeger, p. 87–140.

Stout, S. A. and Spackman, W., 1987, A microscopic investigation of woody tissues in peats: Some processes active in the peatification of ligno-cellulosic cell walls: International Journal of Coal Geology, v. 8, p. 55–68.

Taylor, G. H., and Liu, S. Y., 1987, Biodegradation in coals and other organic-rich rocks: Fuel, v. 66, p. 1269–1273.

Teichmuller, M., 1982, Origin of petrographic constituents in coal, *in* Stach, E., Mackowsky, M.-Th., Teichmuller, M., Taylor, G. H., Chandra, D., and Teichmuller, R., eds., Stach's textbook of coal petrology: Berlin, Gebruder Borntraeger, p. 219–294.

Teichmuller, M., 1989, The genesis of coal from the viewpoint of coal petrology: International Journal of Coal Geology, v. 12, p. 1–87.

Thiessen, R., and Sprunk, G. C., 1936, The origin of the finely divided or granular opaque matter in splint coals: Fuel, v. 15, p. 304–315.

von Post, L., 1924, Das genetische System des organogenen Bildungen Schwedens, *in* Proceedings, International Congress of Pedology, Volume 4, Helsinki, p. 287–304.

White, D., and Thiessen, R., 1913, The origin of coal: U.S. Bureau of Mines Bulletin 38, 390 p.

Whitmore, T. C., 1975, Tropical rain forests of the Far East: Oxford, Clarendon Press, 282 p.

MANUSCRIPT ACCEPTED BY THE SOCIETY JANUARY 14, 1993

Geological Society of America
Special Paper 286
1993

Inorganic constituents from samples of a domed and lacustrine peat, Sumatra, Indonesia

Leslie F. Ruppert, Sandra G. Neuzil, C. Blaine Cecil, and Jean S. Kane*
U.S. Geological Survey, 956 National Center, Reston, Virginia 22092

ABSTRACT

Two peat cores and one substrate core were collected from the Siak Kanan domed peat deposit in Riau Province, eastern Sumatra, Indonesia. The peat cores consisted of a 9.5-m core from the subaerial part of the domed peat and a 0.51-m core from the bottom of a lake located within the dome. Approximately 1.0 m of substrate material was obtained from a core from the bottom of the lake adjacent to the lacustrine peat core. Samples from selected intervals from the peat cores and from the underlying mineral substrate were low-temperature ashed and examined using light-optical, cathodoluminescent, and scanning electron microscopy, as well as by energy-dispersive X-ray analysis to chemically characterize the inorganic constituents. In addition, some of the ashed samples were analyzed by inductively coupled plasma-atomic emission spectroscopy and Fourier-transform infra-red spectrometry techniques.

The low-temperature ash (LTA) of the peat samples is composed of discrete grains of biologic origin and volcanic eolian origin and aggregates of micrometer to submicrometer-sized particles that are largely artifacts of low-temperature ashing. Most of the aggregate material, which makes up approximately 90% of the LTA, probably results from reactions between organically bound cations and S and N which produce sulfate and nitrate compounds. All the Na, Ca, and Mg, and most of the K and the Al in the peat are concentrated in the aggregate fraction. The sulfate and the nitrate salts are readily soluble in water.

Different suites of siliceous organisms that may have potential for environmental indicators are present in the lacustrine and domed peat intervals. In addition, rare but identifiable volcanic constituents are present in the samples of all the peat intervals. The volcanic constituents include glass shards, feldspars, zircons, and blue luminescent quartz. Most of the quartz grains from intervals of samples from the lake peat luminesce, however, in the orange plus blue range, which is indicative of a metamorphic origin. Phytoliths and siliceous organisms from samples of both the peats and the substrate intervals did not luminesce in the visible range, which is a characteristic of authigenic quartz and, apparently, biogenic silica. Biogenic silica grains, quartz grains, and other silicate minerals observed in the peat have etching features that may be caused by simple dissolution or interactions with multiple types of multiprotic organic acids. Dissolution and precipitation features are visible on quartz grains from the mineral substrate samples immediately below the peats.

*Present address: NIST, Office of Standard Reference Materials, Building 202, Room 215-A, Gaithersburg, Maryland 20899.

Ruppert, L. F., Neuzil, S. G., Cecil, C. B., and Kane, J. S., 1993, Inorganic constituents from samples of a domed and lacustrine peat, Sumatra, Indonesia, *in* Cobb, J. C., and Cecil, C. B., eds., Modern and Ancient Coal-Forming Environments: Boulder, Colorado, Geological Society of America Special Paper 286.

INTRODUCTION

Tropical, ombrogenous, domed peat deposits in Indonesia have been proposed as possible modern analogues for Lower through mid-Middle Pennsylvanian low-ash coal beds in the Appalachian Basin (Neuzil and Cecil, 1984; Cecil et al., 1985 and this volume). Although some information on Indonesian peats and peat-water chemistry is available (Anderson, 1964, 1973; Esterle et al., 1989a, 1989b; Neuzil et al., 1988 and this volume; Cameron et al., 1989; Moore, 1990), very little data on the composition and texture of the inorganic constituents of the peat is available. A systematic examination of mineral matter in a domed peat and associated substrate material from the tropics would, therefore, add to our knowledge of the mineral precursors in peats and the genesis of the minerals in coal.

A series of peat cores was collected in the Siak Kanan peat deposit of Indonesia during summer 1987 (Neuzil et al., this volume); three cores from two sites were examined in this study. One of these sites was located in a small lake within the peat dome (Fig. 1). Cores from the lake bottom consisted of (1) a 0.51-m thin bed of peat and (2) approximately 1.0 m of the underlying mineral substrate. The lake is located at 00°42′N,

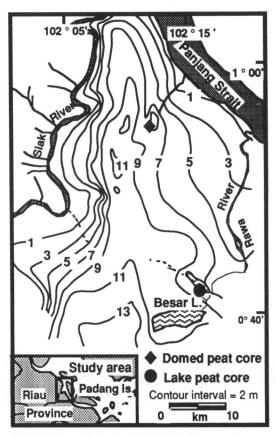

Figure 1. Peat thickness isopachs of the Siak Kanan peat deposit, north-central Sumatra, Indonesia. The thin, or lacustrine peat core (solid circle) was collected in a lake just north of Besar Lake and the thick, or domed peat core (solid diamond) was collected approximately 12 km southwest of the Panjang Strait.

102°17′E, just north of Besar Lake in the Siaksriindrapura quadrangle of Riau Province on the eastern coastal plain of Sumatra (Fig. 1). This lake covers an area of approximately 5 km², has a water depth of approximately 4.5 m, and is of unknown origin (Cameron et al., 1982). The lake, which is highly acidic (pH = 3.3), is supplied exclusively by rainfall and runoff from the surrounding Siak Kanan peat dome (Supardi et al., this volume). The peat at the bottom of the lake appears to be forming by sedimentation of fine particles of organic matter carried into the lake from the surrounding peat dome (Supardi et al., this volume). Samples from Little Lake are referred to as LL in this paper.

A core from the second site was collected from the thick portion of the peat dome approximately 17 km southwest of Panjang Strait (Fig. 1). The core was 9.5 m in length and contained approximately 8.5 m of peat and 1.0 m of underlying mineral substrate. The pH of the interstitial water from an adjacent piezometer site in the domed peat ranged from 3.7 near the top of the peat to 6.5 at the base (Fig. 2). The pH increase was correlated to an observed increase in mineral matter below 8 m. Samples from the domed peat are designated D in this paper.

SAMPLE COLLECTION

The peat core from Little Lake was collected by Pflagger corer in a rigid, split plastic tube and capped on the ends. (Note: Any use of trade, product, or firm names in this publication is for descriptive purposes only and does not imply endorsement by the U.S. Government.) The second core containing peat and approximately 1.0 m of substrate material was obtained from an adjacent site with a piston corer. The substrate core was stored in a sealed, capped, rigid, split plastic tube. The cores were collected in the eastern portion of the lake where water depth is approximately 4.5 m.

The peat from the domed deposit was sampled by using a Macaulay-type auger. The core was described in the field and divided into three parts on the basis of texture (Fig. 2). The three parts were (1) the upper 8.0 m of the core, which contained hemic, fibric-hemic, and sapric peat; (2) a 0.50-m transitional portion of the peat (located at depths from 8.0 to 8.5 m), which contained sapric peat and ooze, and clayey-peat to peaty clay; and (3) the lowermost part of the core, the substrate, which was located at depths from 8.5 to 9.5 m and was composed of organic-rich clay. The top 8 m of peat was subdivided into 1-m intervals except for the uppermost meter, which was divided into two 0.25-m and one 0.50-m intervals. All of the intervals were stored in plastic bags. The lower parts of the core, from 8.0 to 8.5 and 8.5 to 9.5 m, were stored in capped, rigid, split plastic core tubes to maintain stratigraphic integrity.

Methods

The peat core from the lake bottom was described and divided into seven intervals on the basis of the megascopic and

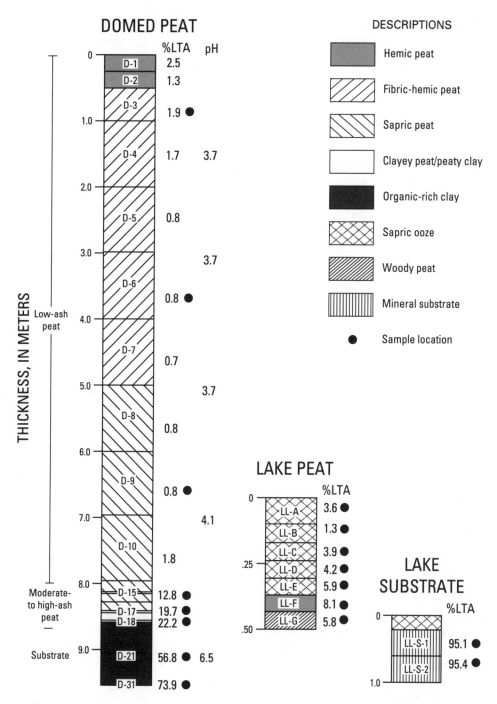

Figure 2. Representations of the cores showing peat types collected for this study from the Siak Kanan peat deposit, Indonesia. The intervals examined in this study are marked with a solid circle. They include three intervals composed of fibric and sapric peat from the low-ash part of the domed peat (D-3, D-6, and D-9); three intervals composed of sapric and clayey peat from the transitional base, or the moderate- to high-ash part, of the domed peat (D-15, D-17, and D-18); and two substrate intervals composed of organic-rich clay (D-21 and D-31). All the intervals from the lacustrine peat (LL-A through LL-G) and two lacustrine substrate intervals (LL-S-1 and LL-S-2) were also examined. Water depth above the lake cores was approximately 4.5 m. The pH of interstitial water was available for some of the domed peat intervals only and was measured in adjacent peat boreholes. Note differences in scale between the cores.

the textural descriptions, and the intervals were designated as LL-A through LL-G (Fig. 2). Representative subsamples, or splits, of each interval were obtained by cutting the core in half lengthwise. Two substrate intervals from depths of 0.22 to 0.38 m and 0.52 to 0.60 m were removed from the second lake core, which contained both sapric ooze and substrate material. The sapric ooze from the second core was not examined in this study. The lake substrate intervals were designated with sample numbers LL-S-1 and LL-S-2. All the lake peat interval splits were freeze-dried; the two lake substrate intervals were oven-dried to remove excess water. Splits were low-temperature ashed by the procedure outlined by Gluskoter (1965) for optical and chemical characterization. Low-temperature ashing methods were used exclusively for all of the samples examined in this study because the method minimizes alteration of mineral matter in coal and peat.

The transitional base and substrate parts of core from the domed peat (Fig. 2) were subdivided into 2-cm intervals in the lab. Selected intervals of the transitional base and the substrate and the peat from the top 8 m of the domed deposit were freeze-dried and then low-temperature ashed. It was presumed that the eight samples (Fig. 2) chosen for further analyses were representative of the low-ash peat (0–8.0 m), the peat at the transitional base (8.0–8.5 m), and the substrate (8.5–9.5 m) parts of the domed peat. Three interval samples from near the top (sample designation D-3), the middle (sample designation D-6), and near the bottom (sample designation D-9) of the low-ash portion of the domed peat were selected for analyses. Three additional samples from the moderate- to high-ash portion of the domed peat were also chosen. These samples are designated in this paper as samples D-15, D-17, and D-18, respectively. Two substrate samples, designated as D-21 and D-31, were also chosen for analyses.

Analyses of the LTA were performed on selected intervals from the domed and the lake cores (Fig. 3). To obtain chemical data on LTA, splits of four intervals were removed, prepared according to the methods documented in Kane and Neuzil (this volume), and analyzed by inductively coupled plasma atomic emission spectroscopy (ICP-AES) (B in Fig. 3). These intervals included those from the low-ash part of the domed peat (intervals D-3, D-6, D-9) and one from the lacustrine peat (interval LL-E). In addition, splits of the LTA from the intervals from the low-ash part of the domed peat were removed, prepared into KBr pellets, and analyzed by using Fourier-transform infra-red (FTIR) spectrometry (C in Fig. 3).

All of the LTA was water-washed to separate discrete mineral grains from soluble constituents for optical and scanning electron microscopic (SEM) analyses. The decantant was collected from samples D-3, D-9, and LL-E and filtered through a 0.22 μm polycarbonate filter. The filtrate was analyzed directly by using ICP-AES (Fig. 3). In addition, the discrete grains and the evaporated residue from the decantant from interval D-9 were analyzed individually by using ICP-AES (D in Fig. 3).

Only the LTA of the samples was analyzed. Mineralogic

identifications of the discrete grains are based on crystal habit, chemical composition as determined by energy-dispersive analysis of X-ray (EDAX), cathodoluminescent (CL) characteristics and when possible, polarized light extinction characteristics. In addition, some of the larger grains were mounted on a fiber and analyzed using single-crystal X-ray diffraction. Qualitative terms such as "common," "rare," and "trace" amounts are used to describe the relative abundance of discrete grains in the LTA.

Results

Low-temperature ash. LTA yields can be used to group the intervals from the domed portion of the peat deposit into three parts (Fig. 2). Intervals in the top 8 m are very low in ash; LTA yields are less than 2.5%, and the mean ash yield is 1.3%. Ash yields for the examined interval samples, D-3, D-6 and D-9, are 1.9%, 0.8%, and 0.8%, respectively (Fig. 2). The transitional base samples, henceforth called the moderate- to high-ash samples, at 8.0 to 8.5 m generally contain less than 25% ash. The three analyzed peat samples (D-15, D-17, and D-18) contain 12.8%, 19.7%, and 22.2% ash, respectively. Ash contents for the lowermost part of the core, from 8.5 to 9.5 m, designated as D-21 and D-31, contain 56.8% and 73.9% LTA.

Low-temperature ash yields show that the peat intervals in the lacustrine portion of the peat deposit contain less than 8.5% LTA. The mean ash yield for the peat intervals is 4.6% and the lake substrate intervals contain more than 95% LTA (Fig. 2).

Initial examination by SEM showed that the LTA from the peat intervals is composed of (1) discrete grains that are predominantly isotropic and range in size from approximately 5 to 250 μm and (2) aggregates of micrometer- to submicrometer-sized particles that cling to the discrete grains, presumably by static charges, and obscure the surface characteristics of the larger grains (Fig. 4). This aggregate material does not appear to be composed of distinct crystal phases, even at the highest magnifications of the SEM, although some SEM images show some rare extremely fine grained (<2 μm) clay minerals. Chemical composition, as determined by EDAX, varies from analytical point to analytical point. The aggregate material is easily disturbed by gentle air movement and constitutes a low-density fraction of the LTA. The low-density material was separated from the discrete grains in the water-washing procedure in order that the two distinct phases in the LTA could be examined and analyzed separately.

Discrete Grains: Domed-peat–low-ash intervals (D-3, D-6, D-9). Discrete grains in the LTA of the low-ash intervals of the domed peat are both biogenic and nonbiogenic in origin. Visual estimates of the grains in these intervals show that the majority (>~75%) are of biologic origin (Fig. 5A). Examination of these grains using optical microscopy and single crystal X-ray diffraction indicate that most of the grains are isotropic and amorphous. ICP-AES analyses of the separated discrete grains (D in Fig. 3) from interval D-9 show that silicon is dominant (Table 1). SEM images show that the biogenic grains from all three intervals are similar (Table 2). Abundant siliceous chryso-phyte cysts, phytoliths (Figs. 5A and 5B), and organisms that resemble filose testacid amoebae (Fig. 5C) are present in these samples. Surface pitting and etching features were observed on some of the grains (Fig. 5A).

Quartz is the dominant phase in the nonbiogenic fraction of the discrete grains. Quartz grains appear to be etched (Fig. 5D);

V-shaped pits and honeycomb-like textures are observed. Although most of the quartz grains are too small ($<10~\mu$m) to obtain cathodoluminescent (CL) spectra, the larger grains (>10 μm) show emission in the blue range (Fig. 6).

In addition to quartz, glass shards, identified by morphol-ogy, extinction characteristics, and chemistry as determined by EDAX, are also observed in the nonbiogenic fraction of the

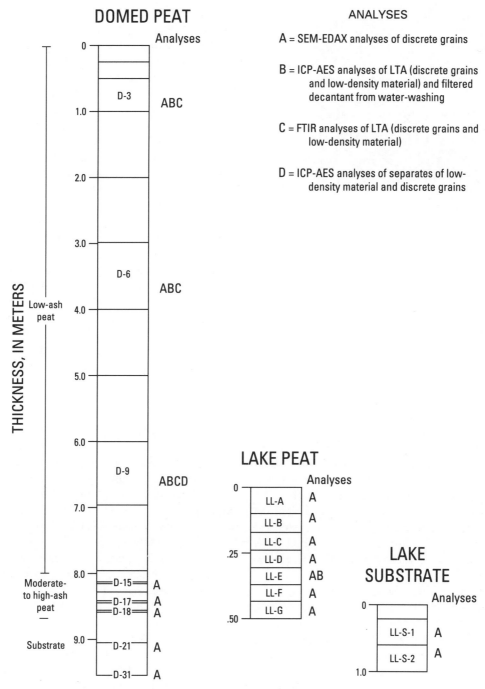

Figure 3. Analyses scheme for LTA interval samples examined. ICP-AES = inductively coupled plas-ma–atomic emission spectroscopy; SEM-EDAX = scanning electron microscopy–energy-dispersive X-ray; FTIR = Fourier-transform infra-red spectrometry; LTA = low-temperature ash. Care was taken to make each split representative of the entire sample. Note differences in scale between the cores.

Figure 4. SEM photomicrograph of submicrometer-sized aggregates of low-density material (LDM) clinging to discrete grain of volcanic glass (G). Because the low-density material partially obscured the discrete grains, the two types of material were separated by water-washing and examined individually. Scale bar is 10 μm.

LTA (Fig. 5E). They are most abundant near the top of the peat (interval D-3). The shards are composed of Si, Al, K and minor amounts of Ti, Mg, and Fe. Potassium feldspar (K-feldspar) grains are also observed in the LTA and appear to be etched.

Heavy minerals are present only in trace amounts in intervals D-3, D-6, and D-9 (Table 2). These minerals include iron oxides, barite, and an unidentified Pb-Mn-Ca-alumino silicate. Only two grains of pyrite (Fig. 5F), from the LTA of interval D-3, were identified by using SEM-EDAX.

Discrete grains: Domed-peat–moderate- to high-ash intervals (D-15, D-17, D-18). SEM examination of the discrete grains from the moderate- to high-ash intervals of the domed peat show that the biogenic grains are not as abundant as nonbiogenic grains, which constituted more than approximately 60% of the discrete grains. The tests of organisms that resemble filose testacid amoebae are absent in the LTA from the moderate-to high-ash part of the domed peat. Some siliceous phytoliths and chrysophyte cysts, however, are present.

In the nonbiogenic fraction, dissolution features on silicate minerals are more abundant in these intervals than they are in the overlying low-ash intervals. Compositionally, the nonbiogenic grains are similar to the overlying intervals (Table 2). The relative abundance of different types of nonbiogenic grains, however, is quite different. Quartz was rare. One of the few quartz grains observed is euhedral and highly etched (Fig. 7A-B); it luminesces in the blue range. Etched grains of amorphous silica, mica, glass shards, and K-feldspar (Fig. 7D) make up the bulk of the LTA (Table 2). Zircons are present but not common, and some of the

grains show surface textures that could be the result of etching. Pyrite is observed in trace amounts (Fig. 7C) and is more abundant than in the overlying low-ash intervals. Rutile or anatase, ilmenite, mica, and Mn- and Fe-Ti-Mn oxides are also present in trace amounts. Two grains composed of calcium only (calcite or Ca-oxalate?) were observed with boring-like features (Andrejko et al., 1983).

Discrete grains: Lacustrine peat (intervals LL-A through LL-G). There are approximately equal porportions of nonbiogenic and biogenic discrete grains in the lacustrine peat samples. In contrast to the domed peat, siliceous sponge spicules (Fig. 8A) dominate the biogenic fraction of the LTA (Table 2). Organisms that resemble filose testacid amoebae are absent in these intervals; chrysophyte cysts (Fig. 8B) and phytoliths (Fig. 8C) are present. Chrysophyte cysts commonly showed mace-like shapes, and apertures or pores (Adam and Mahood, 1981) are observed on some of them (Fig. 8B). The overall abundance and size of the biogenic grains decrease towards the base of the lacustrine peat. Boring-like features (Andrejko et al., 1983) that measure 1 to 2 μm in diameter are also observed on the sponge spicules, but not on chrysophyte cysts and phytoliths.

In the nonbiogenic fraction, quartz grains predominate and are larger in size than those found in the domed peat. The mean size of the quartz grains decreases from approximately 30 to 50 μm at the top to approximately 10 to 30 μm at the base of the lacustrine peat core. CL spectra of quartz in intervals LL-A, LL-C, LL-F and LL-G show two types of emission—grains that emit light in the orange range and have a minor blue component (Fig. 6), and those that emit light only in the blue range (Fig. 6), which is similar to the quartz in the domed peat. Blue luminescent grains are rare, can show conchoidal fracture, and sometimes appear to be less etched than orange plus blue luminescent grains. Like the quartz in the domed peat, V-shaped and honeycomb-like textures are observed on grain surfaces (Fig. 8D).

Other nonbiogenic discrete grains include glass shards, K-feldspar, and mica. All are less common than quartz, but together they constitute as much as 5% of the LTA in some of the sample intervals. K-feldspar grains are deeply etched. Iron oxide phases dominate the heavy mineral suite. Also present are trace amounts of pyrite, ilmenite, rutile or anatase, zircon, and amphiboles. Other heavy minerals, such as apatite, an Al-Ce phosphate (florencite?), pyroxene, and compounds that include Ni-Fe, Cr-Fe, Pb and Bi oxides are observed in the lacustrine peat intervals but not the domed peat intervals. Rare micron-sized flakes of gold and copper are also observed in the lacustrine peat.

Substrate samples. Major minerals identified in the substrate material from the lacustrine substrate core and the domed peat core are similar; however, they differ in composition from the overlying peat intervals (Table 2). Biogenic grains are rare. Sponge spicules and organisms that resemble filose testacid amoebae are absent, but siliceous diatoms and chrysophyte cysts were observed from both substrate sets (Fig. 9A). Etched quartz (Fig. 9A), amorphous silica, K-feldspar, kaolinite(?), and muscovite (Table 2) dominate the discrete grain fraction of both sub-

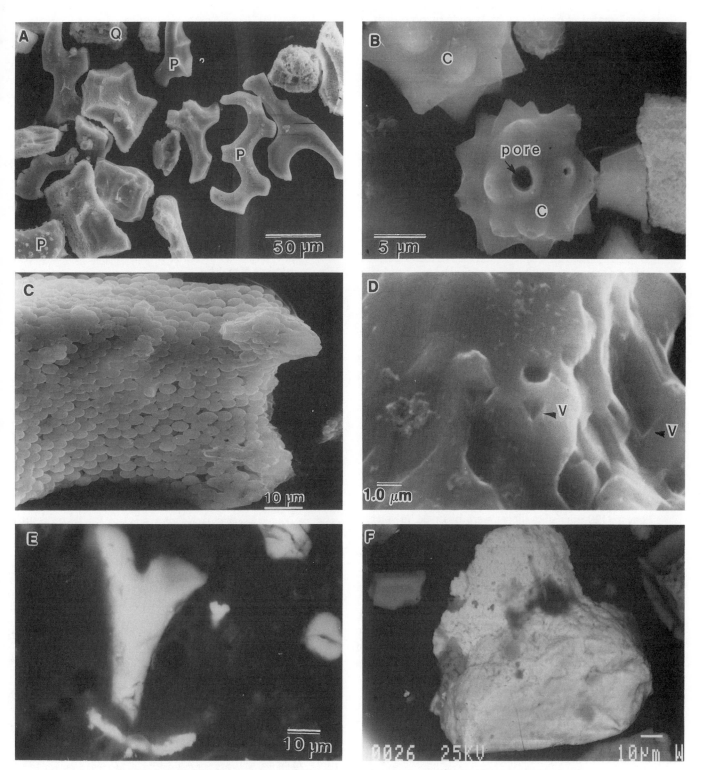

Figure 5. SEM secondary electron (SE) and backscatter electron (BSE) photomicrographs of the discrete grains in the LTA from low-ash intervals D-3, D-6, and D-9 of the domed peat. Unless indicated below, all grains are composed of silica only (based on SEM-EDAX). (A) SE image of discrete grains from interval D-9. The majority of these grains are biogenic in origin. P = phytolith and Q = etched quartz. (B) SE image of chrysophyte cyst, C, showing pores. (C) SE image of organism that resembles a filose testacid amoeba. (D) SE image of close-up of a quartz grain showing V-shaped etch pits, V. (E) BSE image of glass shard composed of Si, Al, Fe, and minor Ti and Mg. (F) One of two pyrite grains observed in the low-ash portion of the domed peat (BSE image).

**TABLE 1. RESULTS OF ICP-AES ANALYSES OF DISCRETE
GRAIN FRACTION FROM INTERVAL D-9**

Oxide	Weight Percent
SiO_2	81.5
Al_2O_3	3.7
Fe_2O_3	6.9
CaO	2.3
MgO	1.6
Na_2O	2.1
K_2O	<0.5
TiO_2O	0.4
P_2O_5	0.8
MnO	0.3

The discrete grains represent about 7 weight percent of the LTA.

strates. CL spectra were obtained on selected quartz grains, and the majority luminesce in the blue plus orange range. Only a few quartz grains luminesce in the blue range.

The LTA from the substrate intervals from Little Lake (intervals LL-S-1, LL-S-2) contain mostly quartz, K-feldspar, and amorphous silica. Pyrite (Fig. 9C); ilmenite, gold, Fe oxide, zircon, and amphibole are also present. Spinel and barite are rare. In addition, a rare-earth alumino-silicate phase that sometimes contains phosphate (Fig. 9B) was present in both lacustrine substrate intervals but not in the domed peat substrate. Substrate samples from the domed peat (intervals D-21, D-31) were dominated by quartz, K-feldspar, amorphous silica, abundant pyrite (Fig. 9C), an impure silica phase that has minor amounts of Fe and Mg, zircon, amphibole, mica, and a rare-earth silicate phase.

Long (~120-μm) silica fibers were observed in interval sample LL-S-1. The fibers appear to have replaced plant trachea cells (Fig. 9D). The fibers are nonluminescent and are, therefore, authigenic in origin (Zinkernagel, 1978).

Low-density material from the domed and lacustrine peats. EDAX analyses of the low-density fraction of the LTA samples from the domed and the lacustrine peat show varying amounts of Si, Al, K, Ca, S, Mg, and, sometimes, Fe. With the exception of Si and Fe, these elements are concentrated only in the low-density fraction. Most of this fraction is submicrometer in size, and although rare micrometer-sized clay minerals do occur, they do not comprise a significant proportion of the material.

The low-density material from interval D-9, from near the bottom of the low-ash interval of the domed peat, constitutes 93 weight percent of the LTA. ICP-AES analyses of the material (Fig. 3) shows that the residue primarily contains Al, Ca, Mg, and Na (Table 3). Only minor amounts of Si and Fe are present. The sum of the concentrations of oxides in the residue, however, does not equal 100%; 80% of the low-density material (Table 3) is unaccounted for in the ICP-AES analyses. Qualitative FTIR analyses, which were performed on the LTA from the low-ash intervals of the domed peat core (C in Fig. 3) by John Kovach (Department of Energy, 1989, written communication), show

**TABLE 2. IDENTIFICATION OF DISCRETE GRAINS IN THE LTA
FROM THE DOMED AND LACUSTRINE PEAT SAMPLES**

Domed Peat

Low-ash intervals

D-3 Testacid amoebae(?), chrysophyte cysts, phytoliths, quartz, glass shards, mica, kaolinite(?), impure silica with Al, barite, pyrite, iron oxides.

D-6 Testacid amoebae(?), chrysophyte cysts, phytoliths, quartz, glass shards, mica, kaolinite(?), impure silica with Al, iron oxides.

D-9 Testacid amoebae(?), chrysophyte cysts, phytoliths, quartz, glass shards, hematite, Pb-Mn-Ca silicate.

Moderate- to high-ash intervals

D-15 Amorphous silica, quartz, glass shards, mica, zircon, barite, pyrite, plagioclase, anatase or rutile, amphiboles, ilmenite, Ca (calcite or oxalate).

D-17 Quartz, amorphous silica, mica, ilmenite, rutile or anatase, pyrite, spinels(?), impure silica phases (minor Al and Fe), K-feldspar, glass shards.

D-18 Quartz, amorphous silica (some with Fe and Mg), illite, kaolinite(?), K-feldspar, Mn oxide.

Substrate

D-21 Kaolinite(?), mica, quartz, amorphous silica, K-feldspar, pyrite, zircon, amphibole.

D-31 Quartz, amorphous silica, pyrite, ilmenite, Fe-oxides, zircon, mica, amphibole, diatoms, K-feldspar, rare-earth silicates.

Lacustrine Peat

LL-A Sponge spicules, chrysophyte cysts, phytoliths, glass shards, quartz, mica, rutile or anatase, amphiboles, zircon, Fe-oxides.

LL-B Sponge spicules, chrysophyte cysts, mica, quartz, apatite, Fe-oxides.

LL-C Sponge spicules, chrysophyte cysts, quartz, amorphous silica.

LL-D Sponge spicules, chrysophyte cysts, glass shards, wickmanite(?), Fe-oxides, amorphous silica.

LL-E Sponge spicules, quartz, amorphous silica, mica, Fe-oxides, pyrite, K-feldspar, FeNi (oxide?), zircon, jacobsite(?), Au.

LL-F Sponge spicules, amorphous silica, impure silica, mica, zircon, quartz, K-feldspar, glass shards, rutile or anatase, mica, ilmenite.

LL-G Sponge spicules, amorphous silica, quartz, K-feldspar, mica, barite, pyrite, bismuth, amphiboles, spinels, crandallite, Pb and Cu oxides, pyroxene, ilmenite.

Lacustrine Substrate

LL-S-1 Quartz, K-feldspar, ilmenite, mica, rutile or anatase, rare-earth alumino-silicates (some with phosphate), crandallite,

LL-S-2 zircon, diatoms, amorphous silica.
Quartz, K-feldspar, rare-earth alumino-silicates (some with phosphate), pyrite, zircon, diatoms, amorphous silica, spinel, barite.

Mineralogic identifications are based on crystallographic habit, optical microscopy, and chemistry, as determined by an energy-dispersive X-ray unit attached to the scanning-electron microscope. (?) = tentatively identified. Order of occurrence represents relative concentration.

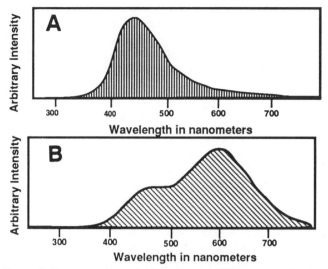

Figure 6. Representative cathodoluminescent (CL) spectra of quartz grains in the LTA samples of the domed peat (A) and the lacustrine peat (B). Peak at approximately 440 nm is in the blue range and is indicative of volcanic quartz (Zinkernagel, 1979). The double peak at around 460 and 610 nm shows emission in the orange plus blue range, which is often not distinguishable to the eye, and which is indicative of metamorphic quartz (Zinkernagel, 1978).

that the ash also contains an abundance of sulfates and minor nitrates.

Direct ICP-AES analyses of the filtered decantant from two intervals in the low-ash portion of the domed peat (D-3, D-9) and one from the lacustrine peat (LL-E) (C in Fig. 3) show that the majority (~90%) of the low-density material is present as a water-soluble fraction (Table 3). Approximately 100% of the Na, Ca, and Mg in the LTA is water soluble. Potassium and Al are also partially soluble in water. In interval D-3 near the top of the low-ash portion of the domed peat, K is 100% soluble, whereas in interval D-9 near the base of the low-ash portion of the domed peat, K is only 50% soluble. Aluminum is only 18% soluble in LTA of interval D-3, but it is 83% soluble in interval D-9. In the LTA from the lacustrine peat sample LL-E, only 8% of the Al is water soluble, and K is not soluble.

DISCUSSION

The LTA yields for the lacustrine and the domed peats are very low, and the inorganic constituents that are present in the LTA can be separated into the following suites: (1) discrete grains of biogenic origin, (2) discrete grains of nonbiogenic, probably volcanic, eolian origin, and (3) a low-density material composed of micrometer to submicrometer aggregates. In the LTA of all peat intervals examined, the low-density material is more abundant than the discrete grain suites.

Relative abundance of the discrete grains that constitute suites 1 and 2 in the previous paragraph appear to vary with peat type and/or ash content. In the low-ash portion of the domed

peat, grains of biogenic origin are much more abundant than those of nonbiogenic origin, but in the basal transitional, or moderate- to high-ash portion of the domed peat, nonbiogenic grains are more abundant than biogenic grains. Nonbiogenic and biogenic grains are roughly equal in abundance in the lacustrine peat.

Four distinct types of biogenic grains are observed in the peat and all are composed solely of silica. Chrysophytes and phytoliths are present in the low- and the moderate- to high-ash portions of the domed and the lacustrine peats, but organisms that resemble filose testacid amoebae are present only in the low-ash portion of the domed peat. The amoebae-like organisms may have some value as an environmental indicator of very low-ash peats if their absence in the moderate- to high-ash part of the domed peat and the lacustrine peat is the result of unfavorable growth conditions (higher pH and inorganic material content at the base of the domed peat or too deep a water cover over the lacustrine peat). Alternatively, their absence may simply be the result of decreased preservation potential in comparison to phytoliths and chrysophyte cysts.

In contrast, sponge spicules are present only in the LTA from the lacustrine peat intervals obtained from Little Lake (Table 2). Standing water may be required for optimum growth of sponges, and their absence in the domed portion of the peat deposit may be the result of a lack of perennial standing water. If the presence of sponge spicules in peats implies standing water, then it may be possible to use their presence to estimate roughly the geometry of paleopeat swamps or to delineate subenvironments within an individual peat swamp; peats with constant standing water cover, such as the lacustrine part of the domed peat deposit in Indonesia, may contain spicules, but domed portions of deposits may not. The preservational potential of the spicules examined in this study is not known; however, freshwater siliceous sponge spicules have been reported in low-rank coal from the Venado Formation coal deposits in Costa Rica (Sanchez et al., 1985); some of the coal beds are interpreted to be lacustrine to restricted lacustrine in origin (Sanchez et al., 1987).

The second discrete grain suite observed in samples from lacustrine and domed peat intervals is nonbiogenic in origin. Because neither sample site is fluvially influenced (Supardi et al., this volume), we suggest that the nonbiogenic, discrete grain fraction of the LTA is eolian in origin. In addition, it is inferred that the eolian material in the domed and, probably, the lacustrine peat is predominantly volcanically derived on the basis of the presence of blue luminescent quartz, which is indicative of volcanic origin (Zinkernagel, 1978), and glass shards, zircon, and euhedral amphibole and pyroxene (Table 2). These mineral suites are typical of volcanic ejecta material. In addition, the peat deposit is near volcanic sources in the Sumatra highlands and the Indonesian archipelago.

Not all the quartz examined was volcanic in origin. The quartz from the substrate material from both cores luminesced in the orange range and had a minor blue component, which is indicative of metamorphic origin (Zinkernagel, 1978). In addition, most, but not all, of the quartz in the LTA from the lacustrine peat is metamorphic in origin, although glass shards and accessory minerals that are typical of volcanic ash-fall material are also present. Burrowing organisms could have mixed metamorphic quartz from the substrate into the lacustrine peat; however, burrows were not observed in the core. Additional work needs to be done to ascertain the actual mechanism of incorporation of orange plus blue luminescent quartz into the lacustrine peat.

The interpreted volcanic origin of the majority of the minerals comprising the nonbiogenic discrete grains in both sets of peat samples may have implications for volcanic material in coal. Some partings in coal beds are the result of volcanic ash falls (Triplehorn and Bohor, 1986). Indonesia is an area of active volcanism, but volcanic ash layers have not been reported from coastal lowland peat deposits in Sumatra and Malaysia. The absence of discrete volcanic ash-fall layers in these peats suggests that predominant wind currents preclude much eolian volcanic material from entering into the Siak Kanan peat deposit. Minor amounts of volcanic minerals, however, are present throughout

the peat, and many of these minerals show signs of chemical dissolution. Dissolution occurs fairly rapidly in the Siak Kanan peat deposit. Etched features were observed on quartz grains from the top interval from the lacustrine peat (LL-A); that interval was estimated to be less than 800 years old based on ^{14}C age dates on a core from another site in the lake (Supardi et al., this volume). This suggests that dispersed volcanic ash material in coal is likely to be obscured by dissolution processes that occurred during peat formation. Triplehorn et al. (1991) and Crowley et al. (1990) were able to identify dispersed volcanic ash partings in the Wyodak-Anderson coal bed in the Powder River Basin only through careful examination of chemical and petrographic data: the dispersed partings were not recognizable megascopically because of extensive chemical alteration and dissolution. Alteration of thicker volcanic parting material can also occur during peat formation but the partings are recognizable in coal (Raymond et al., 1985; Crowley et al., 1989).

Quartz is thought to be relatively stable in sedimentary en-

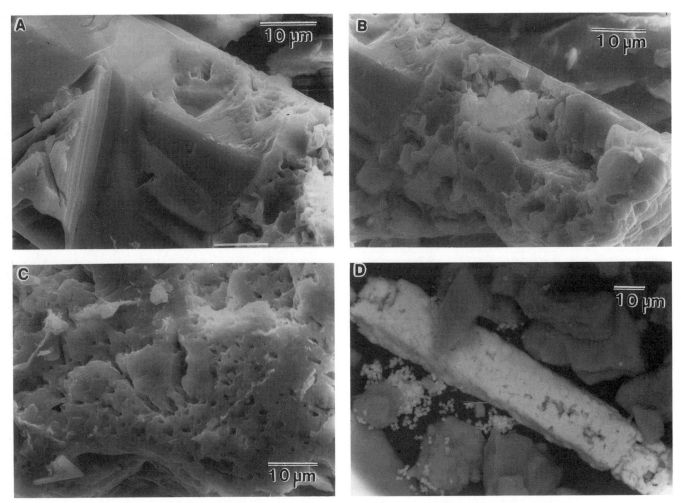

Figure 7. SEM secondary electron (SE) and backscatter electron (BSE) photomicrographs of some of the nonbiogenic discrete grains in the LTA from the moderate- to high-ash intervals of the domed peat; (A) and (B) blue luminescent euhedral quartz showing V-shaped and honeycomb-like etch pits (SE); (C) etched K-feldspar (SE); (D) replacement pyrite composed of individual crystals measuring 1 μm (BSE).

vironments, but most of the quartz examined in the present study shows distinct etching and dissolution features that are not typically observed on quartz grains from modern lake, stream, or marine environments. Quartz can be dissolved by water if the chemical system is undersaturated with respect to silica; however, the process is extremely slow. Although the mechanism for the dissolution features observed is unclear, the high organic acid concentration in the peat may promote very rapid dissolution of quartz and other silicates. Rapid dissolution of silicate minerals by organic acids at near neutral pH is well documented (Tan, 1980; Edman and Surdam, 1986; Kharaka et al., 1986; Seigel et al., 1986; Bennett and Seigel, 1987; Hansley, 1987; Bennett et al., 1988, 1991), but the rate of dissolution for quartz decreases as the pH decreases (Bennett et al., 1988, 1991). Indeed, etching features are most prominent near the interface between the low- and the moderate- to high-ash parts of the domed peat where the pH increases (Fig. 2). Dissolution features, however, are observed on quartz grains from peat intervals that have a low pore

water pH (<4). Experimental dissolution studies at low pH have been carried out by using either single multiprotic or multifunctional monoprotic organic acids (Bennett et al., 1988), but the interstitial peat water from other areas in the Siak Kanan peat deposit contains a mixture of complex organic acids (W.H. Orem, USGS, 1989, oral communication). Dissolution features that we observed on quartz and other silicates in peat at low (<4) pH may be a function of the multiple types of complex organic acids present in the peat water. The decrease in grain size of quartz grains from top to bottom in the lacustrine peat may be the result of increased quartz dissolution through time. Dissolved silica from the dissolution of discrete grains and silica removed from the substrate and recycled by living plants (Ruppert et al., 1985) could be the source of silica for sponges, chrysophytes, phytoliths, and for organisms that resemble testacid amoebae.

Excess silica and other elements released from the dissolution of silicate mineral phases may be precipitating in the substrate of the peats. Nonluminescent silica fibers are observed in

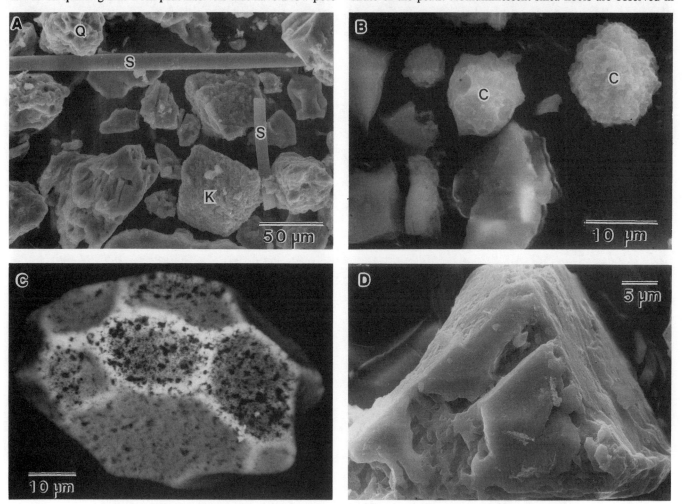

Figure 8. SEM secondary electron (SE) and back-scatter electron (BSE) photomicrographs of both biogenic and nonbiogenic discrete grains from the LTA of the lacustrine peat intervals: (A) siliceous sponge spicules, S; etched K-feldspar, K; and etched quartz, Q (SE; scale bar = 50 μm); (B) chrysophyte cysts, C, with etched appearance (SE); (C) siliceous plant phytolith (BSE); and (D) honeycomb-like etch features in quartz (SE).

94 *L. F. Ruppert and others*

the lacustrine substrate (Fig. 9D). Rare-earth crandallite group minerals and other rare-earth, alumino-silicates appear to be precipitating in the lacustrine substrate samples. Although more work must be done to verify their origin, rare-earth elements can be leached from volcanic minerals by organic acids at low pH (McLennan, 1989) and may accumulate in the substrate in humid tropical climates that have high rainfall (Mariano, 1989) in authigenic mineral phases.

The third suite of inorganic constituents in the samples is a low-density material that accounts for more than 93% of the total ash in interval D-9 from the domed peat. Although the percentage of low-density material was not measured for other intervals in either the domed or the lacustrine peat, the LTA of all the samples was usually dominated by the low-density phase. The material, although it does contain rare micrometer-sized clay minerals which are probably the result of incomplete separation of the discrete grains from the low-density material during the water-washing procedure, is primarily composed of water soluble Al, Mg, Na, K, Ca (Table 3), and sulfates and minor nitrates. Because the ICP-AES data does not detect S and N and the FTIR data is qualitative, the oxides comprising the low-density material could not be quantified. Kane and Neuzil (this volume), however, did verify the presence of abundant sulfates in high-temperature ashes of other Indonesian peat samples. The presence of sulfates and nitrates would be expected to lower the oxide compositional totals, as would incomplete low-temperature ashing and hydrous mineral phases which are present in similar types of material (Foscolos et al., 1989).

The sulfates and nitrates are formed from the oxidation and fixation of organic S (Miller et al., 1979) and N (Painter et al., 1980) during the low-temperature plasma ashing. Cations, which include Al^{+3}, Mg^{+2}, Na^+, K^+, and Ca^{+2}, are probably present in the peat as metal-organic complexes (Kharaka et al., 1986; Neuzil et al., this volume) and other organic-cation complexes. The ca-

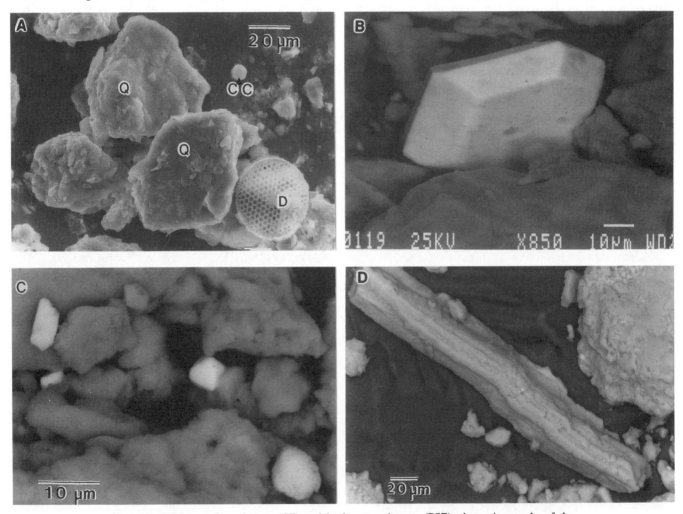

Figure 9. SEM secondary electron (SE) and backscatter electron (BSE) photomicrographs of the discrete grains in the LTA from the substrate intervals from the domed and lacustrine peat: (A) etched quartz, Q, with siliceous diatom, D, and chrysophytes, CC, from the domed peat substrate sample D-21 (SE); (B) Ce-Nd-Cs-La-Ba alumino-silicate from the lacustrine sample LL-S-1 (BSE); (C) pyrite framboids and dodecahedrons (bright areas) from the domed peat substrate sample D-21 (BSE); and (D) nonluminescent silica fiber from the lacustrine substrate sample LL-S-1 (BSE).

TABLE 3. RESULTS FROM ICP-AES ANALYSES OF LOW-DENSITY FRACTIONS

Oxide	Weight Percent D-9[*]	Percent in Filtered Decantant[†]		
		D-3	D-9	LL-E
SiO_2	0.4	0	0	0
Al_2O_3	2.4	18	83	8
Fe_2O_3	0.1	0	0	0
CaO	4.6	100	100	100
MgO	8.5	100	100	100
Na_2O	2.1	100	100	100
K_2O	0.4	100	100	100
TiO_2	<0.01	100	50	0
P_2O_5	0.02	......	......	......
MnO	0.08	......	......	......

Low-density material represents about 93 weight percent of the LTA.
[*]Weight Percent D-9 = weight percent of the oxide in the low-density fraction of interval D-9.
[†]Percent in Filtered Decantant = percentage of the oxide from the whole LTA in the filtered decantant.

tions apparently react with the sulfate and the nitrate that are produced during the low-temperature ashing procedure to create the water-soluble compounds detected in the low-density fraction of the LTA. These reactive cations continue to be partially water and/or ammonium acetate soluble in some lignites, subbituminous, and bituminous coals (Finkelman et al., 1990).

One of the purposes of examining minerals and mineral matter in modern peat is to understand mineral-matter compositions and variations in coal. All of the elements necessary to generate authigenic quartz, kaolinite, smectites, and other common silicate minerals observed in low-ash coals are present in this Indonesian domed peat deposit within either discrete grains of biologic and eolian origin and/or as metal-organic complexes. If, however, all of the inorganic constituents present in the peat crystallized into mineral phases during coalification, the resulting mineral matter would vary in composition from that of most U.S. low-ash coal beds. Specifically, concentrations of kaolinite and other clays that contain abundant Al would be lower in the coal produced from this Indonesian peat. Authigenic and diagenetic processes that either add Al and Si or, alternatively, leach Mg, Ca, Na, Mn, and K, which are inferred to be organically bound, could result in mineralogic compositions similar to those observed in low-ash coal.

CONCLUSIONS

Low-temperature ash contents of interval samples from the domed and lacustrine portions of the Siak Kanan peat deposit are extremely low. Discrete minerals are very rare in the LTA: discrete minerals constitute approximately 7 weight percent of the LTA in the domed peat deposit. The presence of blue luminescent quartz, zircon, amphiboles, and possibly pyroxenes, and the ab-

sence of observed fluvial influence indicate that a significant portion of the discrete minerals in both peats is derived from eolian material that is primarily volcanic in origin.

Measured pH of interstitial water in the peat is low (<4 in the low-ash portion of the domed peat to 6.5 in the substrate material), and eolian-derived quartz and silicates in the LTA from both types of peat are etched. Etching features are thought to be a result of simple dissolution or interactions with multiple types of complex organic acids that are present in the peat pore water. Because dissolution occurs in the peat stage, dispersed volcanic material in coal that is not present as distinct layers may be extremely difficult to recognize.

Some of the dissolved silica is probably precipitated in mineral phases as evidenced by authigenic silica fibers infilling plant material and unidentified rare-earth alumino-silicates in the substrate. Still more silica is utilized by and incorporated into sponges, chrysophytes, organisms that resemble filose testacid amoebae, and plants as phytoliths. Although chrysophyte cysts and phytoliths are present in both peat environments, organisms that resemble the amoebae are present only in the domed peat and sponge spicules are present only in the lacustrine portion of the domed peat deposit. Both of these organisms may have potential as environmental indicators in coals if they are preserved.

The bulk of the inorganic constituents (~90%) in the LTA of the peat samples is composed of aggregates of micrometer- to submicrometer-sized particles of low-density material that is mostly water soluble. This low-density material in ash is probably derived from organically bound and dissolved cations in the peat that are taken up directly by organisms and plants or adsorbed onto decayed organic material. The cations form metal-organic compounds that are released during the low-temperature ash procedure and form water-soluble sulfates and nitrates which are artifacts of the LTA procedure. The metal-organic compounds are expected to be highly reactive during diagenesis and may either be leached from the peat during coalification or crystallized to form authigenic minerals.

It is postulated that if this peat were to be buried and coalified, silicate minerals would form from the inorganic constituents already present in the peat. Authigenic and diagenetic processes, however, which include the addition of Al or the leaching of Ca, Mg, and Na, would be necessary to form the suite of silicate minerals that are observed in low-ash coals from the Appalachian Basin and in other U.S. coals.

ACKNOWLEDGMENTS

We express our appreciation to Howard Evans for his expert help with single crystal X-ray diffraction analyses, Tim Moore for insightful discussions, and James Pontolillo for LTA analyses. Don Triplehorn kindly showed the senior author washing techniques which allowed for the separation and analyses of the LTA phases. In addition, we are grateful to Philip Bennett, Bob Finkelman, Brenda Pierce, and Robert Raymond for providing critical and helpful reviews to this manuscript.

REFERENCES CITED

Adam, D. P. and Mahood, A. D., 1981, Chrysophyte cysts as potential environmental indicators: Geological Society of America Bulletin, v. 92, p. 839–844.

Anderson, J.A.R., 1964, Structure and development of the peat swamps of Sarawak and Brunei: Journal of Tropical Geography, v. 18, p. 7–16.

Anderson, J.A.R., 1973, An introduction to the ecology of the peat swamps of Sarawak and Brunei, in Proceedings, International Peat Society: Classification of peat and peatlands: International Peat Society, p. 44–51.

Andrejko, M. J., Raymond, R., Jr., and Cohen, A. D., 1983, Biogenic silica in peat: Possible source of chertification in lignites, in Raymond, R., Jr., and Andrejko, M. J., eds., Mineral matter in peat: Its occurrence, form, and distribution: Los Alamos, New Mexico, Los Alamos National Laboratory, p. 25–37.

Bennett, P., and Siegel, D. I., 1987, Increased solubility of quartz in water due to complexing by organic compounds: Nature, v. 326, p. 684–686.

Bennett, P. C., Melcer, M. E., Seigel, D. I., and Hassett, J. P., 1988, The dissolution of quartz in dilute aqueous solutions of organic acids at 25°C: Geochimica et Cosmochimica Acta, v. 52, p. 1521–1530.

Bennett, P. C., Siegel, D. I., Hill, B. M., and Glaser, P. H., 1991, Fate of silicate minerals in a peat bog: Geology, v. 19, p. 328–331.

Cameron, C. C., Esterle, J. S., and Palmer, C. A., 1989, The geology, botany, and chemistry of selected peat-forming environments from temperate and tropical latitudes: International Journal of Coal Geology, v. 12, p. 105–156.

Cameron, N. R., Ghazali, S. A., and Thompson, S. J., 1982, The Geology of the Bengkalis and Siak Sri Indrapjura–Tanjungpinang quadrangles, Sumatra: Bandung, Indonesia, Geological Research and Development Centre, 26 p.

Cecil, C. B., Stanton, R. W., Neuzil, S., Dulong, F. T., Ruppert, L. F., and Pierce, B. S., 1985, Paleoclimate controls on late Paleozoic sedimentation and peat formation in the Central Appalachian Basin: International Journal of Coal Geology, v. 5, p. 195–230.

Crowley, S. S., Stanton, R. W., and Ryer, T. A., 1989, The effects of volcanic ash on the maceral and chemical composition of the C coal bed, Emery Coal Field, Utah: Organic Geochemistry, v. 14, p. 315–331.

Crowley, S. S., Stanton, R. W., Triplehorn, D. M., and Ruppert, L. F., 1990, Origin and distribution of inorganic elements in the Wyodak-Anderson coal bed, Powder River Basin, Wyoming, in U.S. Geological Survey Research on Energy Resources—1990: Programs and Abstracts: U.S. Geological Survey Circular 1060, p. 19–21.

Edman, J. D., and Surdam, R. C., 1986, Organic-inorganic interactions as a mechanism for porosity enhancement in the Upper Cretaceous Ericson Sandstone, Green River Basin, Wyoming, in Gautier, D. L., ed., Roles of organic matter in sediment diagenesis: SEPM Special Publication 38, p. 85–109.

Esterle, J. S., Ferm, J. C., and Tie, Y. L., 1989a, A test for the analogy of tropical domed peat deposits to "dulling up" sequences in coal beds—Preliminary results: Organic Geochemistry, v. 14, p. 333–342.

Esterle, J. S., Staub, J. R., Raymond, A. L., and Tie, Y. L., 1989b, Geochemistry of Domed Peat Deposits from Micro- and Mesotidal Deltaic Systems in Sarawak, Malaysia: Geological Society of America Abstracts with Programs, v. 21, no. 6, p. 25.

Finkelman, R. B., Palmer, C. A., Krasnow, M. R., Aruscavage, P. J., Sellers, G. A., and Dulong, F. T., 1990, Combustion and leaching behavior of elements in the Argonne Premium Coal Samples: Energy and Fuels, v. 4, p. 755–767.

Foscolos, A. E., Goodarzi, F., Koukouzas, C. N., and Hatziyannis, G., 1989, Reconnaissance study of mineral matter and trace elements in Greek lignites: Chemical Geology, v. 76, p. 107–130.

Gluskoter, H. G., 1965, Electronic low-temperature ashing of bituminous coal: Fuel, v. 44, p. 285–291.

Hansley, P. L., 1987, Petrologic and experimental evidence for the etching of

garnets by organic acids in the Upper Jurassic Morrison Formation, northwestern New Mexico: Journal of Sedimentary Petrology, v. 57, p. 666–681.

Kharaka, Y. K., Law, L. M., Carothers, W. W., and Goerlitz, D. F., 1986, Role of organic species dissolved in formation waters from sedimentary basins in water diagenesis, in Gautier, D. L., ed., Roles of organic matter in sediment diagenesis: SEPM Special Publication 38, p. 111–122.

Mariano, A. N., 1989, Economic geology of rare earth minerals, in Lipin, B. R., and McKay, G. A., eds., Geochemistry and mineralogy of rare earth elements: Mineralogic Society of America Reviews in Mineralogy, v. 21, p. 309–334.

McLennan, S. M., 1989, Rare earth elements in sedimentary rocks: Influence of provenance and sedimentary processes, in Lipin, B. R., and McKay, G. A., eds., Geochemistry and mineralogy of rare earth elements: Mineralogic Society of America Reviews in Mineralogy, v. 21, p. 169–200.

Miller, R. N., Yarzab, R. F., and Given, P. H., 1979, Determination of the mineral-matter contents of coals by low-temperature ashing: Fuel, v. 58, p. 4–10.

Moore, T. A., 1990, Petrographic methods for the comparison of modern peat in a Miocene age lignite, Kalimantan, Indonesia: Calgary, Alberta, Canada, Society for Organic Petrology, Seventh Annual Meeting, p. 1–3.

Neuzil, S. G., and Cecil, C. B., 1984, A modern analog of low-ash, low-sulfur, Pennsylvanian-age coal: Geological Society of America Abstracts with Programs, v. 16, no. 3, p. 184.

Neuzil, S. G., and 6 others, 1988, Peat deposits on coastal Sumatra—A modern analog of coal formation, in U.S. Geological Survey Research on Energy Resources—1988: Program and Abstracts: U.S. Geological Survey Circular 1025, p. 37–38.

Painter, P. C., Youtcheff, J., and Given, P. H., 1980, Concerning the putative presence of nitrate in lignites: Fuel, v. 59, p. 523–525.

Raymond, R., Jr., Cohen, A. D., and Bish, D. L., 1985, Ash contents of Costa Rican peats reflect depositional conditions: International Peat Society Symposium on Tropical Peat Resources: Prospects and Potential, Kingston, Jamaica, Feb. 25-March 1, 1985, Abstracts, p. 11.

Ruppert, L. F., Cecil, C. B., Stanton, R. W., and Christian, R. P., 1985, Authigenic quartz in the Upper Freeport Coal Bed, west-central Pennsylvania: Journal of Sedimentary Petrology, v. 55, p. 334–339.

Sanchez, J. D., Coates, D. A., Bradbury, J. P., and Bohor, B. F., 1985, Diatoms in coal: Miocene Venado Formation, Limon Basin, Costa Rica, Central America: Geological Society of America Abstracts with Programs, v. 17, no. 7, p. 706.

Sanchez, J. D., Bradbury, J. P., Bohor, B. F., and Coates, D. A., 1987, Diatoms and tonsteins as paleoenvironmental and paleodepositional indicators in a Miocene coal bed, Costa Rica: Palaios, v. 2, p. 158–164.

Seigel, D. I., Baedecker, M. J., and Bennett, P., 1986, The effects of petroleum degradation on inorganic water-rock reactions, in Fifth International Symposium on Water Rock Interactions: Extended abstracts: Reykjavic, Iceland, International Association of Chemistry and Cosmochemistry, p. 524–527.

Tan, K. H., 1980, The release of silicon, aluminum, and potassium during decomposition of soil minerals by humic acid: Soil Science, v. 129, p. 5–11.

Triplehorn, D. M., and Bohor, B. F., 1986, Volcanic ash layers in coal: Origin, distribution, composition, and significance, in Vorres, K. S., ed., Mineral matter and ash in coal: American Chemical Society Symposium Series 301, p. 90–98.

Triplehorn, D. M., Stanton, R. W., Ruppert, L. F., and Crowley, S. S., 1991, Volcanic ash dispersed in the Wyodak-Anderson coal bed, Powder River Basin, Wyoming: Organic Geochemistry, v. 17, p. 567–575.

Zinkernagel, U., 1978, Cathodoluminescence of quartz and its application to sandstone petrology, in Fuchtbauer, H., Lisitzn, A. P., Milliman, J. D., and Seibold, E., eds., Contributions to Sedimentology, no. 8: Stuttgart, E. Schweizerbart'sche Verlagsbuchhandlung, 69 p.

MANUSCRIPT ACCEPTED BY THE SOCIETY JANUARY 14, 1993

Geological Society of America
Special Paper 286
1993

Geochemical and analytical implications of extensive sulfur retention in ash from Indonesian peats

Jean S. Kane* and Sandra G. Neuzil
U.S. Geological Survey, 956 National Center, Reston, Virginia 22092

ABSTRACT

Sulfur is an analyte of considerable importance to the complete major element analysis of ash from low-sulfur, low-ash Indonesian peats. Most analytical schemes for major element peat- and coal-ash analyses, including the inductively coupled plasma atomic emission spectrometry method used in this work, do not permit measurement of sulfur in the ash. As a result, oxide totals cannot be used as a check on accuracy of analysis. Alternative quality control checks verify the accuracy of the cation analyses. Cation and sulfur correlations with percent ash yield suggest that silicon and titanium, and to a lesser extent, aluminum, generally originate as minerals, whereas magnesium and sulfur generally originate from organic matter. Cation correlations with oxide totals indicate that, for these Indonesian peats, magnesium dominates sulfur fixation during ashing because it is considerably more abundant in the ash than calcium, the next most important cation in sulfur fixation.

INTRODUCTION

Interest in the inorganic constituents found in coal is prompted largely by the detrimental environmental and technological effects which are attributable to their presence. Because peat is the precursor of coal, it is natural to study these inorganic constituents in peat as a means of understanding those in coal.

Quantitative inorganic chemistry studies of peat date back as far as 1895, and an abundance of data from many studies of peat from numerous locations is available (Shotyk, 1988). The vast majority of references, however, focus on the geochemistry of one or more cations in peat, but not on total inorganic chemistry. Moreover, in most studies, little is said about the analytical methodology used.

Studies of inorganic constituents in both peat and coal are frequently done on ashed samples, rather than on the original material. The ashing isolates the mineral fraction from the bulk sample, concentrating it to provide better detectability of elements. This concentration is especially necessary in samples having a low inorganic content (low percent ash). Once the peat has been ashed, schemes for whole-rock analysis should be applicable for total chemistry of the ash. Such methods are routinely used successfully to analyze coal-ash standard reference materials (SRMs) (Brown and Smith, 1989). The inorganic matter that results from ashing coal and peat, however, may contain a number of minerals which differ from the principal rock-forming silicate minerals. In that case, the convention of reporting all major element abundances as oxides may lead to a low oxide total, so that the practice of verifying acceptability of analysis by summation to $100 \pm 1\%$ (Abbey, 1983) will be inappropriate.

This problem was noted in our study of Indonesian peats. The average oxide total for the high-temperature ash (HTA) of 93 peat samples is 85%, demonstrating that summation of major cations as oxides fails to verify accuracy for the peat ashes analyzed. The data we present for HTA supports the conclusion that inclusion of sulfur as SO_3 in the totals eliminates most of that error. We also present evidence that this correction is even more important for the low-temperature ash (LTA) samples of peat.

For rock analysis, omission of sulfur as SO_3 from the oxide total only rarely introduces significant error. The omission of sulfur as a major constituent of both peat HTA and peat LTA, however, may be highly significant. It may also be of considerable importance in the major element analysis of ash from low-ash coals, particularly of subbituminous or lower rank (O'Gorman

---

*Present address: NIST, Office of Standard Reference Materials, Building 202, Room 215-A, Gaithersburg, Maryland 20899.

Kane, J. S., and Neuzil, S. G., 1993, Geochemical and analytical implications of extensive sulfur retention in ash from Indonesian peats, *in* Cobb, J. C., and Cecil, C. B., eds., Modern and Ancient Coal-Forming Environments: Boulder, Colorado, Geological Society of America Special Paper 286.

and Walker, 1971; W. C. Grady, unpublished data), whose oxide totals are significantly less than 100%.

This paper presents results of analysis for major cations in HTA of 93 samples from three low-ash and low-sulfur Indonesian peat deposits. Based on the analyses reported here, other aspects of the inorganic geochemistry of these peats are also discussed. Other reports in this volume on peat from these Indonesian deposits (Neuzil et al., Supardi et al., Grady et al., Ruppert et al.) provide information on geology, mineralogy, and the like.

EXPERIMENTAL

Peat sample collection

A Macaulay-type auger was used to core three peat deposits in Sumatra and Kalimantan, Indonesia (Neuzil et al., this volume; Supardi et al., this volume; Supardi and Priatna, 1985). The samples analyzed in this study were collected as discrete depth interval segments from 17 locations and were stored wet under refrigeration until moisture determinations were made and the samples were further processed, as described in the following sections.

Figure 1 shows a schematic diagram of the complete analytical sequence for all studies on these peats and indicates sample

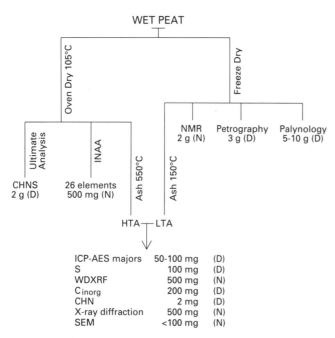

To have sufficient sample to perform all listed analyses, 115 g dry peat or 1.15 kg wet peat is required, assuming 2 percent ash in peat. Requirements increase at lower percent ash, and decrease at higher percent ash.

Figure 1. Schematic of sample splitting for complete sequence of analyses. Minimum sample weights for several indicated analyses are given in mg. N = nondestructive analysis (sample can be recovered after analysis). D = destructive analysis (sample is totally consumed).

size requirements (Neuzil et al., this volume; Grady et al., this volume; Ruppert et al., this volume).

Peat sample drying, sieving, and ashing

For high-temperature ashing, a split of wet peat was oven-dried to constant weight at 105 °C (or, in some cases, at 60 °C). Moisture was calculated as weight lost in drying. The dry sample was ground in a blender for 30 sec and sieved repeatedly, until all of the sample passed a 40-mesh screen (screen opening 420 μm). The ground peat sample was split for ultimate analysis, instrumental neutron activation analysis, and ashing. The split for ashing was redried at 105 °C, and reweighed before ashing. The samples were placed in a muffle furnace, and the temperature was gradually increased over 3 hours to the HTA ashing temperature of 550 °C, and that temperature was maintained for 16 hours.

For low-temperature ashing, splits of wet peat were freeze-dried to constant weight. Moisture was determined as the weight lost in drying. The freeze-dried peat from most sampling sites was ground in a blender to pass only the 10 mesh screen (screen opening 2,000 μm). For one sampling site, SK11, the dried samples (10 discrete depth segments) were sieved to give three size fractions, >2,000 microns, 420 to 2,000 microns, and <420 microns. Splits were taken for petrographic analysis, palynology, and nuclear magnetic resonance studies as well as for low temperature ashing. The peat was oven-dried at 105 °C before ashing to remove any moisture adsorbed during splitting and grinding and to assure a more accurate ash content. The ashing was performed in a low-temperature plasma asher at 35–40 watts per chamber and approximately 1 torr pressure to constant weight within 1 mg or 0.1% of initial peat weight, whichever was larger. Percent ash was determined as ash weight/dry peat weight × 100 for both HTA and LTA.

Determination of major elements in ash

Ninety-three HTA samples were analyzed by inductively coupled plasma atomic emission spectrometry (ICP-AES) following mixed lithium metaborate/lithium tetraborate fusion (Shapiro and Brannock, 1962). Optimum sample size for the analysis is 100 mg; for some analyses, however, only 25 mg was available. The ash is primarily a low-density "fluff" in which a small number of dense mineral grains occur (Ruppert et al., this volume); the 25-mg sample is not necessarily representative of this material, and its use could introduce sampling errors in the analysis.

Indium was used as an internal standard to correct for instrumental drift during the analysis. Following optimized routine rock analysis procedures (Kane and Dorrzapf, 1987; Leary et al., 1982), all analytes were measured and reported as percent oxide in the sample, and totals were calculated.

Twenty LTA samples were also analyzed. One LTA sampling site was also an HTA sample location; samples from three depths at that site were analyzed both as LTA and as HTA samples.

Quality control (QC) checks for HTA and LTA analyses

were based on concurrent analysis of National Institute of Standards and Technology (NIST) standard reference coal ashes SRM 1633 and SRM 1633a. Analysis of the five U.S. Geological Survey rock reference samples used as calibration standards also provide QC data (only two of five were used in calibration for each element). Routine QC checks through duplicate analysis of individual samples, were not possible because of sample quantity limitations. Summation of cations reported as oxides is an independent approach to QC verification of analyses. This verification of analytical accuracy is inappropriate for these samples, which contain abundant sulfur in ash (see Results), because of the inability to measure S by ICP-AES.

RESULTS

Analytical results were obtained for HTA samples representing complete or partial depth profiles from 12 sites in three peat deposits. Table 1 gives the data obtained, with major cations reported as oxides, for HTA samples from three of those sites, one from each peat deposit. It also gives percent ash yield and sulfur content for the unashed peats, and the oxide total for the ICP-AES analyses. Similar data for all LTA samples is in Table 2. The QC sample data is in Table 3. Several things are readily apparent from the data.

1. Based on acceptable totals (Abbey, 1983), representation of major cations as oxides is accurate for the QC analyses, but not for the peat ash samples.

2. The magnitude of error is greater for LTA than for HTA, and for either ash it is generally greatest for samples with low Si and high Ca concentrations. The error varies inversely with percent ash and with sulfur in peat.

3. Cation analyses are accurate, based on the QC data. Erroneous totals result from incorrect stoichiometries used to represent those analyses.

Our principal difficulty in proceeding from this point to a full understanding of the correct stoichiometry for the cations in the ash is lack of sufficient sample for further analyses. Seven different splits were taken from each peat for analysis (Fig. 1). Additionally, the ash content for most samples was considerably less than the suite average of 2.5% (Fig. 2). As a result, we typically had 50 mg or less of HTA, LTA, or both for this study. Therefore, we obtained limited experimental evidence for the role of sulfur fixation in accounting for the shortfall in oxide totals. After presenting our data in the following section, subsequent discussion will depend on supporting evidence drawn from the literature.

To demonstrate that incomplete ashing was not a cause of the low totals for the HTA samples, CHN analyses were performed on the one HTA sample, SK9-P6, giving the lowest oxide total (44%), which we assumed to represent the worst case with respect to residual organic matter. Results are shown in Table 4. The carbon content of 0.57% indicates that the high-temperature ashing process, was complete within 1%–2%, based on reasonable assumptions of organic matter to carbon ratios (organic matter = $C_{TOT} \times 1.3$).

An independent indication of completeness of the high temperature ashing is given by the comparison of ICP-AES and instrumental neutron activation analysis of HTA and whole peat, respectively. ICP-AES determinations of Fe for 36 HTA samples, when back-calculated to a whole-peat basis, agree with INAA analyses on whole peat for these same samples (Fig. 3). A general low bias for the ICP-AES analyses in comparison to the INAA results would indicate incomplete ashing. A wide scatter of individual data about the theoretical correlation line would indicate variable ashing efficiency from one sample to another. Neither bias of the ICP-AES results nor wide scatter in the data is found.

For the LTA, incomplete ashing is likely to be a greater factor in causing low totals. Evaluation of completeness of ashing for our LTA samples is based solely on data for three samples, SK11-P3, SK11-P6, and SK11-P9, included in both the HTA and LTA sample suites. On an absolute basis for dry peat, the LTA ash yields are greater than the HTA ash yields by 0.42%, 0.45%, and 0.49%, respectively. On a relative basis, the LTA ash yields are 126%, 215%, and 209% of HTA ash yields (Fig. 4). This difference could result from incomplete low-temperature ashing. Essential water remaining after low-temperature ashing and/or a greater degree of sulfur retention in LTA could also contribute to the observed differences in LTA and HTA yields. The CHNS analyses, however, needed to establish the presence of residual organic or essential water, or to quantify sulfur retention, could not be performed on any LTA sample because of insufficient sample.

Either residual organic or essential water, or both, particularly in LTA, might be important in obtaining accurate oxide totals. Retention of sulfur as SO_3 in ash, however, seems to be the dominant cause of error in the oxide total for both HTA and LTA. Because of sample limitations, only nine HTA samples, from two of the three peat deposits studied, have been analyzed for sulfur by combustion methods (Rait and Aruscavage, 1989). The data for sulfur in ash is given in Table 5. It shows wide variation in the extent of sulfur retention in the ash. For the Siak samples (coastal lowland), only 14% to 35% of the sulfur in peat is retained in ash. For the Bengkalis (island) samples, 47% to 96% of the original sulfur is retained.

SEM studies on Indonesian peat LTA (Ruppert et al., this volume) and wavelength dispersive X-ray fluorescence (WDXRF) examination of two HTA and six LTA samples (Robert G. Johnson, 1989, unpublished data) identified major amounts of sulfur in all samples examined. Also, X-ray diffraction studies of 11 HTA (Frank T. Dulong, 1989, unpublished data) identified anhydrite in some of those samples. Sample quantity limitations restricted these efforts to just a few samples; the work on the LTA samples is presented elsewhere in this volume (Ruppert et al.).

DISCUSSION

Peats are defined as having 25% or less inorganic material on a dry-weight basis (Standard Method D-2607, ASTM, 1988). The average HTA contents of 2.5% for our 93 samples (Neuzil

TABLE 1. ANALYTICAL DATA FOR HTA FROM ONE SAMPLING LOCATION IN EACH OF THREE PEAT DEPOSITS

Identification Number	In Peat		In Ash										Oxide Total (%)
	Ash Yield (%)	S (%)	SiO$_2$ (%)	Al$_2$O$_3$ (%)	Fe$_2$O$_3$ (%)	MgO (%)	CaO (%)	Na$_2$O (%)	TiO$_2$ (%)	P$_2$O$_5$ (%)	MnO (%)	K$_2$O (%)	
SK9-P1	2.40	0.15	64.00	7.40	4.00	4.30	7.30	0.16	0.32	4.00	0.08	0.64	92.40
SK9-P2	1.6	0.16	66.70	5.10	5.10	5.40	3.60	0.27	0.14	3.40	0.07	0.89	90.67
SK9-P3	1.62	0.12	61.40	9.30	4.70	5.10	5.70	0.43	0.33	3.30	0.33	0.60	91.19
SK9-P4	1.23	0.15	54.20	9.70	5.90	5.90	5.60	0.72	0.31	3.50	0.11	0.96	86.90
SK9-P5	0.75	0.15	29.20	18.40	14.50	14.60	3.10	2.20	0.34	2.30	0.26	1.20	86.10
SK9-P6	0.71	0.10	13.40	8.60	8.30	9.10	1.30	1.50	0.17	0.99	0.19	0.29	43.84
SK9-P7	1.00	0.14	9.20	16.70	24.60	22.70	2.30	3.60	0.18	1.40	0.60	0.75	82.03
SK9-P8	1.30	0.14	4.40	9.00	26.50	21.50	2.30	3.70	0.13	1.15	0.71	0.59	69.98
BK7-P1	0.66	0.12	38.10	5.10	4.80	13.90	5.00	4.00	0.55	12.70	0.15	4.90	89.20
BK7-P2	0.40	0.14	39.90	9.50	4.10	14.70	2.10	5.70	0.57	11.60	0.09	4.20	92.46
BK7-P3	0.44	0.14	40.20	6.80	3.70	13.50	2.60	5.10	0.48	11.20	0.12	4.10	87.80
BK7-P4	0.29	0.13	18.70	8.80	4.50	33.60	2.40	7.90	0.56	4.80	0.10	2.40	83.76
BK7-P5	0.32	0.11	12.00	8.70	5.40	36.60	2.70	7.50	0.44	3.10	0.12	1.40	77.96
BK7-P6	0.46	0.12	5.70	6.00	4.60	41.00	4.00	6.60	0.30	2.20	0.13	0.99	71.52
BK7-P7	0.56	0.10	5.30	3.50	4.90	39.60	6.50	6.30	0.34	1.60	0.15	0.76	68.95
BK7-P8	0.73	0.11	9.90	7.00	5.20	41.10	7.80	7.40	0.27	1.30	0.15	1.40	81.52
BK7-P9	1.54	0.12	1.50	3.20	9.20	35.70	15.10	7.80	0.12	0.86	0.25	2.40	76.13
BK7-P10	2.29	0.11	0.98	4.00	9.00	29.70	18.10	9.60	0.07	0.66	0.36	1.00	73.47
BK7-P11	2.14	0.19	3.10	8.40	8.80	25.40	14.40	10.90	0.16	0.88	0.41	0.96	73.41
BK7-P12	1.22	0.12	22.30	9.10	8.70	26.20	7.30	6.50	0.38	0.67	0.19	0.69	82.03
WK3-P1	0.69	0.10	22.50	4.50	2.80	30.80	8.50	6.20	0.29	7.60	0.11	4.50	87.80
WK3-P2	0.47	0.10	14.40	5.40	3.90	41.90	1.70	8.50	0.27	5.90	0.07	2.70	84.74
WK3-P3	0.40	0.11	13.50	4.90	3.40	40.40	3.30	9.30	0.27	5.90	0.07	1.80	82.84
WK3-P4	0.37	0.10	11.40	4.50	3.70	42.40	4.10	10.20	0.28	3.40	0.09	1.60	81.67
WK3-P5	0.69	0.10	5.90	1.40	2.20	46.70	9.40	8.10	0.14	1.40	0.05	1.30	76.59
WK3-P6	1.31	0.10	2.44	0.10	3.10	42.50	16.10	6.40	0.11	1.10	0.06	0.94	72.85
WK3-P7	2.44	0.10	0.48	0.10	0.00	42.80	26.10	4.10	0.06	0.64	0.06	0.74	75.08
WK3-P8	3.65	0.12	0.74	0.10	0.77	37.30	30.60	6.60	0.06	0.38	0.14	0.72	77.41
WK3-P9	6.05	0.46	10.40	2.60	2.20	26.80	24.50	7.00	0.20	0.24	0.19	1.20	75.33

TABLE 2. ANALYTICAL DATA FOR LTA ANALYSES

Identification Number	In Peat Ash Yield (%)	In Ash										
		SiO_2 (%)	Al_2O_3 (%)	Fe_2O_3 (%)	MgO (%)	CaO (%)	Na_2O (%)	TiO_2 (%)	P_2O_5 (%)	MnO (%)	K_2O (%)	Oxide Total (%)
SK1-P1	3.63	28.66	5.48	9.44	3.48	11.05	0.38	0.27	1.63	0.25	0.51	61.14
SK1-P2	2.32	1.07	6.81	9.29	5.14	6.44	0.77	0.17	0.78	0.30	0.23	30.99
SK1-P3	4.4	30.58	14.56	7.86	2.65	1.82	0.39	0.78	0.28	0.15	1.49	60.56
SK2-P10	1.22	34.22	3.97	1.86	7.63	2.94	1.00	0.25	5.28	0.04	5.42	62.59
SK2-P30	2.05	1.07	1.83	0.81	12.77	7.00	2.43	0.07	0.39	0.03	0.61	27.01
SK2-P32	9.8	1.07	1.40	4.72	9.45	6.72	1.47	0.05	0.21	0.06	0.36	25.50
SK2-P33	5.67	33.15	4.16	1.72	6.80	14.69	0.66	0.45	2.11	0.01	0.86	64.60
FebBK1-P1	2.45	27.37	3.59	1.57	5.97	13.15	0.49	0.93	1.83	0.01	0.83	55.76
FebBK1-P4	2.02	5.77	6.81	1.27	10.28	5.32	3.51	0.03	0.85	0.05	0.71	34.59
FebBK1-P5	1.62	5.13	5.67	1.26	13.10	4.90	5.93	0.40	0.89	0.06	0.84	38.19
FebBK1-P6	2.72	1.07	5.67	1.39	18.07	7.84	3.78	0.27	0.39	0.11	0.66	39.23
SK11-P3	2.05	39.30	1.79	2.17	3.57	2.84	0.10	0.15	2.18	0.07	0.49	52.66
SK11-P6	0.84	23.80	7.36	2.98	4.45	2.61	1.56	0.19	1.86	0.06	0.10	44.97
SK11-P9	0.94	10.90	4.91	3.97	7.04	4.11	1.35	0.18	0.96	0.09	0.71	34.22
SKLL-DC1	5.68	46.39	12.10	0.99	2.28	2.37	0.88	0.98	0.30	0.04	1.64	67.97
FebBK2-P23/24	0.87	1.91	1.31	1.87	12.50	4.82	2.74	0.16	0.92	0.07	1.36	27.66

TABLE 3. ANALYTICAL DATA FOR QC SAMPLES

Identification Number	Reference*	SiO_2 (%)	Al_2O_3 (%)	Fe_2O_3 (%)	MgO (%)	CaO (%)	Na_2O (%)	TiO_2 (%)	P_2O_5 (%)	MnO (%)	K_2O (%)	Oxide Total (%)
NBS 1633	1	48.85	25.67	8.94	2.23	6.55	0.33	1.19	0.26	0.067	1.66	95.747
	2	47.10	23.80	8.81	2.50	6.51	0.42	1.19	0.23	0.064	2.04	92.65
NBS 1633a	1	49.36	29.05	13.12	0.82	1.61	0.14	1.37	0.42	0.027	1.93	97.847
		47.7	29.4	13.8	0.85	1.6	0.15	1.38	0.4	0.026	2.31	97.616
	2	48.8	27.2	13.4	0.75	1.59	0.23	1.37	0.42	0.02	2.2	
AGV-1	1	60.02	17.48	6.76	1.65	5.09	4.3	1.03	0.54	0.1	2.885	99.855
		56.39	17.42	6.83	1.51	4.76	4.31	1.03	0.47	0.095	2.78	95.595
		58.71	17.13	6.69	1.52	4.93	4.19	1.03	0.53	0.099	2.77	97.599
	3	59	17.25	6.76	1.46	4.9	4.26	1.06	0.49	0.1	2.89	
BIR-1	1	48.44	16.08	11.31	9.95	13.46	1.81	0.975	0.07	0.175	0.1	102.37
		47.51	16.27	11.7	9.41	13.09	1.75	0.97	0.03	0.17	0.1	101
		47.38	15.47	11.2	9.39	13.16	1.77	0.93	0.09	0.171	0.1	99.661
	3	47.77	15.89	11.25	9.59	13.3	1.81	0.98	0.04	0.17	0.03	
BHVO-1	1	50.67	13.92	12.22	7.54	11.56	2.31	2.72	0.33	0.17	0.53	101.97
		49.35	14.1	12.57	7.06	11.25	2.27	2.76	0.3	0.17	0.53	100.36
		49.16	13.39	12.03	6.87	11.21	2.15	2.63	0.32	0.17	....	97.93
	4	49.97	13.86	12.3	7.21	11.35	2.24	2.72	0.33	0.17	0.53	
G-2	1	69.84	15.48	2.78	0.84	2.03	4.1	0.49	0.15	0.04	4.5	100.25
		69.27	16.1	2.8	0.79	1.98	4.25	0.5	0.12	0.041	4.36	100.211
		67.96	15.03	2.67	0.79	1.97	3.88	0.47	0.15	0.041	4.19	97.151
	3	68.91	15.52	2.75	0.81	1.99	4.15	0.49	0.15	0.05	4.45	

*1 = this study; 2 = Gladney, 1987; 3 = Flanagan, 1976; 4 = Abbey, 1983.

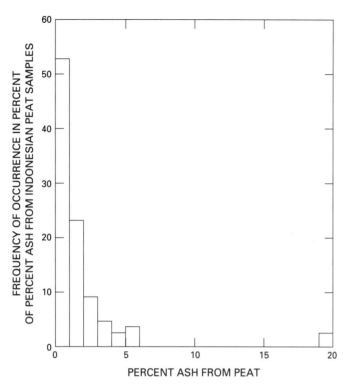

Figure 2. Frequency distribution of percent ash yield of peat for 93 HTA samples.

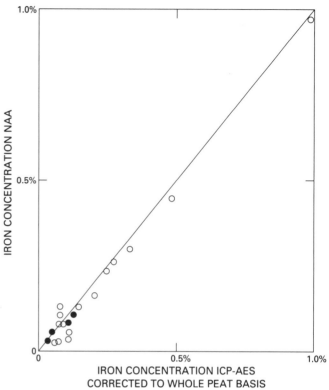

Figure 3. Correlation plot of percent iron in peat, INAA, versus percent iron in ash, ICP-AES, calculated to whole-peat basis. Slope of correlation line = 0.999. Correlation coefficient = 0.994. Diagonal line = theoretical correlation line. Open circle = one sample. Solid circle = several overlapped samples.

TABLE 4. CHN ANALYSIS OF HTA FROM SK9-P6 AND POTENTIAL CONTRIBUTIONS TO DEFICIT IN OXIDE TOTAL

	Percent in HTA Measured	Accounts for a Maximum of
C	0.57	0.74% residual organic
H	0.74	6.7% essential water
N	0.67	2.97% nitrate
Could account for		~10% deficit in totals
Observed deficit in total		56%

et al., this volume), is therefore generally very low for peats. At the same time, for the suite of samples under consideration, it is atypically high as a whole (Fig. 2). This characteristic indicates extremely limited input of minerals during peat accumulation. The low density and water solubility of most of the ash (Ruppert et al., this volume) also indicates that many of the inorganic constituents in these peats are not allogenic minerals.

The inorganic constituents in these peats, then, probably originate largely as essential constituents of, or adsorbed onto, organic matter. The ashing process releases the inorganic cations from organic association, producing either discrete, crystalline phases (Bardin and Bish, 1983; Sawyer and Griffin, 1983) or "amorphous" inorganic matter in which no crystals of micron or larger size were observed (Ruppert et al., this volume), or both.

These phases are almost certainly artifacts created in the ashing process (Painter et al., 1980; Miller et al., 1979). Any detrital mineral component of the ash should be represented accurately by the oxide convention routinely used in rock analysis. The oxide representation, however, is inaccurate for crystalline or amorphous ashing artifacts, which are derived from organic sulfur, nitrogen, and/or carbon fixed during ashing and also include additional oxygen incorporated during fixation. This causes oxide totals to be erroneously low.

One possible stoichiometry which would improve the low totals observed for peat HTA, is the occurrence of Ca as oxalate rather than as oxide. Both whewellite and weddellite, the calcium oxalate mono- and dihydrates respectively, have been identified by X-ray diffraction in LTA of Okefenokee swamp plants (Bardin and Bish, 1983). The occurrence of weddellite in LTA of Everglades peat also has been documented (Sawyer and Griffin, 1983). Calcium oxalate is unlikely in HTA, however, based on differential thermal analysis (DTA) plots, which show decomposition of calcium oxalate monohydrate to $CaCO_3$ and CO between 450 °C and 500 °C (Liptay, 1971).

The occurrence of nitrate in the HTA, similar to that observed in LTA by Painter et al. (1980), resulting from fixation of

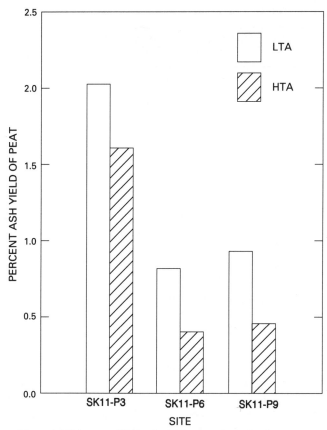

Figure 4. LTA versus HTA ash yield comparisons for three peats.

TABLE 5. SULFUR IN PEAT AND HTA SAMPLES WITH CALCULATED SULFUR RETENTION

Sample Identification Number	Sulfur in Peat* (%)	Sulfur in Ash as SO_3* (%)	Sulfur Retention (%)	Adjusted Total (Oxide Total Plus SO_3)
SK4-P3	0.129	8.0	18.1	86.0
SK4-P7	0.367	1.2	19.2	102.2
SK9-P1	0.152	5.5	34.7	97.9
SK9-P3	0.124	6.5	34.0	97.7
SK11-P2	0.129	6.0	16.4	88.9
BK5-P7	0.112	19.0	46.8	89.5
BK7-P11	0.192	12.3	54.6	85.7
BK10-P1	0.108	6.0	98.0	74.9
BK10-P8	0.359	7.5	49.8	91.2

*Analyst: C. J. Skeen

organic nitrogen, is also possible. The one CHN analysis performed on HTA sample SK9P6 (Table 4) did not show a sufficient quantity of nitrate in the sample to eliminate the deficit in oxide total which had been observed.

We therefore conclude that the presence of sulfur species in the ash is the primary cause of our observed low oxide totals in both the peat HTA and LTA. Literature support for this conclusion comes from the low-temperature coal-ashing studies of Miller et al. (1979) and from low-temperature lignite-ashing studies by Adolphi and Storr (1985). The coal combustion studies of Huffman et al. (1988, 1989) also support our postulated sulfate speciation in peat HTA.

The strongest literature support for that speciation comes from the recent study of mineral matter in 28 Greek lignites (Foscolos et al., 1989). They found as much as 32.72% SO_3 in the ash after ashing at 1,000 °C, and fixation of from <1% to 88% of the sulfur originally in lignite (average, 27%). For LTA, they found fixation of 6% to 100% of the original sulfur (average, 80%). Intermediate degrees of sulfur fixation were observed after 750 °C ashing. Mineralogically, anhydrite ($CaSO_4$), was identified as the only sulfur-bearing species in 1,000 °C ash. Anhydrite, bassanite, and gypsum, however, were identified in all LTA samples. Hydrous sulfate species (e.g., jarosite, meta-aluminite, ammonium iron sulfate, epsomite, hexahydrate) were identified in

one or more LTA samples. Examination of 550 °C and 750 °C ash showed degrees of hydration of sulfate species intermediate between those in the LTA and in the 1,000 °C ash.

DTA data (Liptay, 1971) provides added indications that ashing at 550 °C, as specified by Standard Method ASTM D-2974 (ASTM, 1988) and the temperature that was used for this study, will not expel all water of hydration. The hydrogen measured (Table 4) in SK9-P6 is equivalent to 6.6% water of hydration, assuming total destruction of organic matter. Addition of SO_3 to the initial totals improves them, but accuracy is not achieved within the desired 100 ± 5% limits (Table 5). This suggests the need to include essential water as well, in order to obtain accurate totals.

Pyrite was identified in lignite and in LTA, but not in any of the HTA samples, of Foscoles et al. (1989). They concluded from these observations that pyritic sulfur does not survive high-temperature ashing. They further concluded that organic sulfur is fixed, regardless of ashing temperature. DTA results (Liptay, 1971) also show total decomposition of pyrite at temperatures below 550 °C. Supardi et al. (this volume) have reported that 80% or more of the total sulfur in the suite of Indonesian peats is organic sulfur. The Foscolos et al. (1989) observations, DTA data, and the very low average ash content for these peats support our conclusion that the sulfur being fixed during ashing and the cations fixing that sulfur occur predominantly in organic matter.

Coefficients for the correlations of cation concentrations in ash with percent ash yield of peat and with oxide totals calculated for HTA are tabulated in Table 6. Also included are coefficients for correlations of percent S in peat (Supardi et al., this volume) with percent ash of peat and with oxide totals. Correlations are significant at the 95% confidence level when the correlation coefficient, r, is significantly different from zero (if $-0.35 \leqslant r \leqslant 0.35$, for this sample suite). The strength of the correlation increases as r approaches a value of 1.

A positive correlation between a given cation and percent ash yield of peat indicates that the cation is derived mainly from

TABLE 6. CORRELATION COEFFICIENTS (r), CALCULATED FOR INDIVIDUAL PEAT DOMES

	Correlation Coefficients for HTA Oxide Total			Correlation Coefficients for % Ash in Peat Yield		
	Siak	Bengkalis	W. Kalimantan	Siak	Bengkalis	W. Kalimantan
Si in ash	0.68	0.89	0.89	0.49	0.50	0.75
Al in ash	0.22	0.24	0.46	-0.08	0.35	0.37 (-0.27)
Fe in ash	-0.02 (0.3)	-0.37	-0.05 (0.07)	-0.28	-0.27	-0.22
Mg in ash	-0.58	-0.58	-0.48	-0.28	-0.52	-0.51
Ca in ash	-0.25	-0.57	-0.57	-0.30	-0.03 (0.44)	-0.12 (0.68)
Na in ash	-0.46	-0.44	-0.26 (0.05)	-0.29	0.43	-0.45
P in ash	0.06	0.32	+0.31	-0.33	-0.34	-0.43
T in ash	0.42	0.74	0.72	0.63	0.43	0.76
Mn in ash	-0.22	-0.46	-0.23	0.03	-0.10	-0.17
K in ash	0.17	0.37	0.53	-0.24	-0.21	-0.03
S in peat	0.31	0.36	0.27 (-0.26)	0.11 (-0.14)	0.62	0.78

Correlations are given for all samples from a given dome. If the correlation for all samples differs in sign from that for only those samples having <5% ash, then the latter correlation coefficient is also given, in parentheses
Null hypothesis: no significant correlation at the 95% confidence level; true, if $-0.35 \leq r \leq 0.35$.
Significant correlations (false null hypothesis) are underscored.
Positive correlations of analyte with total: analyte has minimal role in sulfur retention; positive correlations of analyte with percent ash: dominantly detrital input of analyte.

detrital minerals. A negative correlation indicates that the cation originates in organic matter in the peat (McCarthy et al., 1989). A positive correlation between a cation and oxide total indicates that the cation has no role in sulfur fixation. A negative correlation indicates the importance of the cation in sulfur fixation. A positive correlation of sulfur in peat with percent ash yield indicates that the anionic sulfur in ash originates from organic matter in the peat.

Strong positive correlations of silicon and titanium with percent ash yields show that they are associated with minerals in all deposits, and do not originate in organic matter. This does not, however, refute the finding (Ruppert et al., this volume) that some of the silica in the LTA is biogenic. Strong negative correlations for cationic magnesium, and strong positive correlations for anionic sulfur, with percent ash yields indicate that these elements originate in organic matter, for the Bengkalis and West Kalimantan peats. Weaker correlations indicate that aluminum is allogenic in the Bengkalis and West Kalimantan peat, that sodium is organically associated in these same peats, and that phosphorus is organically associated in the West Kalimantan peat. No other correlations are significant, so conclusions regarding other cation origins can not be drawn from this data.

For ash from all deposits, silicon and titanium correlate positively with the oxide totals, and therefore, these cations are not important for sulfur fixation in the peat ash. For ash from Bengkalis and West Kalimantan, potassium also correlates positively with oxide totals, and therefore has a minimal role in sulfur fixation. This implies an absence of jarosite, which was identified in ash of Greek lignites (Foscolos et al., 1989), in ash from the Indonesian peats. The strong negative correlation of calcium and oxide totals for the Bengkalis and West Kalimantan samples is consistent with work by Ibarra et al. (1989), who found that

calcium played a major role in sulfur fixation during ashing and charring of lignites. It is also consistent with the identification of gypsum, bassanite, and anhydrite in lignite ash by Foscolos et al. (1989). The very high degree of sulfur retention for both the Bengkalis and West Kalimantan samples (Table 5), however, requires more calcium than is available in our samples. This explains the dominant magnesium role in sulfur fixation which we have observed.

Magnesium appears from the correlation data to be the most important element for sulfur fixation in the ash from all deposits. It shows a strong negative correlation with oxide total in all three deposits. The Foscolos et al. study (1989) identified two magnesium sulfates, epsomite and hexahydrite, in their study. The approximately 10:1 MgO:CaO ratio for the Siak samples having the lowest oxide totals explains the dominance of magnesium over calcium in sulfur fixation for the Siak peats. Magnesium is also more abundant than calcium in the other two deposits.

Our data identify an additional role for sodium in sulfur fixation for both Siak and Bengkalis peats, and for iron in Bengkalis peat. Foscolos et al. (1989) did not identify any sodium-bearing sulfate in the ash from Greek lignites The jarosite that they identified, however, has a sodium-bearing equivalent, natrojarosite, which may form in the Indonesian peat ash. Its formation would explain the observed correlation of sodium with oxide total for Siak and Bengkalis. Its occurrence should also show a role for iron in sulfur fixation, which is found in our data for Bengkalis ash, but not Siak ash.

There are several mathematical solutions which are consistent with our cation analyses, e.g., the potential stoichiometries for the ash suggested by the Foscoles et al. study (1989) and accurate oxide totals. Without sufficient sample to permit experimental verification, however, we cannot select the best model.

CONCLUSIONS

Acceptable totals for the analysis of both HTA and LTA from low-ash peats by conventional rock analysis schemes can be achieved only by including sulfur as SO_3 and also essential water in the summation of analyses. Omitting these constituents from oxide totals cause only trivial deviations from theoretical values for most rock analyses. Deviations of as much as 60%, however, have been observed in the HTA of our peat sample suite; the average oxide total without these constituents for 93 HTA samples of Indonesian peats analyzed is 85%. The magnitude of error varies inversely with percent ash yield for the peat. It is considerably greater for LTA than for HTA. The 20 LTA we have analyzed yielded an average oxide total of only 30% without SO_3 and water.

Our data show that, even at very low ash yields, correlations of cation concentration in ash versus percent ash yield for peat may distinguish between organic and allogenic/detrital origin of the individual inorganic cations in the ash. Further, because the low oxide totals for ash (omitting sulfur and water) are likely to indicate that organic sulfur in peat was fixed during ashing, the plot of oxide total versus percent ash yield may also be highly informative regarding mode of sulfur occurrence in the peat.

None of the established analytical protocols for major element analysis of peat or coal ash take into sufficient account the importance of sulfur as a major constituent of the ash from low-ash peats or coals. The omission of sulfur from major element analytical protocols leads to low totals and makes it impossible to use summation of oxides to $100 \pm 1\%$ as a check on internal consistency of the analysis.

ACKNOWLEDGMENTS

We wish to thank the Directorate of Mineral Resources, Bandung, Indonesia, and their personnel for providing peat samples for this study and J. Pontolillo and C. Skeen (U.S. Geological Survey) for assistance with sample preparation and analyses.

REFERENCES CITED

Abbey, S., 1983, Studies in "standard samples" of silicate rocks and minerals 1969–1982: Ottawa, Geological Survey of Canada Paper 83-15, 114 p.

Adolphi, P., and Storr, M., 1985, Glow discharge excited low temperature ashing—A new technique for separating mineral matter of coals: Fuel, v. 64, p. 151–155.

American Society for Testing and Materials (ASTM), 1988, Annual book of ASTM standards, v. 5.05 and v. 4.08.

Bardin, S. W., and Bish, D. L., 1983, The occurrence of calcium oxalate minerals within aquatic macrophytes from Okefenokee Swamp, *in* Raymond, R., Jr., and Andrejko, M. J., Mineral matter in peat—Its occurrence, form, and distribution: Los Alamos, New Mexico, Los Alamos Scientific Laboratory, p. 53–62.

Brown, F. W., and Smith, H., 1989, Analysis of coal ash by atomic absorption spectrometric and spectrophotometric methods, *in* Golightly, D. W., and Simon, F. O., eds., Methods for geochemical analyses: U.S. Geological Survey Bulletin 1823, p. 41–45.

Flanagan, F. J., editor, 1976, Descriptions and analyses of eight USGS rock standards: Washington, D.C., Geological Survey Professional Paper 840, p. 132–134.

Foscolos, A. E., Goodarzi, F., Koukouzas, C. N., and Hatziyannis, G., 1989, Reconnaissance study of mineral matter and trace elements in Greek lignites: Chemical Geology, v. 76, p. 107–130.

Gladney, E. S., O'Malley, B. T., Roelandts, I., and Gills, T. E., 1987, Compilation of elemental concentration data for NBS clinical, biological, geological, and environmental standard reference materials: Washington, D.C., NBS Special Publication 260-111, p. 1633-2.

Huffman, G. P., Lytle, T. W., Greegor, R. B., and Jenkins, R. G., 1988, In situ XAFS investigation of Ca and K catalytic species during pyrolysis and gasification of lignite chars: Fuel, v. 67, p. 1662–1667.

Huffman, G. P., and 6 others, 1989, Investigation of atomic structures of calcium in ash and deposits produced during the combustion of lignite and bituminous coal: Fuel, v. 68, p. 236–242.

Ibarra, J. V., Palacios, J. M., and de Andres, A. M., 1989, Analysis of coal and coal ashes and their ability for sulfur retention: Fuel, v. 68, p. 861–867.

Kane, Jean S., and Dorrzapf, A. F., Jr., 1987, Current atomic absorption and inductively coupled methods for geochemical investigations, *in* Elliot, I. L.,
and Smee, B. W., eds., Geoexpo/86 exploration in the North American Cordillera, p. 184–188.

Leary, J. J., Brookes, A. E., Dorrzapf, A. F., Jr., and Golightly, D. W., 1982, An objective function for optimization techniques in simultaneous multiple-element analysis by inductively coupled plasma spectrometry: Applied Spectroscopy, v. 36, p. 37–40.

Liptay, G., editor, 1971, Atlas of thermoanalytical plots: Amsterdam, Hayden and Sons, v. 1, 116 p.; v. 2, 161 p.; v. 3, 162 p.; see especially v. 1, curve 21.

McCarthy, T. S., and 5 others, 1989, The inorganic chemistry of peat from the Maunachira channel–swamp system: Geochimica et Cosmochimica Acta, v. 53, p. 1077–1089.

Miller, R. N., Yarzab, R. F., and Given, P. H., 1979, Determination of the mineral-matter contents of coals by low-temperature ashing: Fuel, v. 58, p. 4–10.

O'Gorman, J. V., and Walker, P. L., Jr., 1971, Mineral matter characteristics of some American coals: Fuel, v. 50, p. 135–151.

Painter, P. C., Youtcheff, J., and Given, P. H., 1980, Concerning the putative presence of nitrate in lignites: Fuel, v. 59, p. 523–526.

Rait, N., and Aruscavage, P. J., 1989, The determination of forms of sulfur in coal, *in* Golightly, D. W., and Simon, F. O., eds., Methods for sampling and inorganic analysis of coal: U.S. Geological Survey Bulletin 1823, p. 63–68.

Sawyer, R. K., and Griffin, G. M., 1983, The source and origin of the mineralogy of the northern Everglades, *in* Raymond, R., Jr., and Andrejko, M. J., eds., Mineral matter in peat—Its occurrence, form, and distribution: Los Alamos, New Mexico, Los Alamos Scientific Laboratory, p. 189–198.

Shapiro, L., and Brannock, W. W., 1962, Rapid analysis of silicate, carbonate, and phosphate rocks: U.S. Geological Survey Bulletin 1144A, 56 p.

Shotyk, W., 1988, Review of the inorganic geochemistry of peats and peatland waters: Earth-Science Reviews, v. 25, p. 95–176.

Supardi and Priatna, 1985, Endapan gambut didaerah Siakkanan, Riau (The peat deposit in the Siakkanan area, Riau): Bandung, Indonesia, Directorate of Mineral Resources, Department of Mines and Energy, Project for coal inventory, 20 p. (in Indonesian).

MANUSCRIPT ACCEPTED BY THE SOCIETY JANUARY 14, 1993

Geological Society of America
Special Paper 286
1993

Detrital peat formation in the tropical Mahakam River delta, Kalimantan, eastern Borneo: Sedimentation, plant composition, and geochemistry

Robert A. Gastaldo
Department of Geology, Auburn University, Auburn, Alabama 36849-5305
George P. Allen
TOTAL, Compagnie Français des Pétroles, Route de Versailles, 78470 St. Remy Les Chevreuse, France
Alain Y. Huc
Institut Français du Pétrole, 1 & 4 avenue de Bois-Préau, B.P. 311, 92506 Rueil Malmaison Cédex, France

ABSTRACT

The Holocene Mahakam River delta, Kalimantan, Indonesia, is a complex fluvial- and tidal-influenced regime prograding into the Makassar Strait. Three plant communities are segregated and controlled by salinity and edaphic conditions. Salt-tolerant mangrove taxa pioneer the lower delta plain tidal flats and are quickly replaced by *Nypa* palms. These palms comprise the vast majority of dense swamps in the delta plain. A more diverse hardwood forest replaces *Nypa* swamps when sediments have accumulated to subaerial heights where better drainage is possible. The interior of the island supports well-established primary tropical forest. Autochthonous peat deposits have not been identified in the region.

Fluvial distributary channels are the principal conduits through which plant parts originating in these communities are transported to the delta front. Plant parts that reach the delta front may remain resident at the sediment-water interface until they are reworked into accumulations, as thick as 2.5 m, onlapping the interdistributary lower delta plain tidal flats. These allochthonous peat bodies are composed of fragmented canopy detritus from various sources, but mainly from dicotyledonous angiosperms. Plant parts include leaves, cuticles, wood fragments, petiole parts (both dicotyledonous angiosperm and monocotyledonous *Nypa*), damar (dipterocarp resins), fruits, and seeds. Deposits occur as high-tide beach ridges. Peat beach ridges alternate with tidal mud flats as the headlands aggrade into the Makassar Strait.

Geochemical analyses have been conducted on the peat recovered in bulk samples and from core. These include ^{14}C dating, Total Organic Content (TOC), bulk sulfur content, Rock-Eval pyrolysis, Py-GC and Py-GCMS. Peat is Holocene (Recent) in origin, with ^{14}C dates of bulk samples as old as 1050 yr B.P. Damars recovered from the beaches are also Recent in origin (2645 ± 215 yr B.P.; 930 ± 205 yr B.P.). TOC ranges from 27.5%–39.4% with accompanying hydrogen indice (HI) varying from 250-450. Sulfur ranges from 0.65%–2.75%. H/C ratios average 1.43, and O/C ratios average 0.44. The occurrence of relatively high values of the HI are probably due to the incorporation of pieces of Recent damar which exhibit HI as high as 1,130.

Gastaldo, R. A., Allen, G. P., and Huc, A. Y., 1993, Detrital peat formation in the tropical Mahakam River delta, Kalimantan, eastern Borneo: Sedimentation, plant composition, and geochemistry, *in* Cobb, J. C., and Cecil, C. B., eds., Modern and Ancient Coal-Forming Environments: Boulder, Colorado, Geological Society of America Special Paper 286.

INTRODUCTION

The accumulation of terrestrial plant parts as peat has long been recognized as being either autochthonous (in situ) or allochthonous (see Stevenson, 1912; McCabe, 1984). Autochthonous peat may be widespread geographically and attain considerable thickness. Its occurrence and distribution have been used as an environmental indicator in assessing ancient global conditions (e.g., Parrish and Barron, 1986). Detrital peat, those buildups composed of plant parts transported considerable distances from their site of growth, have received little attention (G. P. Allen and Pizon, 1986; Gastaldo et al., 1987). This peat accumulates in specific depositional context and may act as a precursor substrate for autochthonous peat swamp forestation (J. Staub, September 1989, personal communication).

Accumulations of detrital peat are restricted in lateral distribution and abruptly pinch out into adjacent facies. These peats are intercalated within other coastal sediments. The plant constituents of this peat may or may not be related to vegetational communities adjacent to the deposit. The occurrence of detrital peat appears to be a common phenomenon at the interface between fluvial and marine processes operating in coastal regimes. These "coffee ground" accumulations have been recognized in the deltas of the Mississippi River (Coleman and Prior, 1980; S. Penland, April 1987, personal communication), the Mobile-Tensaw River (Gastaldo et al., 1987; Gastaldo, 1989), the Niger River (J.R.L. Allen, 1965), and the Mahakam River (G. P. Allen et al., 1979). Similar deposits have been recognized in the stratigraphic record (G. P. Allen, 1987). Recognition of such deposits provides insights into the sedimentological processes occurring within the regime, affording additional data to assist in the interpretation of coastal terrestrial systems.

Detrital peat is a common feature in the interdistributary zones, lower delta plain, of the tropical Mahakam River delta, Kalimantan, Indonesia (G. P. Allen et al., 1979). Peat occurs in a series of chenier-like accumulations along the coastal interface. Accumulations may extend as a continuous lateral deposit for distances greater than 7 km and are recognized inland for distances of approximately 3 km. Such deposits appear to have some potential for lithification. The object of this paper is to describe the occurrences of detrital peat in the Recent Mahakam River delta, to detail the plant components and community representatives that comprise these peat deposits, and present geochemical data of the principal contributing macrodetritus. These studies are made in an effort to characterize the petroleum potential of these shoreline deposits.

SETTING

The Neogene-Holocene Mahakam delta, Kalimantan, is located on the island of Borneo (Fig. 1). The island is approximately 1,300 by 990 km and straddles the equator. Six principal rivers drain the island, the headwaters of which originate in the central highlands of Kalimantan. The Mahakam delta is located at the eastern edge of the Kutei Basin, between lat 0°21' and 1°10' and long 117°40'. Deltaic sedimentation began in the middle Miocene (LaLouel, 1979). Three major deltaic complexes, separated by two marine transgressions, have been recognized in the basin (Loired and Mugniot, 1982) and the Quaternary history of the delta has recently been detailed (Carbonel and Moyes, 1987).

The eastward prograding sedimentary sequence is between 6,000 and 8,000 m thick. The modern Mahakam is a thin accumulation of 50–60 m overlapping the paleodeltas (Roux, 1977). The Mahakam River drains a 75,000 km^2 basin, and the sediment transported through the equatorial basin forms an interactive fluvial- and marine-controlled delta complex (G. P. Allen et al., 1977, 1979; Combaz and De Matharel, 1978). Presently, the deltaic system is approximately 50 km in "length," as measured from the delta front to the first bifurcation of the Mahakam River in the hinterland, and is distributed along the coast for nearly 100 km. It is composed of approximately 2,000 km^2 of wetlands of the subaerial delta plain and 1,800 km^2 of delta front and prodelta.

The Mahakam has two active fluvial distributary systems directed northeast and southeast. An intervening area, termed interdistributary (G. P. Allen et al., 1979), consists of a series of tidal channels practically unconnected to the fluvial regime (Fig. 2). River depths average 7–10 m, with maximum depths of 15 m. Waters have a high suspension load of mud, whereas fine to medium sand and silt are transported in bedload. Tidal channels are similar in depth to river channels, with the base of tidal channels often at the contact with delta front sands. Medium- to fine-grained sand and silt are localized in the distributary channels and delta front (G. P. Allen et al., 1979; Gayet and Legigan, 1987), whereas the remainder of the delta is characterized by mud. Identifiable plant macrodetritus accompanies organic mud and clay distributed throughout the various deltaic environments of deposition.

The interior subzone (equivalent to the upper delta) is characterized by a hardwood tropical forest. No systematic botanical survey has ever been conducted in eastern Kalimantan. Based on vegetational surveys of other Indonesian coastal zones (Anderson, 1964, 1983; Andreisse, 1974), the vegetation is probably dominated by Dipterocarp angiosperms, Pandans (Pandanaceae), Palms (Palmae), and other angiosperm families. The subtidal zone (equivalent to the lower delta) is forested by *Nypa* palm swamps. The shoot apex of the palm is at ground level and from this develop large erect leaves. The leaves of *Nypa* attain heights of 8 m. Coastal mangroves (*Avicennia, Rhizophora,* and *Sonneratia*) colonize newly formed tidal flats and aggradational headlands. Glades of *Acrostichum* (mangrove fern) are also present. Mangrove taxa may be found adjacent to channels toward the interior of the delta where salinity gradients are generated by tides (Fig. 2).

The delta front fringes the delta plain and is an intertidal to subtidal platform 8–10 km in width (Combaz and De Matharel, 1978). Localized sand bars interrupt a monotonous mud sequence in which a marine fauna is often preserved. At the

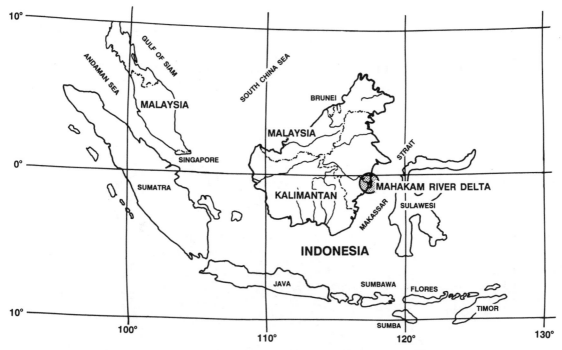

Figure 1. Location map of the Mahakam River delta, Kalimantan, eastern Borneo.

boundary of the delta front and the delta plain, laterally extensive detrital peats accumulate as beach ridges (G. P. Allen et al., 1979). These beach ridges may be as much as 2.5 m in thickness, are distributed as much as 3 km inland, and may cover a total surface area of approximately 50 km^2 (G. P. Allen and Pizon, 1986).

Prodeltaic sediments accumulate on the outer limit of the delta front where water depth begins deepening to 35 m within a kilometer of the delta front (Kartaadiputra et al., 1975). The prodelta is characterized by a homogenous massive mud in which beds of carbonaceous clay and silt occur. Although phytoplankton would be expected to accumulate within this area, none have been observed in the sediment (Combaz and De Matharel, 1978). Decayed organic debris originating from the delta plain has been noted to occur. This includes degraded wood fibers and other aerial plant macrodetritus, but few spores and pollen have been recovered (Combaz, 1964; Bellet, 1987).

Setting of detrital peat accumulations

Detrital peats occur as tidal flats and high-tide beach ridges along the delta front of the Mahakam delta. They are composed of allochthonous plant parts of various systematic affinities and sizes (see following discussion) along with an admixture of mud. Accumulations occur on the southern side of the Handil distributary and along several headlands of the interdistributary zone (Tandjung Bayor, Tandjung Timbanglugan, Tandjung Terantang, and Tandjung Ayu to Tandjung Kaeli; Fig. 2). The position of these accumulations appears to be adjacent to "abandoned" flu-

vial channels that are undergoing erosion by increased tidal domination. Where they occur along headlands, detrital peat accumulations alternate with tidal mudflats in a chenier-like sequence. This alternation is easily recognizable from aerial photographs because these deposits are accentuated by the distribution of vegetation along the headlands (Fig. 3). Detrital peat is colonized toward land by dense herbaceous growth (principally a dicotyledonous ground creeper with 3–4 stolons overlying each other), whereas mudflats are vegetated by mangrove trees and ferns (predominantly *Avicennia* and *Acrostichum* sp., respectively). The peat beach investigated in this study is on the aggrading headland of Tandjung Bayor.

Tandjung Bayor is the most prominent interdistributary headland in the delta. It has a present width of 7 km and a length of 10 km (as measured from the first well-developed tidal channel to the north). It is subjected to daily tides, with spring tides exceeding 2 m in range. Several small tidal channels, originating on its northern and southern tide, penetrate into the interior of the headland. These are not as well developed as tidal channels found in the interior of the delta (Fig. 2). Tidal channels in their distal part are blackwater, clastic-free, waterways. Small waves, with an amplitude generally less than 0.5 m, are generated in the Makassar Strait and directed toward the south. This marine influence is responsible for the transport and deposition of both clastic and organic sedimentary particles onto Tandjung Bayor.

The headland is vegetated mainly by a dense *Nypa* palm swamp, disturbed by the construction of aquaculture ponds. The 3 km of the headland that are most seaward, however, are characterized by alternating vegetational zones of mangrove (*Avicen-*

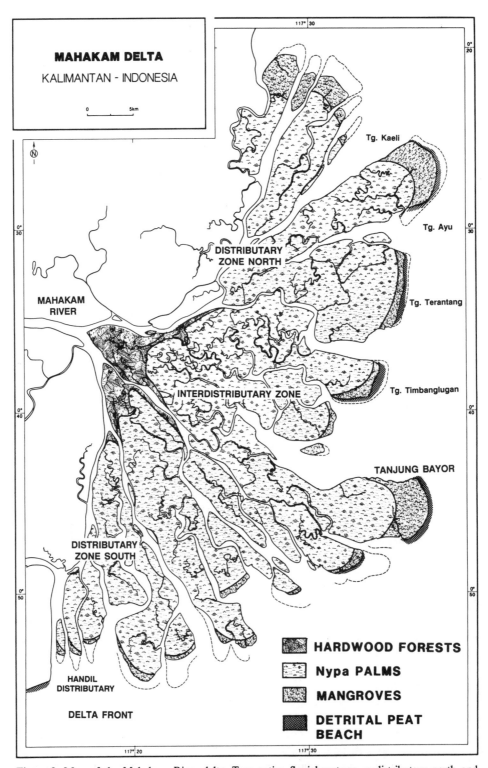

Figure 2. Map of the Mahakam River delta. Two active fluvial systems, a distributary north and distributary south zone, bound an interdistributary zone that is dominated by tidal activity. The delta is vegetated by hardwood, *Nypa* palm, and mangrove swamps along a salinity and topographic gradient. Detrital peat is found accumulating along headlands of the interdistributary zone.

nia) and herbaceous plants. Locally, the mangrove *Sonneratia* is established on the edges of aquaculture ponds. This tree is not a principal component in the community. The distribution of plants reflects the differences in soil substrata noted previously. The vegetational zones have developed in an arcuate pattern corresponding with the prevailing direction of littoral transport of sediment (Fig. 3).

The present detrital peat beach occurs at the seaward margin of the headland. It is composed of a supratidal, intertidal, and subtidal zone (Fig. 4). The supratidal zone extends for at least 100 m from the high-tide beach ridge toward the interior of the headland, where it is stabilized by herbaceous vegetation. It is a relatively flat area that is bounded seaward by the beach ridge. An escarpment at the seaward edge of the beach ridge, resulting from high-tide erosion, is slightly greater than 0.5 m in height (0.65 m as measured on October 21, 1988). The top of the beach ridge is littered with heterogeneously sized plant parts including drift-

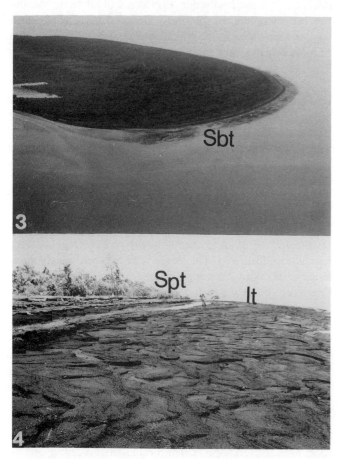

Figure 3. Oblique aerial photograph of Tandjung Bayor with a subtidal (Sbt) zone characterized by a slurry of detrital plant parts and suspension-load mud. The arcuate distribution of mangrove and herbaceous vegetation reflects differences in mineral and peat soils. Photograph taken October 15, 1988.

Figure 4. Photograph of the peat beach at low tide with supratidal (Spt) and intertidal (It) zones indicated. The high-tide beach escarpment may be as much as 0.65 m in height. Photograph taken October 22, 1988.

wood (tree trunks and branches as much as several meters in length), seeds of *Nypa* (10 cm in length), entire leaves of various dicotyledonous angiosperm taxa (as much as 30 cm in length), and fragmented canopy litter (see following discussion). The intertidal zone extends from the high-tide beach ridge to the shoreline and is characterized by a cuspate morphology caused by reworking of the peat. The peat beach is composed of fragmented aerial plant detritus that is reworked by tide and wave activity. These plant parts are principally less than 1 cm^2 in area and are sorted into coarser-fraction (approximately 1 cm) and finer-fraction (approximately 0.5 cm) accumulations. The distribution of sorted plant parts appears to be random with respect to sedimentological processes during tide cover. Delicate gastropod shells, some exceeding 10 cm in length, occur within the intertidal zone. The shells are complete, preserved because each is packed with detrital peat. Lenticular accumulations of mud are found within the peat beach and reflect suspension-load sedimentation in localized protected depressions within the peat body. Linear mud deposits accumulate behind spits composed of peat transported by littoral currents. The subtidal peat zone extends approximately 200 m into the shallow water surrounding the headland (Fig. 3). These waters contain a slurry of concentrated, fragmented plant parts mixed with suspension-load mud.

METHODOLOGY

Peat beach sediment was sampled by two procedures. It was not possible to transport a vibracoring apparatus onto the beach, thus requiring that cores had to be driven into the sediment by hand. This resulted in the acquisition of a compressed core of 90 cm length. Immediately after splitting, pH and Eh values were measured throughout the length of the core. One half of the core was logged and photographed; the other half was subsampled. Four horizons were bulk-sampled for phytodebris (depths of 5, 25, 76, and 90 cm), and six horizons were sampled for geochemical analyses (5 [IFP88694], 20 [IFP88695], 35 [IFP88696], 50 [IFP88697], 70 [IFP88698], and 90 cm [IFP88699] depths). In addition, bulk core samples, varying in the amount of recovered peat, were taken. The second sampling method involved grab samples (plastic zip-lock bag full) recovered from the exposed supratidal and intertidal zones of the beach.

All samples containing recoverable plant macrodetritus were sieved (100 μm mesh) to remove most of the clastic component. Sieving was conducted either on the barge by Indonesian assistants or at TOTAL's Handil base. Samples containing highly fragmented macrodetritus were maintained for processing to recover dispersed cuticles. All samples were dried under heat lamps at 60 °C at the end of each field day. Detrital damar (dipterocarp resin) was removed, where recognized, and separate plant samples were recovered for ^{14}C dating. Whole leaves, comprised solely of xylary (conducting) tissue surrounded by a cuticular sheath, and allochthonous woods were collected from the supratidal zone. Sieved macrodetritus has been examined using a Wild M-8 stereoscope and a Philips 505 scanning electron microscope

(SEM) (Université de Paris-Sud, Orsay). Component plant parts from each sample have been identified, categorized (*Nypa* laminae, leaf laminae of dicotyledonous angiosperms, *Nypa* petiole parts, dicot petiole parts, dispersed cuticles, reproductive propogules [seeds], woods and fibrous materials, roots and rootlets, mosses, and damar [resins]), and quantified.

In an effort to characterize the principal plant parts of the litter beds, a quantification procedure used to characterize palynomorph assemblages has been applied to each sieved sample. Plant parts from each sample, or sample suite from an identified sedimentary facies, are counted in five sets of 300. Each plant-part category from each sample examined has been placed into a glass vial and labeled according to locality and part category. The five tallies are then averaged to provide a percentage of contribution for each plant part category.

Geochemical samples were transferred from the central parts of cores to glass jars (10 cc), bulk samples of the peat beach also were placed in glass jars. All samples were dried at 60 °C. These have been subjected to TOC analyses, ^{14}C dating, and Rock-Eval pyrolysis. TOC was determined by several methods including Carmhograph, Rock-Eval module, Leco carbon analyzer, and elemental analysis. Individual plant parts recovered from phytological samples and identified in the laboratory were subjected to preparative pyrolysis (according to the procedure of Vandenbrouke and Behar, 1988). These included wood, hardwood dicot leaf fragments (as differentiated from mangrove taxa), dispersed dicot cuticle (separated from its parent leaf by mechanical fragmentation during reworking), mangrove cuticle (removed from degraded leaves collected on the surface of the peat beach; see Figs. 5, 6), and damar (resin) picked from the surface of the peat deposit. A small quantity of each identified plant part (20–30 mg) was deposited on a gold rod, placed in a furnace (Delsi/Girdel) under a flow of argon and an initial oven temperature of 320 °C, and pyrolyzed to 550 °C (60 °C/min). The C14+ pyrolyzate was trapped in a small cooled reaction vessel. The saturated and unsaturated hydrocarbons were isolated by liquid chromatography and analyzed by gas chromatography and gas chromatography–mass spectroscopy (R10C Quadripole Nermag) at the Institut Français du Pétrole.

RESULTS

Peat beach composition: Biological components

Examination of the detrital peat beach along Tandjung Bayor revealed that this accumulation is composed of a collection of plant parts of various systematic affinities and sizes. The small fraction, those parts less than or equal to 1 cm, predominated. Large tree trunks of dicotyledonous angiosperms, as well as bases of *Nypa* palm, were obvious on top of the accumulation in the supratidal zone. Few hardwood logs were seen to be incorporated in the beach. No *Nypa* bases were seen incorporated into the beach sediments. The residence time of large woody trunks within beach deposits may be short lived. For example, on Oc-

tober 21, 1988, we observed a large part of a prostrate tree trunk embedded in the peat. On October 22, 1988, during the low neap tide cycle, we observed the same tree trunk completely exposed. All peat surrounding the trunk had been eroded from around the wood. Most woods, however, are within the pebble-sized fraction and incorporated in the peat. Fruits of *Nypa* were found lying on and incorporated into the peat. Accumulations of *Nypa* fruits were both concentrated and scattered. None were found germinating.

Entire leaves of dicotyledonous angiosperms, some of which were mangroves, also were distributed on and incorporated into the beach (Fig. 5). Both entire leaves (with parenchymatous tissues intact) and degraded leaves were found. In the case of mangrove leaves, the mesophyll was commonly degraded, leaving only the interior xylary architecture and the surrounding cuticle (Figs. 6, 7). In most instances, the cuticle completely enveloped the xylary architecture. Degradation appeared to be solely through bacterial action, as evidence of fungal degradation was extremely rare. In specimens examined using SEM, fungal hyphae were uncommon. Where hyphae were encountered, one or more bridge xylary elements. No fungal reproductive bodies have been seen within any cuticular sheath.

The most common condition for leaves was fragmented. Leaf parts averaged a little more than 1 cm in size in the most coarse fraction of the peat beach. In the fine fraction, leaf size was comparable to coffee grounds. Beds of fine and coarse plant detritus sometimes alternated, reflecting the conditions under which they were deposited. Dispersed cuticles, separated from their source leaves, might have been the result of the mechanical fragmentation during bedload transport by wave and tide agitation.

The contributions of hardwood leaves and woods were approximately equal in frequency in the recovered phytological assemblages (Table 1). The leaves were overwhelmingly dominated by dicotyledonous angiosperms (those originating from the upper delta plain hardwood forests and interior of the island). There was little evidence for an important *Nypa* leaf (laminar) contribution. A small admixture of *Nypa* petioles occurs but is not significant. A single leafy moss fragment and a few seeds have been recovered to date. This depositional environment has a larger contribution from dispersed cuticle and damar (resin) than any other in the delta (Gastaldo and Huc, 1992). A bryozoan colony was recovered on one fragment of dicot leaf.

There appears to be a significant concentration of damar (resin) within the peat, as local fishermen collect it in 20–30 kg sacks for sale in Samarinda. The local fisherman collect only the damar that occurs at the surface, although it is mixed with the plant detritus. Damar was not observed to be concentrated into zones but, rather, dispersed throughout the peat. Pieces attained lengths of as much as several centimeters and diameters of at least 1 cm. They were generally rounded at the ends and might also be rounded into a cylindrical shape. None were similar in morphology or size to *in situ* resin ducts collected from allochthonous wood in fluvial sediments. Resin ducts in wood are small and cylindrical, attaining a maximum diameter of less than 300 μm (Gastaldo, 1993). Damar may be resident offshore prior to transport to the peat beach, as is evidenced by the attachment of bivalves and bryozoa to collected specimens. It is common to find the damar burrowed.

In addition to the phytological elements, other biological components have been recovered from the peat. Insects made up a small part of recovered biotic parts. Most of these specimens were ant heads, beetle elytra, and isolated legs that were difficult to assign taxonomically (W. Shear, May 24, 1989, written communication). Bivalve and gastropod shells were present

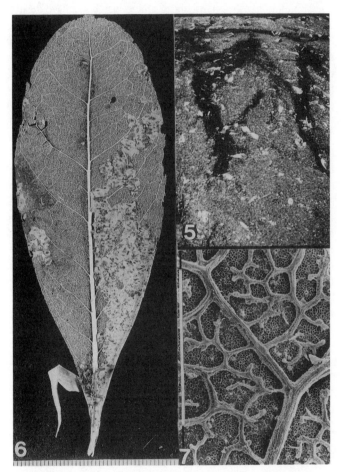

Figure 5. Photograph of peat beach supratidal zone with degraded mangrove leaves scattered over the surface.

Figure 6. Recovered mangrove leaf as depicted in Figure 5. The upper cuticle has been partially removed by mechanical processes, the remainder of it can be seen in the lower right of the leaf. All parenchymatous tissues have been degraded leaving the veins (xylary elements) and the lower cuticle. Scale is in mm.

Figure 7. Scanning electron micrograph of lateral venation of these mangrove leaves showing complete degradation of intervascular parenchymatous tissues. The lower cuticle with stomata is below the venation. Note that there are no fungal hyphae or reproductive bodies within the leaves. Scale equals 100 μm.

within the detrital peat. Bivalve shells generally were small and similar to those found in other lower delta-plain/delta-front environments. The gastropod shells, however, were much larger. Thin-walled gastropod shells were packed with detrital peat that maintained the integrity of the wall (personal observation, October 22, 1988).

Peat beach composition: Geochemical components

As measured from within the split core, pH values of the peat ranged from 5.47 and 5.64 near the top to 7.35 near the base. Eh values decreased down core. The TOC, as determined from peat beach core samples, ranged from 27.5 to 39.4, with corresponding ash contents of 18% to 27% (Table 2). Bulk samples, representing sample sizes ranging from 15 to 25 cm of core length, varied in ash content as much as 48%. Sulfur values (organic and inorganic fractions) obtained from bulk sediment samples ranged from a low of 0.65% to a high of 2.75% (sample IFP 88698 where mudflat overlies detrital peat). Hydrogen indices of core samples ranged from a low of 250 to a high of 450. Hydrogen:carbon atomic ratios averaged 1.43 (range from 1.39 to 1.51); oxygen:carbon ratios averaged 0.44 (range from 0.46 to 0.61).

J. L. Oudin and G. P. Allen (associated with TOTAL, Bordeaux) conducted ^{14}C analyses on bulk peat samples taken from core recovered in the early 1980s. Detrital damar (resin) samples collected during October 1988 were sent to Krueger Enterprises for dating. [14]C dates on subsurface peat samples varied depending upon their stratigraphic and geographic position. Peat recovered from the base of a 1.5 m compressed core taken 400 m inland was dated at 1050 yr B.P., whereas a sample taken at a depth of approximately 0.7 m was dated at 880 yr B.P. Peat recovered 50 m inland from the interface with the ocean, at a depth of 0.3 m, was dated at 640 yr B.P. Two samples of damar picked from the surface of the peat beach were analyzed. The first sample appeared "fresh." The damar was amber (dark yellowish orange, 10 YR 6/6) in coloration and translucent. The second sample appeared "weathered." It was bluish white (5 B 9/1) and opaque. The transluscent specimen was ^{14}C dated at 2645 ± 215 yr B.P., whereas the opaque specimen was ^{14}C dated at 930 ± 205 yr B.P.

To determine the geochemical contribution of each major plant component deposited in the peat, selective plant parts were analyzed. The Rock-Eval data for these plant parts are displayed on a hydrogen index/oxygen index (HI/OI) diagram (Fig. 8). The data are widespread with dicotyledonous wood of a HI low of 210 and the damar exhibiting an exceptionally high HI of 1,130. Intermediate values belong to the other samples: 870 for the mangrove cuticle (see: Figs. 5, 6), 560 for dispersed dicot cuticle fragments, and 315 for hardwood dicot leaf pieces. The Py-GC traces of the saturated + unsaturated hydrocarbon fraction for these plant parts are illustrated in Figure 9. With the exception of the damar (resin), the distribution of saturates and unsaturates is dominated by linear hydrocarbons occurring as

TABLE 1. DETRITAL PEAT COMPONENT PLANT AND ANIMAL PARTS

Leafy moss	0.1
Woody and fibrous detritus	43.5
Nypa Petiole	3.8
Leaf Laminae	36.0
Dispersed cuticle	5.1
Fruits and seeds	0.3
Rootlets	6.7
Damar (resin)	4.4
Millimeter-sized resistant parts	P*
Insect parts	P*
Bivalve shells	P*
Gastropod shells	P*

Frequency of occurrence expressed as a percentage of sample. (N = 1,500)
*Indicated presence in the accumulation.

doublets, with the presence of the n-alk-1-ene just preceeding the n-alkane. In these samples, the Prist-1-ene is often a major peak and the carbon distribution pattern is characterized by a large range of carbon number with a maximum in the high molecular weight (near NC24–nC33). This distribution is characteristic of land-plant derived hydrocarbons. Besides these commonalities, several differences can be pointed out between the various samples. These include a flatter distribution of the n-alkanes/n-alkenes and a generally higher importance of the nC28 fraction for the hardwood dicot parts as compared with the mangrove cuticle. These differences must be considered with care because of the small number of analyses.

The GC trace of the hydrocarbons in the pyrolyzate of the damar (resin) exhibits a clear and specific pattern. This pattern is characterized by two discrete sets of peaks, one corresponding to

C15 sesquiterpenoid hydrocarbons and the other to C30 triterpenoid hydrocarbons. It is interesting to note that the specific GC signature of the damar products can be recognized in the other GC traces to various degrees. The question arises as to the possible contamination of the other plant parts by microscopic damar fragments, although none were noted appressed to plant parts examined using SEM.

DISCUSSION

The recognition of detrital peat accumulations at the terrestrial/marine interface provides insight into several aspects of modern and ancient coastal regimes. First, they provide a definitive shoreline marker, particularly where beach sand would not accumulate, that can be used to evaluate basinal eustatic changes. Second, evaluation of the plant constituents yields information about the regional vegetational patterns. Last, provided a sufficient burial, these organic accumulations are probably a part of the source rock system responsible for the occurrence of hydrocarbons within the Mahakam deltaic sequence.

The deposits accumulating along the aggrading headland of Tandjung Bayor are peats, with ash contents as low as 18% (Table 2). These deposits exhibit sedimentary structures equivalent to any other beach (G. P. Allen et al., 1979). They accumulate only at the shoreline and have not been identified in any other terrestrial depositional environment in the Mahakam delta. The deposits accumulate in direct response to marine processes interacting with allochthonous plant detritus. Fluctuations in the marine processes result in the alternation of mud-flat/delta front and peat beach deposits, establishing a unique signature to the sedimentological sequence.

The recovered phytomacrodetritus is comprised of a wide diversity of plant parts of various systematic affinities. At the present time, constituents representing the fringing *Avicennia*

TABLE 2. GEOCHEMICAL DATA OF PEAT BEACH SAMPLES

IFP Sample Number	Depth in Core	TOC* (%)	TOC† (%)	TOC§ (%)	TOC** (%)	Ash (%)	Sulfur (%)	Hydrogen Index	Oxygen Index	H/C Atomic Ratio	O/C Atomic Ratio	Maximum Temperature (°C)
88694	5 cm	39.0	39.4	38.2	40.8	22.0	0.66	446	101	1.45	0.48	387
88695	20 cm	39.2	36.1	36.7	41.4	18.1	0.84	406	115	1.44	0.46	386
88696	33 cm	30.9	31.8	29.8	32.4	27.4	1.73	267	130	1.4	0.59	377
88698	68 cm	33.2	30.6	30.3	34.0	22.8	2.76	354	114	1.39	0.54	378
88699	86 cm	28.9	27.5	26.8	29.7	24.8	1.54	252	141	1.51	0.61	320
88700	Bulk‡	35.7	36.1	35.9	39.2	22.0	0.65	266	151	NA	NA	318
88701	Bulk§§	32.1	30.5	27.0	32.8	31.5	1.55	262	150	NA	NA	324
88702	Bulk***	23.3	22.5	22.6	22.6	48.2	1.50	420	118	NA	NA	384

*Carmhograph
†Rock-Eval module
§Leco
**Elemental analysis
‡Bulk sample from 10 to 20 cm depth
§§Bulk sample from 25 to 40 cm depth
***Bulk sample from 65 to 85 cm depth

PEAT BEACH PLANT DETRITUS
ROCK–EVAL DATA

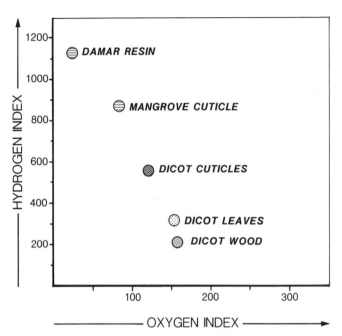

Figure 8. Hydrogen index:oxygen index plot of Rock-Eval data for selected plant parts sampled from the detrital peat beach.

mangrove community, the lower delta plain *Nypa* palm swamps, the upper delta plain hardwood forests, and the interior island vegetation are all represented in differing quantities. A detailed assessment of systematic contribution to the deposit will be forthcoming after completion of dispersed cuticle studies of samples collected within the delta. For the moment, the following remarks can be made concerning the plant constituents.

Leaf and wood fragments, both size-sorted to approximately 1 cm^2 area, dominate the assemblage. Frequently, entire leaves can be found in the peat, and these are either mangrove or hardwood taxa. The leaves of the mangrove *Avicennia* tend to be retained as a cuticular sheath with the internal parenchymatous cells degraded. To date, these leaves have not been recognized in any other state of preservation within the deposit. The transport of these leaves to the beach may reflect slightly different conditions than those of hardwood taxa not living in the immediate vicinity (see following discussion). Several whole leaves of hardwood taxa were recovered both from within and on the peat beach, but most leaves are highly fragmented. Whereas the majority of the lower delta plain vegetation is *Nypa* palm, and it is common to find fruits of the palm concentrated in the peat, quantitatively it is underrepresented in the deposit. This is in contrast to the assertion made by G. P. Allen and Pizon (1986) that the peat beach was comprised mainly of *Nypa*. Pieces of

either *Nypa* laminae or petioles are rare. This is the result of the *Nypa* swamp community structure, its growth habit, and the retention of the nonfunctional leaves on the plant after leaf death. The density of *Nypa* palm in wetland swamps is high. Leaves are produced from the base of the plant and may be functional for several years before death. The leaf laminae of nonfunctional leaves degrade while the laminae are still attached to the fibrous and resistant petiole. The petiole, in turn, is retained on the plant base until weakened by either insect inhabitation (ant colonies thrive within *Nypa* petioles) or fungal degradation (that may be induced by the colonization by ants). When petioles are weakened, they normally fall over but do not reach the ground or adjacent channel. Rather, they are often caught within the leaves of other plants and remain suspended above the soil or water level. Degradation continues and fragments of the petiole, the most resistant part, may eventually end up in the mud or waterway. *Nypa* is transported to the detrital beach site by water. The closest stand of the palm is more than 5 km away from Tandjung Bayor, separated by mangrove swamp toward the interior of the delta. The primary means of transport is probably by tide activity, although fluvial transport cannot be ruled out as this is the mechanism for movement of plant parts from hardwood forest to the delta front.

Hardwood forest leaves, small leafy mosses, woods, and damars (dipterocarp resins) are transported in the river channels for at least 50 km, and probably greater distances. These plant parts occupy different sites of transport in the fluvial system. The leaves and delicate mosses are not resident in bedload during their movement downstream, as bedload transport would result in fragmentation. Rather, they are transported in the suspension load and out into the expanse of the subaerial delta proper. They may either float at the surface or slightly beneath the surface, depending upon their density and bouyancy. As Gastaldo et al. (1987) documented, whole leaf transport is the rule rather than the exception in a river-dominated regime. Scheihing and Pfefferkorn (1984) also reported the transport of leaf rafts through the Orinoco delta in Venezuela. This also seems to be the case in the Mahakam. In macrodetrital litter beds recovered from lateral channel bars in the major distributary channels and the delta front sand, most leaves were recovered whole (Gastaldo and Huc, 1992). The descent from suspension load to the sediment/water interface in the lateral channel bars was precipitated by high tidal velocities (spring tides) moving against river flow. Shallowing, coupled with interactions with wave and tidal currents, cause some of the suspension-load organics to drop out in the delta front sands. Leaves are apparently transported for some distance offshore before they settle out of suspension, if they have not already been trapped in other depositional environments. All leaves recovered are black in color, evidencing chemical alteration of the cellular contents.

Leaves are transported as far as the point where clear-water conditions prevail, evidenced by the leaf fragment recovered with an attached bryozoan colony. Offshore residence time for leaves transported through the delta may be as much as several months

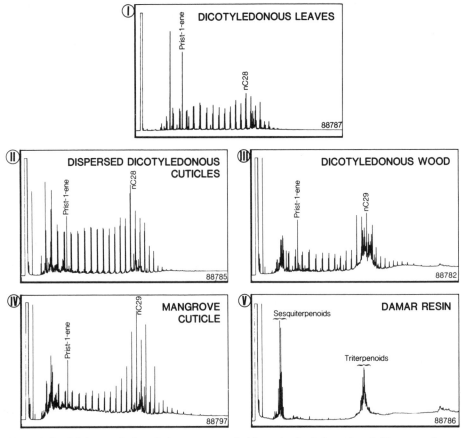

Figure 9. Py-GC traces of selected plant parts sampled from the detrital peat beach. Traces are those of saturates + unsaturates. IFP sample numbers are included in the lower right.

or longer. Fungal infestation does not appear to be the principal cause of degradation once these parts are resident in the water column (Gastaldo and Huc, 1992; Gastaldo, 1993). The ^{14}C dates of peat and other leaf samples demonstrate that the integrity of these parts may be retained for hundreds of years without decay. The deterioration and fragmentation of leaves appears to be related to their resuspension and transport toward shallow water areas where they are subjected to continuous, and often intense, agitation by waves and tides. In their weakened, saturated condition, the whirlpool-like effect operating at the shoreline pulverizes these plant parts. Tidal activity appears to be the principal means by which the plant parts are deposited onto the fronts of the headlands. Those entire or nearly entire leaves, infrequently encountered in the peat, have not been resident in the water column as long as those that are fragmented. The entirety of such leaves is suggestive that these parts were never moved out of the suspension load.

In the case of the mangrove leaves, preserved as cuticular sheaths encasing venation, these may have been transported to the beach by wind from nearby trees. Such leaves were continu-

ally encountered on the peat surface of the supratidal zone. Rarely were they found within the peat deposit. It is also possible that they were transported via water, flushed from the mangrove swamps by tidal currents, and deposited on top of the beach during an overwash event.

Wood may be transported either in suspension load or in bedload depending upon their initial density and/or maintenance of their buoyancy. Dipterocarpaceae and other tropical woods possess a range of physical characters including the ability to float or sink immediately when entering the water (Wood, 1964). Floating may occur either at the water's surface or, most commonly, below water level. This is controlled strictly by the weight and density of the wood in question. Those that float may eventually sink when rotting has proceeded to the point where water infiltration increases their density, and they fall out of suspension. Wood may also shoal in shallow water sites. Wood fragments have been transported principally in bedload in the deepest parts of channels to the delta front. These are rounded in shape, similar to those reported by Gastaldo et al. (1987), but show no signs of fungal degradation. All alteration appears to be mechanical. It is not known how far offshore these small parts may have been transported before being reworked to the delta front principally by waves and tides. No small wood fragments with an epifauna have been recovered. One larger wood piece had been "*Teredo*-bored," indicating that it was resident offshore for a considerable period of time before being picked up and moved back to the headland.

In the case of the damars, these plant parts also appear to have been transported in bedload. Damar density is high, and it sinks in water. The pieces of damar taken from the peat beach are rounded at their ends, and may be cylindrical in shape. They are larger than all other plant parts in the peat, with the exception of woody branches, trunks, and whole leaves. These damars are exudates from dipterocarp trees; they do not represent internal resin duct accumulations. The resins are, then, not produced locally but transported from the dipterocarp forests of the island's interior (a minimum of 100 km). Additionally, these plant parts have been resident offshore for some time, as indicated by the presence of borings in the resin and bivalves attached to the exterior of damar specimens.

The peat accumulations are Recent in origin as indicated by the ^{14}C dates taken from samples within beach cores. They do not represent the reworking of mangrove and *Nypa* swamps within the interdistributary zone as physical processes have changed from fluvial domination to tidal domination in this part of the delta. *Nypa* swamp soils are characterized by scattered leaf petioles and rootlets, while the adjacent tidal channels possess a high concentration of *Nypa* aerial parts. *Avicennia* mangrove soils are dominated by pneumatophores (woody roots) with little incorporation of aerial canopy detritus. Reworking of these sites would not provide the suite of plant parts found in the peat beach. During the enlargement of former fluvial pathways, reworking of lateral channel bar deposits occurs. This reworking mechanically grinds the previously deposited litter to particle sizes equivalent to the sand in which they are found. The plant parts that compose peat beaches are direct contributions from the basinal and extrabasinal communities of the island.

Originally it was difficult to imagine a significant input of modern damar to the peat beach. We had thought that the damar might have been reworked from Miocene coals outcropping in the hinterland. These resins might have been dislodged by fluvial activity as the Mahakam River cut through these rocks near Samarinda. This does not seem to be the case, as ^{14}C dates indicate that the damars are Recent in origin. These parts are generated by the wounding of a tree, the wound being healed by the exudate. When the exudate breaks off from a host tree it is moved to a waterway, sinks to the bottom of the channel, and is transported as bedload. Damar is probably contributed from trees adjacent to clastic pathways, as flooding has not been recorded in this delta. The damars are probably not the result of recent deforestation in eastern Borneo. If these parts were from logging, the ^{14}C dates for their formation should indicate an age no greater than 100 yr B.P. A significant quantity of solidified dipterocarp damar enters the river system and is transported to the delta front.

This quantity, of course, only represents a small fraction of damar biomass. Its geochemical signature is unique and recognition of the signature in the Py-GC traces of the dicot wood is not surprising.

The preservation of the peat appears to be excellent even with the pH change down section from slightly acidic (pH 5.47) near the surface to nearly neutral (pH 7.35) at a depth greater than 1 m (when uncompressing the core). The subsurface pH equals that found in most other cores extracted from throughout the delta plain. This suggests that there is either some groundwater flushing or infiltration of surrounding water through this deposit. In either case, the nearly neutral pH does not provide a microenvironment conducive to degradation at depth by fungal activity. It is not known to what extent, if any, bacterial degradation continues because recovered plant parts at depth maintain their integrity. In addition, sulfur values averaging 1.50% have been recorded from samples taken at depth, indicating little bacterial activity at this point in time.

Because we are dealing with recent sediments, which have not already undergone diagenesis, the reported Rock-Eval data cannot be considered as a direct measurement of the actual petroleum potential of the different organic detritus. This is especially true for those parts exhibiting the highest OI (dicot leaves and woods). As a matter of fact, an initial increase of the HI is often reported during the diagenesis of Type III organic matter (Huc et al., 1986). This increase can be explained by a progressive natural loss of CO_2 with a corresponding absolute decrease of the organic carbon content. This results in a relative increase of the hydrocarbons released by pyrolysis when related to the remaining organic carbon. The observed scatter of HI, however, is an indication of the variability of the chemical properties of the organic detritus and their respective ability to generate hydrocarbons during catagensis.

SUMMARY

Detrital peat of recent origin is presently accumulating as beach deposits at the terrestrial/marine interface of the interdistributary zone in the Mahakam River delta, eastern Kalimantan, Borneo. These accumulations alternate with mud flats in a chenier-like arrangement and can be easily recognized by vegetational differences. Mangrove taxa, principally *Avicennia,* inhabit mudflats (mineral soils), whereas herbaceous ground creepers dominate peats (organic soils). The peat accumulations are composed of *Nypa* laminae, leaf laminae of dicotyledonous angiosperms, *Nypa* petiole parts, dicot petiole parts, dispersed cuticles, reproductive propagules [seeds], woods and fibrous materials, roots and rootlets, mosses, and damar (resins). Contributions to these deposits are mainly allochthonous, derived from vegetation either in the upper delta plain or from the interior of the island. Plant particle size in the deposit is the result of mechanical fragmentation of mainly reworked plant parts from offshore sites of residency. The mechanical breakup of parts occurs as the result of tidal and wave activity in shallow water areas fronting the headlands. Individual plant constituents of the beach have been subjected to geochemical analyses, including Rock-Eval and Py-GC. Damars and cuticles exhibit the highest potential for generating hydrocarbons upon pyrolysis. With the exception of damar resin, the Py-GC traces of the plant parts are similar and the distribution of saturates and unsaturates is dominated by linear hydrocarbons occurring as doublets including the n-alk-1-enes and n-alkanes, Prist-1-ene is often a major peak. The carbon distribution pattern is characterized by a large range of carbon number with a maximum in the high molecular weight (near nC24–nC33). Py-GC traces of the damar exhibit a unique chemical signature that is also seen in the sampled wood.

ACKNOWLEDGMENTS

The authors thank Choppin de Janvry, TOTAL Indonésie, for his cooperation and logistical assistance before and during the field portion of this study. Yves Grosjean, TOTAL Indonésie, is acknowledged for his conscientious efforts to ensure that our mission in the Mahakam delta went as smoothly as possible. Aspani, Sukardi, Hongo, and Hary are gratefully acknowledged for their technical assistance during the course of the field season. William A. DiMichele, U.S. National Museum, is thanked for his tireless efforts in Kalimantan. Bruce Purser, professor of sedimentology, Université Paris-Sud, Orsay, is acknowledged for his support and assistance subsequent to our field mission in Indonesia.

The logistical support for this project was provided by TOTAL Indonésie. This project was supported, in part, by a grant from the National Science Foundation (EAR 8803609) to Gasatldo. The Petroleum Research Fund, as administered by the American Chemical Society, is acknowledged for partial support of this project (ACS PRF 20829-AC8). The Institut Français du Pétrole and the Université Paris-Sud, Orsay, are acknowledged for providing partial support to Gastaldo for continuing studies at the Institut Français du Pétrole following the field season.

REFERENCES CITED

Allen, G. P., 1987, Deltaic sediments in the modern and Miocene Mahakam delta: Pessac, France, TOTAL Exploration Laboratory, 53 p.

Allen, G. P., and Pizon, J., 1986, Detrital peat deposits in the modern Mahakam delta (Indonesia): Environmental significance and fossil analogues in adjacent Miocene delta deposits [abs.]: Canberra, Australia, 12th International Sedimentological Congress.

Allen, G. P., Laurier, D., and Thouvenin, J. P., 1977, Sediment distribution patterns in the modern Mahakam Delta, in Proceedings, 5th Annual Convention Indonesian Petroleum Association, v. 1, p. 159–178.

Allen, G. P., Laurier, D., and Thouvenin, J. P., 1979, Étude sedimentologique du delta de la Mahakam: Paris, TOTAL, Compagnie Françaises des Pétroles, Notes et Mémoires, v. 15, p. 1–156.

Allen, G. P., Pizon, J., and Bordenave, M., 1986, Sedimentology of the subsurface Mahakam Miocene deltaic basin (Indonesia): [abs.]: Canberra, Australia, 12th International Sedimentological Congress.

Allen, J.R.L., 1965, Late Quaternary Niger delta and adjacent areas: Sedimentary environments and lithofacies: American Association of Petroleum Geologists Bulletin, v. 49, p. 547–582.

Anderson, J.A.R., 1964, The structure and development of the peat swamps of Sarawak and Brunei: Journal of Tropical Geography, v. 18, p. 7–16.

Anderson, J.A.R., 1983, The tropical peat swamps of western Malesia, in Gore, A.J.P., ed., Ecosystems of the world, mires: Swamp, bog, fen and moor,

v. 4B, p. 181–199.

Andriesse, J. P., 1974, Tropical lowland peats in South-East Asia: Amsterdam, Department of Agriculture Research, Tropical Institute Report, 63 p.

Bellet, J., 1987, Le sondage MISEDOR, Palynofaciès et analyses élémentaires de la matiére organique: Géochimie organique des sédiments Plio-Quaternaires du delta de la Mahakam (Indonésie): Paris, Éditions Technip, p. 183–196.

Carbonel, P., and Moyes, J., 1987, Late Quaternary paleoenvironments of the Mahakam delta (Kalimantan, Indonesia): Palaeogeography, Palaeocliimatology and Palaeoecology, v. 61, p. 265–284.

Coleman, J. M. and Prior, D. B., 1980, Deltaic sand bodies: American Association of Petroleum Geologists Continuing Education Course Note Series, v. 15, p. 1–171.

Combaz, A., 1964, Les palynofaciés: Revue Micropaléontologie, v. 7, p. 205–218.

Combaz, A. and De Matharel, M., 1978, Organic sedimentation and genesis of petroleum in Mahakam delta, Borneo: American Association of Petroleum Geologists Bulletin, v. 62, p. 1684–1695.

Gastaldo, R. A., 1989, Preliminary observations on phytotaphonomic assemblages in a subtropical/temperate bayhead delta: Mobile Delta, Gulf Coastal Plain, Alabama: Review of Palaeobotany and Palynology, v. 58, p. 61–83.

Gastaldo, R. A., 1993, The genesis and sedimentation of phytoclasts with examples from coastal environments, in Traverse, A., ed., Sedimentation of Organic Particles: Cambridge, England, Cambridge University Press.

Gastaldo, R. A. and Huc, A. Y., 1992, Sediment facies, depositional environments, and distribution of phytoclasts in the Recent Mahakam River delta, Kalimantan, Indonesia: Palaios, v. 7, p. 574–591.

Gastaldo, R. A., Douglass, D. P., and McCarroll, S. M., 1987, Origin characteristics and provenance of plant macrodetritus in a Holocene crevasse splay, Mobile delta, Alabama: Palaios, v. 2, p. 229–240.

Gayet, J. and Legigan, Ph., 1987, Étude sédimentologique du sondage MISEDOR (delta de la Mahakam, Kalimantan, Indonésie): GéGéochimie organique des sédiments Plio-Quaternaires du delta de la Mahakam Indonésie): Paris, Éditions Technip, p. 23–72.

Kartaadiputra, L., Magnier, P., and Oki, T., 1975, The Mahakam delta, Kalimantan, Indonésie, in Proceedings, 9th World Petroleum Congress, v. 2, p. 239–250.

LaLouel, P., 1979, Log interpretation in deltaic sequences, in Proceedings, 8th Annual Convention Indonesian Petroleum Association, v. 1, p. 247–290.

Loired, B. and Mugniot, J. F., 1982, Seismic sequences interpretation: A contribution to the stratigraphical framework of the Mahakam area, in Proceedings, 11th Annual Convention, Indonesian Petroleum Association, v. 1, p. 323–334.

McCabe, P. J., 1984, Depositional environments of coal and coal-bearing strata, in Rahmani, R. A., and Flores, R. M., eds., Sedimentology of coal and coal-bearing sequences: Oxford, International Association of Sedimentologists Special Publication 7, p. 147–184.

Parrish, J. T. and Barron, E. J., 1986, Paleoclimates and economic geology: Society of Economic Paleontologists and Mineralogists Short Course, v. 18, p. 1–162.

Roux, G., 1977, The seismic exploration of the Mahakam delta—or—"Nine years of shooting in rivers, swamps and very shallow offshore," in Proceedings, 6th Annual Convention, Indonesian Petroleum Association, v. 2, p. 109–142.

Scheihing, M. and Pfefferkorn, H. W., 1984, The taphonomy of land plants in the Orinoco delta: A model for the incorporation of plant parts in clastic sediments of upper Carboniferous age in Euramerica: Review of Palaeobotany and Palynology, v. 41, p. 205–240.

Stevenson, J. J., 1912, The formation of coal: III, in Proceedings, American Philosophical Society, v. 51, p. 243–553.

Vandenbrouke, M. and Behar, F., 1988. Geochemical characterization of the organic matter from recent sediments by a pyrolysis technique, in Fleet, A. J., Kelts, K., and Talbot, M. R., eds., Blackwell Scientific Publications, Oxford, Geological Society of London Special Publication 40, Lacustrine petroleum source rocks: p. 91–101.

MANUSCRIPT ACCEPTED BY THE SOCIETY JANUARY 14, 1993

Geological Society of America
Special Paper 286
1993

Palynologic and petrographic characteristics of two Middle Pennsylvanian coal beds and a probable modern analogue

Cortland F. Eble*
U.S. Geological Survey, National Center, Reston, Virginia 22092
William C. Grady
West Virginia Geological and Economic Survey, P.O. Box 879, Morgantown, West Virginia 26507

ABSTRACT

Four compositional groups, based on the independent parameters of ash yield, sulfur forms, palynology, petrography, and low-temperature ash mineralogy are recognized in the Stockton and Fire Clay coal beds (Kanawha Formation, Middle Pennsylvanian) from the central Appalachian Basin. The four compositional groups are (1) an arboreous lycopod-dominant group, defined by high percentages of *Lycospora* and vitrinite (telocollinite), generally low but variable ash yield and sulfur content, and a mixed illite-kaolinite-quartz low-temperature ash mineralogy; (2) a transitional group defined by a mixture of lycopods (arboreous and "herbaceous" forms) and ferns, increased percentages of inertinite and desmocollinite (degraded vitrinite), low ash yield and sulfur content, and an increasingly dominant kaolinite ash mineralogy; (3) an "herbaceous" lycopod and fern group defined by high percentages of *Densosporites* (and related crassicingulate taxa) and ferns, high inertinite and desmocollinite contents, very low ash yield and sulfur content, and a kaolinite-dominant mineralogy; and (4) a mixed group defined by a palynoflora co-dominated by arboreous and "herbaceous" lycopods and ferns, with calamite and cordaites occurring in increased percentages, and variable petrography. Typically, this group is high in ash yield, low in sulfur content, and has an illite-dominant ash mineralogy.

The vertical arrangement of these groups define compositional cycles, which are believed to represent different developmental stages of a domed peat swamp. The commonly observed vertical profile of *Lycospora*/vitrinite-rich coal layers changing upward to fern-, small lycopod-, and inertinite-rich coal layers defines the first half of the cycle. A return, gradual or abrupt, to *Lycospora*/vitrinite-rich coal layers toward the top of the bed defines the second half of the cycle. The first half of the cycle is interpreted to represent a progressive change from more planar to more domed peat conditions. This change is marked by increased exposure to air and oxygenated rain water and the establishment of floral seres. The second half of the cycle is interpreted to represent a change from more domed to more planar conditions. Compositional "half cycles," common in both the Fire Clay and Stockton coal beds, are usually the result of cycle truncation by inorganic partings.

Both planar and domed swamp environments contribute to peat formation in modern domed peat systems. Similarly, both of these environments are recognized in the Fire Clay and Stockton coal beds. A majority of these two coal beds are believed to have been derived from planar, to perhaps moderately domed, peat environments. In contrast, a relatively small part of these beds are thought to have been significantly domed.

*Present address: Kentucky Geological Survey, 228 MMRB, University of Kentucky, Lexington, Kentucky 40506.

Eble, C. F., and Grady, W. C., 1993, Palynologic and petrographic characteristics of two Middle Pennsylvanian coal beds and a probable modern analogue, *in* Cobb, J. C., and Cecil, C. B., eds., Modern and Ancient Coal-Forming Environments: Boulder, Colorado, Geological Society of America Special Paper 286.

INTRODUCTION

Modern and ancient peat-forming environments are complex biologic and chemical systems that involve the changing and interdependent effects of the flora, hydrology, climate, and depositional regime. Therefore, investigations of these environments, modern or ancient, should involve the study of all the interrelated parameters affecting the system. This paper presents a multidisciplinary approach to the recognition of ancient peat-forming ecosystems and environments. The techniques that will be discussed include palynology, coal petrography, proximate analysis (ash yield), sulfur-forms analysis, and low-temperature ash (LTA) X-ray diffraction (XRD) mineralogy. In this study we apply these techniques to the Middle Pennsylvanian Stockton and Fire Clay coal beds.

Paleoecology of Pennsylvanian coal beds

Paleoecological reconstruction of Pennsylvanian peat swamps using coal palynology is primarily facilitated by two factors: (1) the spore and pollen rain in large, forested swamps, thought to be good modern analogues for many Pennsylvanian coal beds, is largely autochthonous and therefore represents the local, contemporaneous flora (Anderson and Muller, 1975); 2) a majority of Pennsylvanian miospore taxa have now been affiliated with the parent plant group and in some cases to the parent plant (see Ravn, 1986, and Eble, 1988, for comprehensive reviews).

Paleoecological reconstructions based on palynomorphs have the advantage of providing a floral record of the peat-forming environment relatively unhindered by taphonomic biases. Other biases, inherent to the science of palynology, must be considered, however. For instance, as palynomorph production quantities of Pennsylvanian plants are unknown, some consideration must be given to the possibility of some plants being under- or overrepresented by their spore or pollen abundances. Likewise, preservation of spore and pollen exines may have been different among the contributing plant groups. Also, miospore analyses do not record the prepollen of Medullosan Pteridosperms (*Zonalosporites*), because of its large size. Consequently, macrofloral reconstructions based on the palynofloral record must be considered in relative and not absolute terms.

In reality, all paleobotanical methods have biases. Paleobotanical analysis of permineralized peats (Phillips et al., 1985) and etched, polished coal blocks (Winston, 1988, 1989), although providing a wealth of morphologic and anatomical data on coal-forming plants, record only the preserved remnants of the original contributing flora. Therefore, data derived from these types of studies must be considered with regard to differential degradation of plant tissues during peat accumulation. Phillips et al. (1985) have noted that, based upon a hypothetical 4:1 shoot:root ratio of a completely preserved lycopod tree, some peat compositions indicate a loss of at least 71%–93% of the aerial biomass. Also, differential preservation of biomass contributors on an intergroup level also should be considered. The general overrepresentation of tree fern miospore taxa in macerated Middle Pennsylvanian coal samples contrasts with much lower percentages of tree fern components relative to lycopod remains in coal balls of the same age (Phillips et al., 1985; Willard, 1989). This is most likely a two-fold function of (1) very great homosporous tree fern miospore production relative to other plants (Phillips et al., 1985) and (2) a probable lower preservation potential of tree fern tissues.

It becomes apparent that most, if not all, paleobotanical analyses have their own inherent biases. Therefore, it seems prudent to integrate as much independently derived data into a paleoecological model as possible. Simply put, several lines of evidence in support of a common theme usually provides the most convincing argument. This is the basis of the multidisciplinary approach used in the present study.

Coal petrography, when employed in an interpretive as well as quantitative manner, provides much insight into the preservational and degradational history of a coal bed. As the formational mechanisms of macerals and maceral groups are better understood, the inferred biologic, geochemical, and hydrologic conditions under which peat accumulates can be better established (Stach et al., 1982; Teichmuller, 1989). This ultimately leads to more accurate paleoenvironmental reconstructions based on coal petrography. The close correlation between maceral and palynofloral composition in Pennsylvanian coal beds was first noted by Reinhardt Thiessen and his co-workers (Thiessen, 1930; Thiessen and Sprunk, 1936; Sprunk et al., 1940), and although many studies have followed along these lines the works of Smith (1962, 1963, 1964, 1968), involving the identification of "miospore phases" in British Carboniferous coal beds are probably the most comprehensive efforts in the application of these corroborative parameters to date.

Ash yield, sulfur content, and XRD mineralogy of low-temperature ash are geochemical parameters that can provide important information on the environments and processes associated with peat accumulation (Cecil et al., 1982). For instance, the occurrence of a low-ash (<10%) coal bed indicates that the original peat formed under conditions that inhibited the amount of inorganic material emplaced and retained through detrital, authigenic, airborne, or diagenetic mechanisms. This is particularly true in light of the approximate doubling of ash yield (and subsequent loss of volatile organic matter) during the transformation from peat to bituminous coal (Cecil et al., 1982). Likewise, the amount of inorganic sulfur (especially early syngenetic pyrite) in many cases is a good index to the amount of anaerobic decomposition that has taken place in the peat (Schopf, 1952). The mineralogy of the ash-forming minerals is also an important paleoenvironmental indicator with a close correlation existing between ash yield and ash composition (Renton and Cecil, 1979; Cecil and others, 1985).

To summarize, the integration of the analytical methods discussed, when applied to coal beds sampled in relatively small vertical increments, provides the researcher with a very powerful tool with which to discern the paleoecology, paleoenvironments of peat formation, and depositional history of a coal-forming peat body. The biases of one type of analysis are lessened by a corroborative data base, derived from several independent disciplines.

Stratigraphic position of the Stockton and Fire Clay coal beds

This study will concentrate on the compositional characteristics of two Middle Pennsylvanian coal beds, the Stockton and the Fire Clay (Fig. 1). The Stockton coal bed occurs at the top of the Kanawha Formation and is equivalent with the basal Westphalian D of western Europe (Gillespie and Pfefferkorn, 1979) and the Atokan of the Illinois basin (Eble and Gillespie, 1986). The stratigraphically lower Fire Clay coal bed occurs in the upper-middle part of the Kanawha Formation and is equivalent with the upper Westphalian B of western Europe (Eble, 1988) and the upper Morrowan of the Illinois Basin (Henry, 1984; Eble, 1988).

The name Fire Clay is used throughout this paper, even though the name "Hernshaw" has been applied to this coal bed in certain portions of West Virginia (Eble, 1988). This is because the type location of the Hernshaw coal bed in Kanawha County, West Virginia, has been shown to be the stratigraphically lower Cedar Grove coal bed (Henry, 1984). As such, the name Hernshaw is technically invalid and should be suppressed.

MATERIALS AND METHODS

Vertically continuous 0.5 ft (0.15 m) increment samples of the Stockton and Fire Clay coal beds were collected from locations in the central Appalachian Basin (Fig. 2). All samples were mechanically stage crushed to –20 mesh (846 μm) and split into two subsamples (American Society for Testing and Material [ASTM], 1992a). One subsample was used for palynological maceration and petrographic pellet preparation. The other was further reduced to –60 mesh (250 μm) for proximate and sulfur-forms analysis and low-temperature ashing for XRD mineralogic analysis.

Isolation of miospores from coal and coal-parting material (shale, siltstone) was accomplished using techniques outlined by Barss and Williams (1973) and Doher (1980), with minor modifications to achieve the most satisfactory results. To determine the relative abundances of miospore taxa in each sample, 250 miospores were counted at a magnification of 400× from Canada balsam mounts. A 250-miospore sample count has been adopted by other workers (Kosanke, 1973, 1984, 1988; Ravn, 1979, 1986) and is based on providing a minimum practical figure for statistical consideration (Patterson and Fishbein, 1989).

Petrographic analyses were performed on polished coal

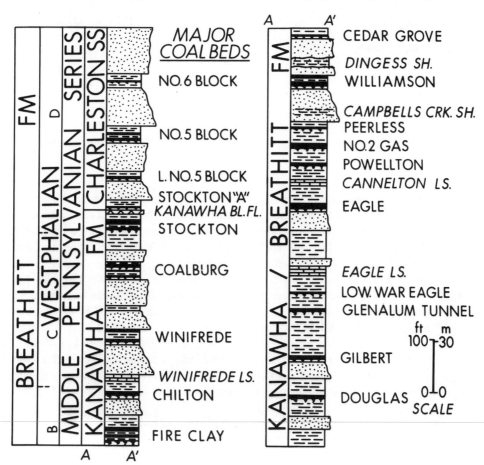

Figure 1. Generalized stratigraphic column of Middle Pennsylvanian strata in the central Appalachian Basin showing the position of major minable coal beds. Modified from Arndt (1979).

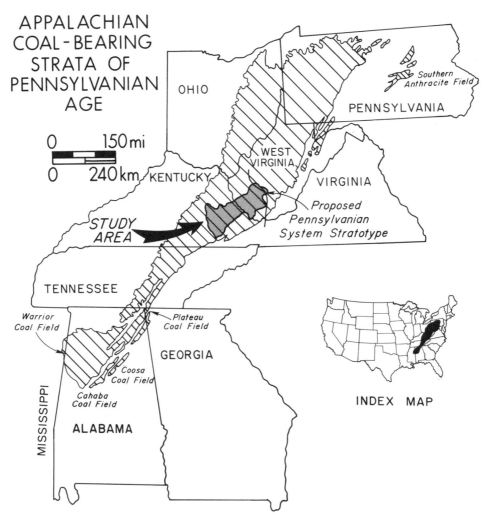

Figure 2. Coal-bearing strata of the Appalachian Basin (diagonal lines). The study area is in the central portion (shaded).

pellets under incident xenon light (oil immersion objective) at 500× magnification in accordance with standard ASTM procedures (1992b). Maceral nomenclature was adopted from Stach et al. (1982) and ASTM (1992c) with minor modifications to the submacerals of the vitrinite group. Vitrinite in bands greater than 50 microns thickness and free of mineral and maceral inclusions was counted as telocollinite. Vitrinite with reflectance equal to that of telocollinite, but occurring in bands less than 50 microns thickness and as a matrix material with maceral and mineral inclusions separated by less than 50 microns, was termed and counted as "matrix" collinite. Desmocollinite, as referred to in this study, is morphologically identical to the "matrix" collinite but has a significantly lower reflectance. Maceral and mineral matter abundances are reported as volume percent of the whole coal and are based on a minimum of 500 points counted from each of two petrographic pellets (ASTM, 1992c).

Six increment columns of the Stockton coal bed were collected from locations in southern West Virginia (Fig. 3). Columns 4, 5, and 6 were analyzed for palynology, coal petrography, ash

yield, sulfur forms, and LTA-XRD mineralogy. Columns 1 and 3 were studied palynologically and analyzed for ash yield and sulfur forms; column 2 was studied only palynologically. Columns 2, 5, and 6 will be used for discussion of the Stockton coal bed in this paper.

Thirteen columns of the Fire Clay coal bed were collected from southern West Virginia and eastern Kentucky (Fig. 4). For this study two columns will be discussed in detail, one from an area of thin coal (column 3; 2.2 ft, 0.67 m), and one from an area of thick coal (column 10; 3.45 ft, 1.05 m). Compositional results of all of these columns are presented in Eble (1988).

RESULTS

Stockton coal bed

The dominant miospore taxa in the Stockton coal, as shown in column 2, (Fig. 5) include representatives of arboreous (*Lycospora* spp. and, to a lesser extent, *Granasporites medius*), and

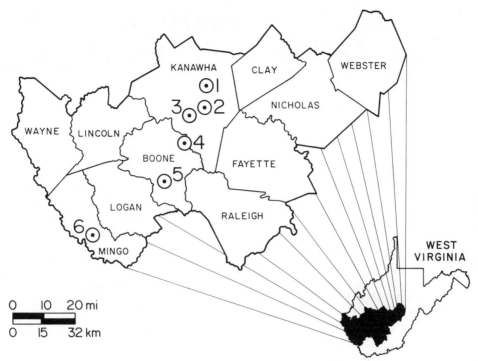

Figure 3. Stockton coal bed sample locations.

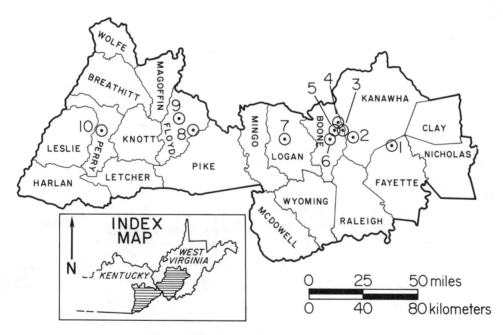

Figure 4. Fire Clay coal bed sample locations.

"herbaceous" (*Densosporites* and related crassicingulate genera) lycopods. Herbaceous is in quotes because Wagner (1989) has demonstrated that these lycopods, once considered to be herbs (Chaloner, 1958; Leisman, 1970), were probably small trees. Ferns are represented (tree-like forms: *Laevigatosporites globosus, Punctatisporites minutus, Punctatosporites minutus,* and *Apiculatisporites saetiger*; and smaller varieties: *Granulatisporites*

and related trilete, sphaerotriangular genera), as are calamites (*Calamospora* and larger species of *Laevigatosporites*), and cordaites (*Florinites*).

The most conspicuous palynomorph abundance patterns in this column are two profiles of upward decreasing *Lycospora*, mirrored by two upward increasing profiles of "herbaceous" lycopods and tree ferns. The first begins at the base of the coal bed

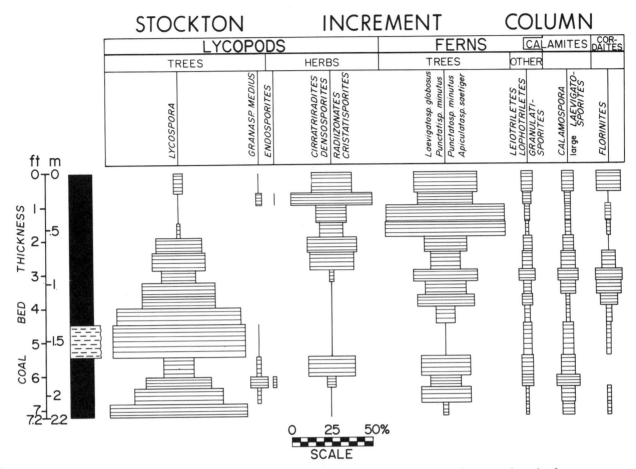

Figure 5. Miospore taxa distribution in Stockton column 2. Note the two decreasing upward trends of *Lycospora*, and two increasing upward trends of "herbaceous" lycopods and tree ferns.

and ends at the shale parting. The second begins in the shale parting and ends near the top of the bed. This upward decreasing pattern of arboreous lycopods (*Lycospora*) is a common feature in the Stockton coal bed (Fig. 6). Palynomorphs related to other ferns, calamites, and cordaites occur consistently throughout the vertical extent of column 2 with *Florinites* being more abundant in coal layers just above the shale parting and also at the top of the bed.

The concept of compositional cycles

The Stockton coal bed at location 6 probably best demonstrates the interrelationships among the different types of data examined. Column 6 shows vertical stratification of ash yield, sulfur content, palynology, petrography, and mineralogy (Fig. 7). From bottom to top, arboreous lycopods, moderately represented in the bottom increment, decrease upward to the increment at 2.0–2.5 ft (0.61–0.76 m), and then increase upward to the top of the bed; the top two increments of this column are dominated by arboreous lycopods. This profile is paralleled by the distributions of total vitrinite, telocollinite (Fig. 7C), petrographic mineral matter, ash yield and illite (Fig. 7D).

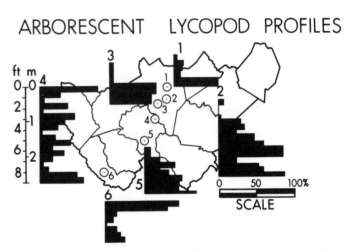

Figure 6. Distribution of arboreous lycopods (*Lycospora* + *Granasporites medius*) in the Stockton coal bed. Columns 1, 2, 3, and 5 all display upward decreasing trends. Column 6 shows an upward decreasing, followed by an upward increasing pattern. Column 4 is a very complex column showing numerous patterns.

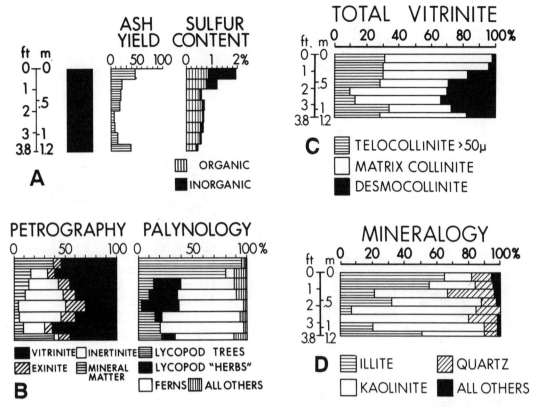

Figure 7. Stockton column 6. (A) ash yield and sulfur content distribution; (B) distribution of macerals mineral matter and miospore taxa; (C) distribution of telocollinite, matrix collinite, and desmocollinite; (D) distribution of major minerals (illite, kaolinite, and quartz) in the low-temperature ash.

An opposite trend is displayed by "herbaceous" lycopods and ferns, desmocollinite (Fig. 7C), inertinite, exinite, and kaolinite. All of these constituents increase upward to the increment at 2.0–2.5 feet (0.61–0.76 m) and then decrease upward to the top of the bed. The sulfur content of the coal bed is low, except for the top two increments, which contain moderate percentages of sulfur (average 1.6%) (Fig. 7A).

These two diametric vertical abundance patterns, one increasing upward to the middle of the bed, then decreasing to the top, and the other decreasing upward to the middle of the bed, then increasing to the top, collectively define, as it will be referred to in this paper, one complete compositional cycle.

Column 5 is split into two benches by a shale parting that occurs near the middle of the coal bed (Fig. 8e). The lower bench displays upward decreasing trends of arboreous lycopods, total vitrinite, telocollinite, ash yield, and illite abundances. Upward increasing trends of tree ferns, "herbaceous" lycopods, inertinite, liptinite, desmocollinite, and kaolinite abundances also are apparent in the lower bench. These trends are identical to those which were observed in the lower half of column 6 and, therefore, represent a half cycle of peat development.

The top bench of column 5 contains another half cycle with trends identical to those seen in the lower bench. This half cycle terminates at the top of the coal bed. To summarize, column 5 exhibits two repetitive half cycles, each terminating in an inorganic unit (shale parting and roof shale). Cycle truncation by partings is a common feature in the Stockton and Fire Clay coal beds.

Fire Clay coal bed

Palynologically, the overall miospore flora of the Fire Clay coal bed closely resembles that of the stratigraphically higher Stockton coal bed. Arboreous and "herbaceous" lycopods and ferns are generally primary in occurrence with calamites and cordaites being secondary in abundance.

In Fire Clay column 3 (Fig. 9), the bed is split into two benches by a flint clay parting, which represents an ancient volcanic ash fall (see Chesnut, 1983, for a review of the work done on this distinctive parting). The bottom bench contains two increments, the basal of which is dominated by arboreous lycopods and vitrinite macerals. This increment is low in ash yield but is moderate in sulfur content. The increment directly subjacent to the flint clay parting, in contrast, shows an increase in ash, inertinite, and desmocollinite content and has a more even distribution of arboreous lycopods and ferns.

The deposition of the flint clay parting represents a significant interruption in peat accumulation. The flint clay palynoflora

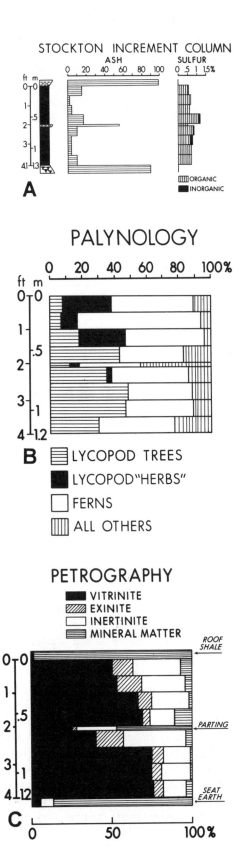

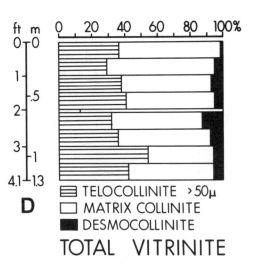

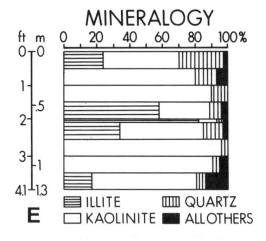

Figure 8. Stockton column 5. (A) ash yield and sulfur content distribution; (B) miospore taxa distribution; (C) petrographic maceral and mineral matter distribution; (D) distribution of telocollinite, matrix collinite, and desmocollinite; (E) distribution of major minerals in the low-temperature ash.

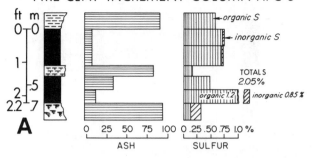

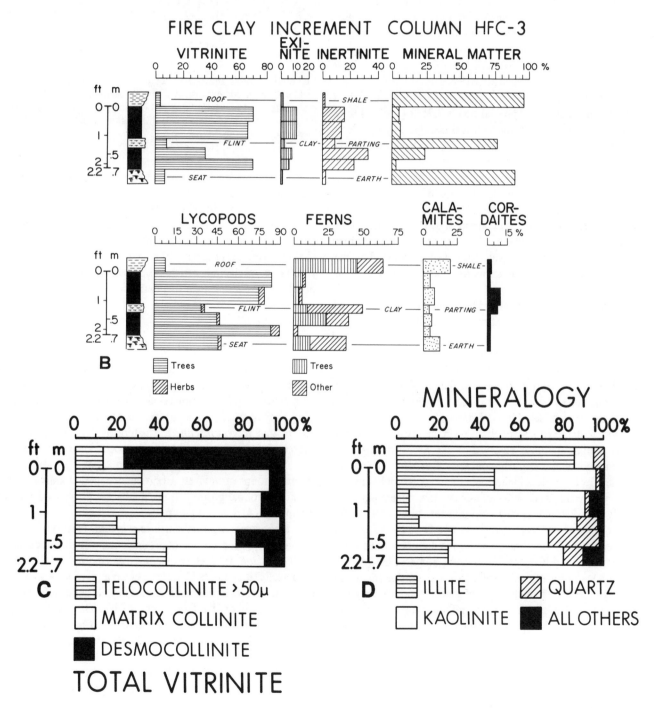

Figure 9. Fire Clay column 3. (A) ash yield and sulfur content distribution; (B) distribution of macerals, mineral matter, and miospore taxa; (C) distribution of telocollinite, matrix collinite, and desmocollinite; (D) distribution of major minerals in the low-temperature ash.

contains a higher amount of small ferns and cordaites, with the latter "carrying over" into the coal increment directly above the flint clay. Mineralogically, the flint clay parting and coal increment above are dominated by kaolinite (Fig. 9D). The increment beneath the flint parting is high in quartz. The remainder of the upper bench is dominated by arboreous lycopods, is vitrinite rich and low in ash yield and sulfur content, and shows an even distribution of kaolinite and illite.

Fire Clay column 3 is a good example of a coal bed with only the initial phases of two compositional cycles, which seems to be common in areas where the Fire Clay coal bed is thin. It also is common to see areas where both coal beds are composed of several thin coal layers, each containing the initial phases of cycle development, repeatedly truncated by inorganic partings. In these cases, the "coal bed" would effectively comprise multiple peat swamps, separated by inorganic deposition events.

Increment column 10A was collected in an area of thick Fire Clay coal (Fig. 10). The coal bed is 3.75 ft (1.14 m) thick at this location, exhibits distinct vertical changes in palynologic and petrographic composition (Fig. 10A), and (except for the basal coal increment) is extremely low in ash yield and sulfur content (Fig. 10B). The flint clay parting and overlying high-ash coal increment are co-dominated by arboreous lycopods and ferns and have increased percentages of cordaite pollen, similar to what was seen in Fire Clay column 3. The occurrence of increased percentages of cordaite pollen, an otherwise uncommon plant group in coal palynofloras, associated with the flint clay parting is a common pattern across the study area (Eble, 1988; Eble and Grady, 1990).

Above the basal high-ash coal increment, the palynoflora changes vertically from one dominated by arboreous lycopods to one showing increased percentages of "herbaceous" lycopods and ferns. Petrographically, this trend is paralleled by decreasing upward total vitrinite content and increasing upward desmocollinite (Fig. 10C), exinite, and inertinite abundances. Collectively, this pattern constitutes one half cycle of peat development, with the half cycle being terminated by roof strata.

Overall, the relationships observed in the Fire Clay coal bed are very similar to those observed in the Stockton coal bed. Increments containing high percentages of arboreous lycopods are marked by high telocollinite contents and commonly low, but variable ash yields and sulfur contents. Ash mineralogies are generally illite-kaolinite-quartz co-dominant. In contrast, increments dominated by "herbaceous" lycopods and ferns contain higher percentages of inertinite, exinite, and desmocollinite. These increments are very low in ash and sulfur and are characterized by kaolinite-dominant, or kaolinite-quartz–co-dominant ash mineralogies. High-ash increments are generally variable in petrographic and palynologic composition, although a general increase in calamites and cordaites and an illite-dominant ash mineralogy in these increments is noteworthy. Sulfur contents of these increments is generally low.

As an independent test to see if the compositional parameters were interrelated on a statistically significant basis, the com-

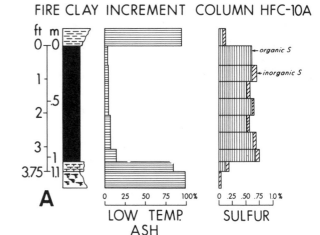

FIRE CLAY INCREMENT COLUMN HFC-10A

LOW TEMP. ASH   SULFUR

Figure 10 (on this and facing page). Fire Clay column 10A. (A) ash yield and sulfur content distribution; (B) distribution of macerals, mineral matter, and miospore taxa; (C) distribution of telocollinite, matrix collinite, and desmocollinite; (D) distribution of major minerals in the low-temperature ash.

bined data sets for both coal beds were analyzed for Pearson Product Moment (parametric statistics), and Spearman Rank (nonparametric statistics) correlation coefficients. The results are shown in Table 1. Three groups of significant positive correlation coefficients, very similar to the interrelationships identified through visual inspection of the data set, can be identified (Fig. 11). These results indicate that the observed compositional relationships have a statistical foundation.

DISCUSSION

Compositional groupings

There are correlations among miospore and maceral abundance, ash yield and sulfur content distribution, and low-temperature ash X-ray mineralogy in both the Stockton and Fire Clay coal beds. In this study we recognize four compositional groups that serve to define the nature of the swamp flora, the degree of peat degradation, and the geochemical conditions that prevailed during the time of peat accumulation. The four compositional groups are (1) an arboreous lycopod dominant group, (2) a transitional group, (3) an "herbaceous" lycopod and fern group, and (4) a mixed group.

Arboreous lycopod dominant group. Palynologically, the arboreous lycopod-dominant group contains high percentages of *Lycospora* and, to a lesser extent, *Granasporites medius*. Petrographically, it is marked by high percentages of telocollinite and low exinite and inertinite contents. Ash yields and sulfur contents commonly are low but tend to be variable. The low-temperature ash mineralogy of this group is characterized by a co-dominance of illite, kaolinite, and quartz, with the latter usually being less abundant.

Transitional group. The transitional group is characterized by a co-dominance of two or more plant groups, the most abun-

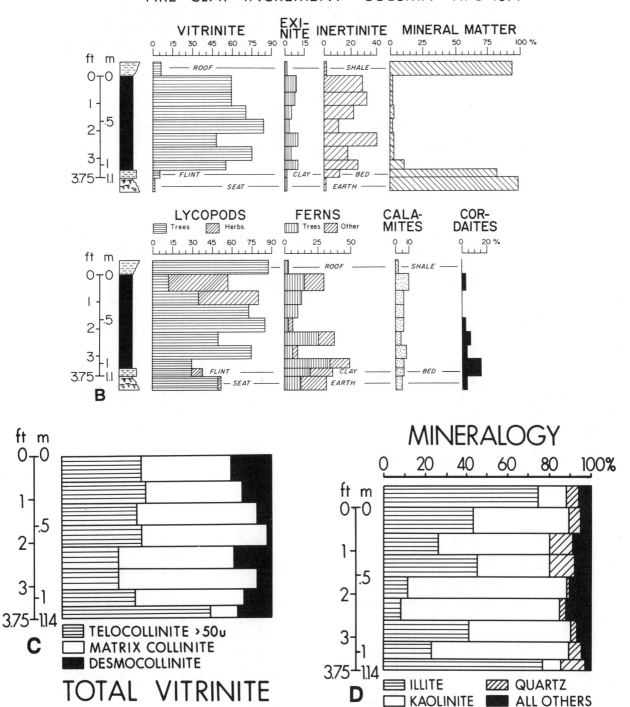

FIRE CLAY INCREMENT COLUMN HFC-10A

dant being arboreous and "herbaceous" lycopods and ferns. Petrographically, increments defined by this group contain increased amounts of desmocollinite, exinite, and inertinite. Ash yields and sulfur contents are low. Mineralogically, an increase in kaolinite over illite and quartz is usually seen.

"Herbaceous" lycopod and fern group. The "herbaceous" lycopod and fern group is characterized by a flora consisting almost exclusively of "herbaceous" lycopods and ferns, and petrographically by high percentages of desmocollinite, exinite, and inertinite macerals. The ash yield and sulfur content of increments containing this flora are both very low. Kaolinite and quartz are the dominant ash-forming minerals in this group.

Mixed group. The mixed group is characterized by a co-dominance of lycopods (trees and "herbs") and ferns, with cala-

C. F. Eble and W. C. Grady

TABLE 1. SIGNIFICANT CORRELATION COEFFICIENTS

	Sulfur	Illite	Kaolin	Quartz	Vitrin	Exinite	Inertin	Minmat	Tc > 50	Matrix	Desmoc	Arblyc	Herblyc	Tfern	Ofern	Calam	Cord
Ash		++	--	+	--			++	-		+			++	-		++
Sulfur	Sulfur	+	-		+												+
Illite		Illite	--	-				++			++	--		--			++
Kaolin			Kaolin	-				--								-	--
Quartz				Quartz		++	++							+	++		
Vitrin					Vitrin	--	--	--	--		--	--		++		+	
Exinite						Exinite	++		--		++	++		--		--	++
Inertin							Inertin				++	--		++			
Minmat								Minmat	-		+	--		++			
Tc > 50									Tc > 50	--	--						
Matrix										Matrix	--	++	-	--	-		
Desmoc											Desmoc	--		++	--	-	
Arblyc												Arblyc	--	--		++	-
Herblyc													Herblyc				
Tfern														Tfern	-	++	
Ofern															Ofern	++	+
Calam																Calam	
Cord																	Cord

The plus (+) sign indicates the variables have a significant positive correlation. The minus (-) sign indicates the variables have a significant negative correlation. The sign on the left reports the relation from the Pearson Product Moment (parametric) correlation, and the sign on the right reports the relation from the Spearman Rank (nonparametric) correlation. Abbreviations: Kaolin = kaolinite; Vitrin = vitrinite; Inertin = inertinite; Minmat = petrographic mineral matter; Tc > 50 = telocollinite >50 microns; Matrix = matrix collinite; Desmo = desmocollinite; Arblyc = arboerous lycopods; Herblyc = "herbaceous" lycopods; Tfern = tree ferns; Ofern = other ferns; Calam = calamites; Cord = cordaites.

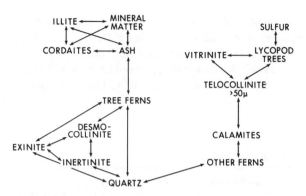

Figure 11. Graphic depiction of significant, positive intervariable correlations based on both Pearson Product Moment and Spearman Rank correlation coefficients.

mites and cordaites generally occurring more frequently in this group than in the others mentioned in the previous paragraphs. Petrographically, increments defined by the mixed compositional group are variable in composition, but they tend to be enriched in inertinite and, to a lesser extent, in exinite. Ash yields typically are high, with illite generally being the dominant component. Sulfur contents, in contrast, are usually low.

Cecil et al. (1985) introduced the terms "planar" and "domed" to describe the types of peat swamps from which Pennsylvanian coal beds in the Appalachian Basin were derived. Planar swamps have relatively flat surfaces in profile, are generally topogenous, mesotrophic to eutrophic, and usually contain spatially complex floras. In contrast, domed peat swamps have convex upward profiles (to varying degrees), are typically ombrogenous and oligotrophic, and contain spatially zoned floras. Cecil et al. (1985) consider these two types of swamps to represent end members of commercial-quality coal formation.

Anderson (1961, 1964) and Romanov (1968) have also indicated that planar and domed swamps represent a continuum in swamp evolution. Swamps start out being planar and, given sufficient moisture to maintain a perched water table, can mature into domed peat swamps. It is this evolution of swamp morphologies which we believe best explains the compositional stratification in the Fire Clay and Stockton coal beds. A summary of the compositional parameters for each group is presented in Table 2. The placement of the four compositional groups in a domed peat swamp model is shown in Figure 12.

Paleoecological interpretations of compositional groups

Arboreous lycopod-dominant group. The arboreous lycopod-dominant group is interpreted to have developed in more planar areas of peat swamps that were poorly drained. These areas have the best chance of maintaining a standing water cover, or at least supersaturated peat conditions (Anderson, 1964), which the arboreous lycopods may have required for reproduction (Phillips, 1979). The large lycopod trees probably dominated these areas because ground cover and homosperous,

free-sporing plants, such as ferns, sphenopsids and small pteridosperms, would have had difficulty becoming established in a periodically flooded environment (DiMichele and Phillips, 1985).

Water cover also would inhibit oxidation of the peat and promote the formation of a high vitrinite content coal. The increased percentages of telocollinite associated with this group is probably a two-fold function of good preservation of plant materials and physically larger plant parts contributed from the dominant lycopod trees. The commonly low, but variable, ash yield and sulfur content of increments containing this flora are consistent with a more planar swamp origin for this compositional group. The generally acid swamp waters present in many swamp environments prevent some mineral matter, notably clays, from being introduced into the swamp as a result of flocculation and fallout at the more acid–more neutral pH water boundary (Staub and Cohen, 1978). Likewise, vegetation baffling (White, 1913; Kravits and Crelling, 1981) may also have been an inhibitive mechanism for large-scale mineral transport into the swamp environment.

On the other hand, planar swamp environments are more susceptible to receiving detritus from flood waters than are domed swamps simply because of their topography. Likewise, the influence of more neutral pH surficial waters could potentially accelerate microbial degradation of the peat (both anaerobic and aerobic) and result in the emplacement of authigenic mineral matter (Cecil et al., 1979), and inorganic sulfur (Schopf, 1952). The co-dominance of illite, kaolinite and, to a lesser extent, quartz in the low-temperature ash supports a detrital/authigenic origin for the mineral matter of this compositional group.

Transitional group. The transitional group is interpreted to have formed in parts of peat swamps that were increasingly domed and better drained. The continued abundance of arboreous lycopods in this flora indicates that a water cover (supersaturated substrates) was being maintained, although the decreased percentages would suggest that its consistency was being jeopardized. Increased fluctuation of the water table would expose the peat to air and oxygenated rain water, creating a situation favorable for inertinite (mainly degradational varieties) production. Higher percentages of desmocollinite in increments containing the transitional, and "herbaceous" lycopod and fern flora discussed in the next section, is interpreted to represent increased degradation of vitrinite precursors. Desmocollinite represents strongly decomposed plant parts that have become homogenized through geochemical gelification (Stach et al., 1982). In this study, a direct relationship is observed between desmocollinite and inertinite, suggesting a similar genesis mechanism.

The low ash yield and sulfur content of increments containing this flora are believed to be the result of peat formation in more domed areas of the swamp, as these areas are protected from flood-water influence. These parts of the swamp are also better drained, and are marked by increased rain water flushing of the peat. This type of environment minimizes the chance of authigenic mineral formation and inorganic sulfur emplacement. The increasingly dominant kaolinite mineralogy of the coal ash is interpreted to be a result of increased rain-water flushing of the peat and the leaching of mineral matter and mineral precursors.

"Herbaceous" lycopod and fern group. The "herbaceous" lycopod and fern group also is interpreted to have developed in more domed parts of swamps that were well drained. The paucity of large lycopod trees (mostly *Lepidophloios* and *Lepidodendron*) in these areas is considered to be a function

of an insufficient water cover, necessary to accommodate the specialized reproductive mechanisms of these plants and also possibly because of extreme nutrient deficiency. The smaller statured nature of this flora (dominantly tree-like ferns and small lycopods) may reflect this (perhaps mimicking the stunted "pole forests" described from similar ares of modern domed swamps).

The increased percentages of inertinite macerals (mainly

TABLE 2. SUMMARY OF COMPOSITIONAL GROUP PARAMETERS

Compositional Group	Dominant Floral Components	Petrographic Characteristics	Ash Yield	Sulfur Content	Ash Mineralogy	Inferred Swamp Environment
Arboreous Lycopsid dominant	Lycopsid trees	High vitrinite (telocollinite)	Low to high (<10% to >15%	Low to high (<1% to >2%)	Illite, quartz, and kaolinite	More planar
Transitional	Lycopsids (trees and "herbs") and tree ferns	Moderate to high inertinite and liptinite; increased desmocollinite	Low (<10%)	Low (<1%)	Kaolinite and quartz	More domed
"Herbaceous" lycopsid and fern	"Herbaceous" lycopsids and ferns	High to very high inertinite and liptinite; increased desmocollinite	Very low (<5%)	Low (<1%)	Kaolinite and quartz	More domed
Mixed	Lycopsids, ferns, calamites, cordaites	Variable	High (>15%)	Low (<1%)	Illite, quartz, and kaolinite	More planar

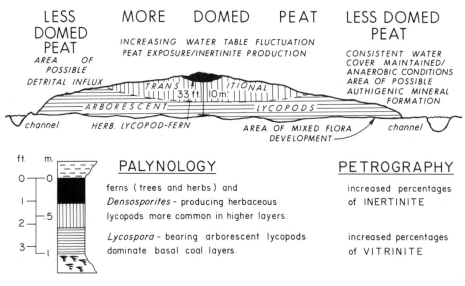

Figure 12. Top portion shows placement of the four compositional groups in a domed peat system model. The mixed and arboreous lycopod dominant groups are inferred to have occupied less raised, more planar, areas of domed peat swamps, whereas the transitional and "herbaceous" lycopod and fern groups are interpreted to have occupied more domed areas. The bottom portion shows a theoretical, 1-m thick compositionally stratified coal bed derived from the 10-m thick domed peat deposit.

degradational varieties) associated with this group is interpreted to be the result of water-table fluctuation and exposure to air and oxygenated rain water. Likewise, the associated high percentages of liptinite macerals found in increments defined by this group are probably attributable to the greater resistance of spore and pollen exines and cuticles to degradation, resulting in a net increase in exinite abundance relative to more easily degraded previtrinite materials. The very low ash yield and sulfur content of increments containing this group is considered to be a two-fold function of development in areas never affected by flood waters and an intense leaching of mineral matter and mineral precursors. This latter scenario is supported by the very high kaolinite, and occasionally quartz, content of the low-temperature ash of this group. Through intense leaching, only the most insoluble elements, aluminum and silicon, are left behind.

Mixed group. In contrast to the transitional and "herbaceous" lycopod and fern compositional groups, the mixed group is considered to have formed in more planar areas of swamps that were susceptible to flood-water influence. This interpretation is supported by the increased ash yield of increments defined by this group and also by the illite-dominant mineralogy of the low-temperature ash. The reduction of arboreous lycopods in this group may indicate well-drained conditions that would inhibit their expansion. Increased percentages of inertinite macerals in this group would support this interpretation.

Calamite and cordaite miospores occur more frequently in the mixed group. These two plant groups may have been more adapted for growth on mineral rather than peat substrates, with at least the cordaites also being able to flourish under conditions of moisture stress (Phillips et al., 1985). An increase in miospores related to "other" ferns also occurs in the mixed group, which may imply an open, perhaps seral/successional vegetation that developed in response to some type of disturbance. These plants tended to be scramblers with some assuming a vine-like habit; a few also may have been lianas (W. DiMichele, 1989, personal communication).

The possibility that some portion of the mixed group palynoflora is of allochthonous origin must be considered, although one might expect increased exine breakage and corrosion from transport (Smith, 1962). As this feature is not readily apparent, the majority of the palynomorphs recovered from increments containing the mixed-group palynoflora are considered to be autochthonous and representative of a new flora that has responded to changes in edaphic conditions.

Spatial distribution of the compositional groups

Where the Stockton and Fire Clay coal beds are thick (>1 m) and uninterrupted by inorganic partings, the compositional groups are arranged as follows: arboreous lycopod or mixed in basal coal layers progressing upward into the transitional, and "herbaceous" lycopod and fern groups in middle to upper coal layers. In some areas, this is followed by a reversal with a change back to the arboreous-lycopod or mixed-compositional groups

occurring toward the top of the coal bed. Where complete, this pattern is believed to reflect a progressive doming of the peat surface, followed by a return to a more planar swamp environment. This sequence defines one complete compositional cycle as was first illustrated in Stockton column 6 (Fig. 13).

In areas where the Fire Clay coal bed is thin (<1 m; all of the Stockton columns were >1 m), the arboreous lycopod-dominant and/or mixed-compositional groups commonly occupy the entire vertical extent of the bed. This is interpreted to represent a persistence of peat development in more planar swamp settings, with peat accumulation being terminated before more domed conditions could evolve. Fire Clay columns 3, 4, and 5 are good examples of this (Fig. 14).

Effect of partings

The presence of inorganic parting, regardless of their origin (detrital, authigenic, or airborne), has a significant effect on the distribution of the compositional groups in both the Stockton and Fire Clay coal beds. The inferred transition from more planar to more domed swamp conditions is often truncated by partings, which result in compositional half cycles. The development of a typical half cycle is well-illustrated in Fire Clay column 8 (Fig. 15).

The bottom coal bench shows arboreous lycopods very abundant in the first two increments, decreasing upward as "herbaceous" lycopods and tree ferns increase in abundance. The introduction of the flint clay parting truncates this progression and establishes the base for another cycle. This second cycle never really develops and is terminated by the overlying sandstone roof strata. Other columns that display one or more half cycles throughout their vertical extent include Stockton columns 1 and 3, and Fire Clay columns 3, 8, and 10. In summary, many of the columns analyzed in this study are marked by multicycle (both half- and complete-cycle) development, encompassing both planar and domed swamp types. In these cases, the coal bed is actually comprised of more than one peat swamp, preserved in different stages of morphologic progression (planar to domed).

Distribution of planar and domed environments in domed peat systems

Examination of Figures 13 and 14 indicates that a substantial portion of the Stockton and Fire Clay coal beds are comprised of the arboreous lycopod-dominant and mixed-compositional groups, two groups that are envisioned to have formed in more planar swamp settings. Therefore, a large part of the "domed peat systems," from which these beds are believed to have been derived, is occupied by planar to perhaps slightly domed environments. In fact, only 35% of the coal increments examined in this study were assigned to the transitional or "herbaceous" lycopod and fern compositional groups, indicating that only a relatively minor portion of the Fire Clay and Stockton "domed peat systems" were significantly domed. Rather, it is the vertical compositional change observed in many of the studied columns that, we believe, warrants a domed peat system origin

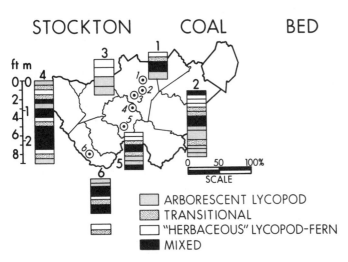

Figure 13. Distribution of the four compositional groups in the Stockton coal bed.

for the Fire Clay and Stockton coal beds. This must be considered, however, with the understanding that both planar and domed peat environments contribute to peat formation in domed peat systems.

A RELATIONSHIP BETWEEN PEAT FORMATION AND SEDIMENT DISPERSAL?

It is clear that the primary determining factor as to whether a swamp will be more planar or more domed is the rainfall/evapotranspiration ratio. Swamp termination mechanisms, however, are less clear. The end of peat accumulation and ultimate covering by sediments has been attributed to many causes, including tectonic lowering of the swamp (subsidence), orogenesis and increased sedimentation, sea-level eustasy, and climate fluctuation. This last factor is considered to be of particular importance. A shift to "drier" conditions during peat accumulation may be an important factor in the deflation of peat domes, allowing them to be more readily covered and preserved by sediments. This concept has been proposed by Cecil (1990), who notes that climate plays an important role in sediment availability. The very wet

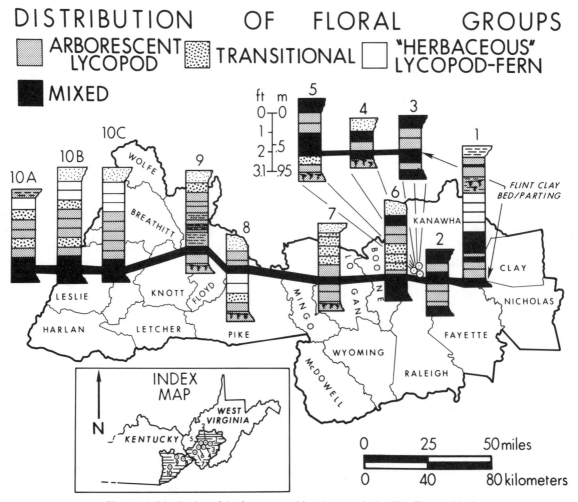

Figure 14. Distribution of the four compositional groups in the Fire Clay coal bed.

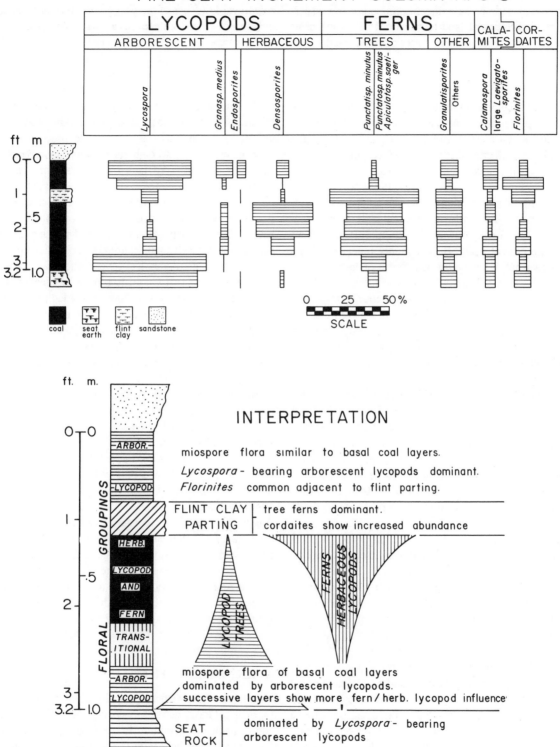

Figure 15. Top portion shows the distribution of miospore taxa in Fire Clay column 8. Note the decreasing upward *Lycospora* and increasing upward "herbaceous" lycopod and tree-fern profiles in the bottom bench, truncation by the flint clay parting containing abundant *Florinites,* and a return to a *Lycospora*-dominant palynoflora. The bottom portion of the diagram is a cartoon depicting these trends.

conditions associated with domed peat development actually restrict sediment dispersal, whereas a shift to "drier" conditions signals the onset of widespread sediment production and dispersal. Hence, the development (and termination) of domed peats and rate of sediment dispersal may be very closely related.

CONCLUSIONS

The inferred existence of the four compositional groups, representing peat accumulation in at least two different types of swamp settings (planar and domed) has been proposed to explain the compositional stratification observed in the Fire Clay and Stockton coal beds. Other studies also have incorporated domed peat models to explain similar compositional variation in Carboniferous coal beds. Smith (1962, 1963, 1964, 1968) developed the concept of "miospore phases" to describe palynologic and petrographic variation in British Carboniferous coal beds. These "miospore phases" are analogous to the compositional groups identified in the present study. He commonly observed an upward transition from a *Lycospora*-rich clarite or vitrite, which represented his "lycospore" phase, through an intermediate "transition" phase, to a durite that contained abundant *Densosporites*, which he called the "densospore" phase. Smith believed this pattern to be the result of peat doming and was the first to draw an analogy between the domed, forested peat swamps of coastal Malaysia (Borneo), and the Carboniferous coal beds of Great Britain.

Other palynologic studies by Fulton (1987) and Bartram (1987), as well as petrographic studies by Esterle and Ferm (1986) and Littke (1987), also have incorporated domed peat models to explain compositional variation in Carboniferous coal beds. Collectively, these results suggest that peat formational processes, namely the development of domed peat systems, may have been similar in the central Appalachian Basin and in the coal basins of Great Britain and western Europe during the Early through mid-Middle Pennsylvanian.

Comparative studies of modern equatorial domed peats as modern analogues for Pennsylvanian coal beds are not new (Potonie and Koorders, 1909; White, 1913; Polak, 1933), but recently they have been the focal point of many investigations. Factors such as their large size, great thickness, and very low ash yield and sulfur content have been cited as making them attractive analogues for Pennsylvanian coal beds of the Appalachian Basin (Neuzil and Cecil, 1984). Anderson (1961, 1964) noted that the floras of these swamps tend to be arranged in concentric seres, the result of progressive doming and increased nutrient deprivation. Likewise, a palynological investigation by Anderson and Muller (1975) detected vertical palynofloral change within a domed peat deposit in Malaysia, which was interpreted to represent the floral response to changing edaphic conditions as the peat became more domed. In the present study, similar palynofloral stratification in the Fire Clay and Stockton coal beds also has been ascribed to peat doming.

Preliminary peat petrographic studies (Esterle et al., 1989; Cohen et al., 1989; Esterle, 1990) have indicated that the peats from this area are dominated by previtrinite macerals (huminites), with the highest concentration of these constituents occurring toward the tops of the peat domes. Inertinite macerals occur only rarely. On the surface, this appears to contradict the interpretations presented in this and other studies that employ domed peat models, as they predict a decrease in vitrinite and an increase in inertinite in higher areas of domed swamps. In making such a comparison, however, the effects of coalification must be considered. Smith and Cook (1980) point out that the high reflectance of most of the inertinite macerals in bituminous coal ("degradoinertinites") is achieved only during the process of coalification over the brown coal to high volatile bituminous coal rank range. Therefore the observed lack of inertinite macerals in modern peats (and most low rank coal) may be primarily a function of rank, and not one of occurrence. The small percentages of inertinites reported from domed peats are probably attributable to primary fusinite or pyrofusinite, both of which are also rare constituents of Pennsylvanian bituminous coal (Teichmuller, 1989).

The peat petrographical investigation by Grady et al. (this volume) suggests that a transition from less degraded peat in the bottom and middle layers to more degraded peat toward the top of a modern domed peat deposit may be analogous to the increase in desmocollinite commonly observed toward the tops of the Stockton and Fire Clay coal beds. They further suggest that the highly degraded peat layers may contain degradosemifusinite and degradofusinite maceral precursors that are not being recognized optically because of the very low rank of the material (Smith and Cook, 1980).

Were all Pennsylvanian coal beds derived from domed peat systems? Probably not. Cecil et al. (1985) consider that although most Lower through mid-Middle Pennsylvanian coal beds probably resulted from domed peat systems, most Upper Pennsylvanian coal beds were derived from planar peats. Upper Middle Pennsylvanian coal beds may have been transitional in character. Again, this needs to be considered with the understanding that both planar and domed peat environments contribute to domed peat systems. Whereas the findings of the present study suggest a domed peat system origin for the Fire Clay and Stockton coal beds, a similar study of the Redstone coal bed (Upper Pennsylvanian, Stephanian) determined this coal to have formed from an entirely planar peat body (Grady and Eble, 1990). These results support the contention (Cecil et al., 1985) that economical coal beds can be derived from both domed and planar peat swamps.

ACKNOWLEDGMENTS

The authors wish to thank Tom Phillips and Bill DiMichele for their critical reviews of this paper. Both offered several helpful comments and suggestions. Frank Dulong and Ron Stanton are thanked for their help with the statistical treatment of the data. Much of this paper was researched and written while the senior author was a National Research Council post-doctoral appointee in residence at the U.S. Geological Survey, Reston, Virginia. Both the NRC and USGS are thanked for financial and logistical support.

REFERENCES CITED

American Society for Testing and Materials (ASTM), 1992a, Standard method of preparing coal samples, D-2013-86: Annual book of ASTM standards, volume 05.05: Philadelphia, American Society for Testing and Materials, p. 239–246.

American Society for Testing and Materials (ASTM), 1992b, Standard practice for preparing coal samples for microscopical analysis by reflected light, D-2797-85: Annual book of ASTM standards, Volume 05.05: Philadelphia, American Society for Testing and Materials, p. 299–302.

ASTM, 1992c, Standard test method for microscopical determination of volume percent of physical components of coal, D-2799-92: Annual book of ASTM standards, Volume 05.05: Philadelphia, American Society for Testing and Materials, p. 307–308.

Anderson, J.A.R., 1961, The ecology and forest types of peat swamp forests of Sarawek and Brunei in relation to their silvaculture [Ph.D. thesis]: Edinburgh, University of Edinburgh, 191 p.

Anderson, J.A.R., 1964, The structure and development of the peat swamps of Sarawek and Brunei: Journal of Tropical Geography, v. 18, p. 7–16.

Anderson, J.A.R., and Muller, J., 1975, Palynological study of a Holocene peat deposit and a Miocene coal deposit from NW Borneo: Review of Paleobotany and Palynology, v. 19, p. 291–351.

Arndt, H. H., 1979, Middle Pennsylvanian Series in the proposed Pennsylvanian System stratotype, *in* Englund, K. J., Arndt, H. H. and Henry, T. W., eds., Proposed Pennsylvanian System stratotype, Virginia and West Virginia, Ninth International Congress for Carboniferous Stratigraphy and Geology, guidebook for field trip number 1: Falls Church, Virginia, American Geological Institute Selected Guidebook Series 1, p. 73–80.

Barss, M. S., and Williams, G. L., 1973, Palynology and nannofossil processing techniques: Canadian Geological Survey Paper 73–26, 25 p.

Bartram, K. M., 1987, Lycopod succession in coals: An example from the low Barnsley Seam (Westphalian B), Yorkshire, England, *in* Scott, A. C., ed., Coal and coal-bearing strata: Recent advances: Geological Society of London Special Publication 32, p. 187–200.

Cecil, C. B., 1990, Paleoclimate controls on stratigraphic repetition of chemical and clastic rocks: Geology, v. 18, p. 533–536.

Cecil, C. B., Stanton, R. W., Dulong, F. T., and Renton, J. J., 1979, Some geologic factors controlling mineral matter in coal, *in* Donaldson, A. C., Presley, M. W., and Renton, J. J., eds., Carboniferous coal guidebook: West Virginia Geological and Economic Survey Coal Geology Bulletin B-37-3, p. 43–56.

Cecil, C. B., Stanton, R. W., Dulong, F. T., and Renton, J. J., 1982, Geologic factors that control mineral matter in coal, *in* Filby, R. H., Carpenter, S. B., and Regaini, R. C., eds., Atomic and nuclear methods in fossil energy research: New York, Plenum, p. 323–335.

Cecil, C. B., Stanton, R. W., Neuzil, S. G., Dulong, F. T., Ruppert, L. F., and Pierce, B. S., 1985, Paleoclimate controls on Late Paleozoic sedimentation and peat formation in the central Appalachian basin (U.S.A): International Journal of Coal Geology, v. 5, p. 195–230.

Chaloner, W. G., 1958, A Carboniferous *Sellagenellites* with *Densosporites* microspores: Paleontology, v. 1, p. 245–253.

Chesnut, D. R., 1983, Source of the volcanic ash deposit (flint clay) in the Fire Clay coal of the Appalachian basin, *in* Compte rendu, Madrid, Tenth International Congress for Carboniferous Stratigraphy and Geology: v. 1, p. 145–154.

Cohen, A. D., Raymond, R., Jr., Ramirez, A., Morales, Z., and Ponce, F., 1989, The Changuinola peat deposit of northwestern Panama: A tropical, back-barrier peat (coal)-forming environment: International Journal of Coal Geology, v. 12, p. 157–192.

DiMichele, W. A., and Phillips, T. L., 1985, Arborescent lycopod reproduction and paleoecology in a coal-swamp environment of late Middle Pennsylvanian age (Herrin coal, Illinois, U.S.A.): Review of Paleobotany and Palynology, v. 44, p. 1–26.

Doher, L. I., 1980, Palynomorph preparation procedures currently used in the paleontology and stratigraphy laboratories: U.S. Geological Survey Circular 830, 29 p.

Eble, C. F., 1988, Palynology and paleoecology of a Middle Pennsylvanian coal bed from the central Appalachian basin [Ph.D. dissert.]: Morgantown, West Virginia University, 495 p.

Eble, C. F., and Gillespie, W. H., 1986, Palynological studies of the upper Kanawha Formation (Pottsville Pennsylvanian) in West Virginia: Compass, v. 63, no. 2, p. 58–65.

Eble, C. F., and Grady, W. S., 1988, Palynologic, petrographic, and coal-quality characterisitcs of Middle and Upper Pennsylvanian coal beds: A comparison [abs.]: American Association of Petrology Geology Bulletin, v. 72, p. 906.

Eble, C. F., and Grady, W. C., 1990, Paleoecological interpretation of a Middle Pennsylvanian coal bed from the Central Appalachian Basin, U.S.A.: International Journal of Coal Geology, v. 16, p. 255–286.

Esterle, J. S., 1990, Trends in petrographic and chemical characteristics of tropical domed peats in Indonesia and Malaysia as analogues for coal formation [Ph.D. dissert.]: Lexington, University of Kentucky, 270 p.

Esterle, J. S., and Ferm, J. C., 1986, Relationship between petrographic and chemical properties and coal seam geometry, Hance seam, Breathitt Formation, southeastern Kentucky: International Journal of Coal Geology, v. 6, p. 199–214.

Esterle, J. S., Ferm, J. C., and Tie, Y.-L., 1989, A test for the analogy of tropical domed peat deposits to "dulling up" sequences in coal beds—Preliminary results: Organic Geochemistry, v. 13, p. 333–342.

Fulton, I. M., 1987, Genesis of the thick Warwickshire Coal: A group of long residence histosols, *in* Scott, A. C., ed., Coal and coal-bearing strata: Recent advances: Geological Society of London Special Publication 32, p. 201–218.

Gillespie, W. H., and Pfefferkorn, H. W., 1979, Distribution of commonly occurring plant megafossils in the proposed Pennsylvanian stratotype, Virginia and West Virginia, *in* Englund, K. J., Arndt, H. H., and Henry, T. W., eds., Proposed Pennsylvanian System stratotype, Virginia and West Virginia (Ninth International Congress for Carboniferous Stratigraphy and Geology, Washington, D.C.): Falls Church, Virginia, American Geological Institute Guidebook Series 1, p. 87–97.

Grady, W. C., and Eble, C. F., 1990, Relationships among macerals, minerals, miospores and paleoecology in a column of Redstone coal (Upper Pennsylvanian) from north-central West Virginia (U.S.A.): International Journal of Coal Geology, v. 15, p. 1–26.

Henry, T. W., 1984, Geologic map of the Mammoth quadrangle, Kanawha and Clay counties, West Virginia: U.S. Geological Survey Quadrangle Map GQ-1576.

Kosanke, R. M., 1973, Palynological studies of the coals of the Princess Reserve district in northeastern Kentucky: U.S. Geological Survey Professional Paper 839, 24 p.

Kosanke, R. M., 1984, Palynology of selected coal beds in the proposed Pennsylvanian stratotype in West Virginia: U.S. Geological Survey Professional Paper 1318, 44 p.

Kosanke, R. M., 1988, Palynological studies of Middle Pennsylvanian coal beds of the proposed Pennsylvanian System stratotype in West Virginia: U.S. Geological Survey Professional Paper 1455, 81 p.

Kravits, C. M., and Crelling, J. C., 1981, Effects of overbank deposition on the quality and maceral composition of the Herrin (No. 6) coal (Pennsylvanian) of southern Illinois, International Journal of Coal Geology, v. 1, p. 195–212.

Leisman, G. A., 1970, A petrified *Sporangiostrobus* and its spores from the Middle Pennsylvanian of Kansas: Paleontographica, v. 129B, p. 166–177.

Littke, R., 1987, Petrology and genesis of Upper Carboniferous seams from the Ruhr region, West Germany: International Journal of Coal Geology, v. 7, p. 147–184.

Neuzil, S. G., and Cecil, C. B., 1984, A modern analog of low-ash, low-sulfur, Pennsylvanian age coal: Boulder, Colorado, Geological Society of America Abstracts with Programs, v. 16, no. 2, p. 184.

Patterson, T. R., and Fishbein, E., 1989, Re-examination of the statistical methods used to determine the number of point counts needed for micropaleontological quantitative research: Journal of Paleontology, v. 63, p. 245–248.

Phillips, T. L, 1979, Reproduction of heterosporous arborescent lycopods in the Mississippian-Pennsylvanian of Euramerica: Review of Paleobotany and Palynology, v. 27, p. 239–289.

Phillips, T. L., Peppers, R. A., and DiMichele, W. A., 1985, Stratigraphic and interregional changes in Pennsylvanian coal-swamp vegetation: Environmental inferences: International Journal of Coal Geology, v. 5, no. 1–2, p. 43–109.

Polak, E., 1933, Ueber Torf und Moor in Niederlandisch Indien: Koninklijke Akademie Wetenschappen, Verhandelingen Deel, v. 30, no. 3, p. 1–85.

Potonie, H., and Koorders, S. H., 1909, Die Tropen—Sumpfflaachmoor—Natur der Moore des produktiven Karbons: Geologisches Jahrbuch Landenstatt, v. 30, no. 1, p. 389–443.

Ravn, R. L., 1979, An introduction to the stratigraphic palynology of the Cherokee Group (Pennsylvanian) coals of Iowa: Iowa Geological Survey Technical Paper 6, 117 p.

Ravn, R. L., 1986, Palynostratigraphy of the Lower and Middle Pennsylvanian coals of Iowa: Iowa Geological Survey Technical Paper 7, 245 p.

Renton, J. J., and Cecil, C. B., 1979, The origin of mineral matter in coal, *in* Donaldson, A. C., Renton, J. J. and Presley, M. W., eds., Carboniferous coal guidebook: West Virginia Geological and Economic Survey Bulletin B-37-1, p. 206–223.

Romanov, V. V., 1968, Hydrophysics of bogs: Jerusalem, Israel Program for Scientific Translations, 299 p.

Schopf, J. M., 1952, Was decay important in the origin of coal?: Journal of Sedimentary Petrology, v. 22, no. 2, p. 61–69.

Smith, A.H.V., 1957, The sequence of microspore assemblages associated with the occurrence of crassidurite in coal seams of Yorkshire: Geological Magazine, v. 94, p. 345–363.

Smith, A.H.V., 1962, The paleoecology of Carboniferous peats based on miospores and petrography of bituminous coals, *in* Proceedings: Yorkshire Geological Society, v. 33, p. 423–463.

Smith, A.H.V., 1963, Paleoecology of Carboniferous peats, *in* Nairn, A.E.M., ed., Problems in paleoclimatology: London, John Wiley and Sons, p. 57–66.

Smith, A.H.V., 1964, Zur Petrologie und Palynologie der Kohlenfloze des Karbons und ihrer Begleitschichten: Fortschrite Geologisches Rheinland Westfalia, v. 12, p. 285–302.

Smith, A.H.V., 1968, Seam profiles and seam characteristics, *in* Murchison, D., and Westoll, T. S., eds., Coal and coal-bearing strata: Edinburgh, Oliver and Boyd, p. 31–40.

Smith, G. C., and Cook, A. C., 1980, Coalification paths of exinite, vitrinite and inertinite: Fuel, v. 59, p. 641–646.

Sprunk, G. C., Ode, W. H., Selvig, W. A., and O'Donnell, H. J., 1940, Splint coals of the Appalachian region: Their occurrence, petrography and comparison of chemical and physical properties with associated bright coals: U.S. Bureau of Mines Technical Paper 615, 59 p.

Stach, E., Mackowski, M.-Th., Teichmuller, M., Taylor, G. H., Chandra, D., and Teichmuller, R., 1982, Stach's textbook of coal petrology: Berlin, Stuttgart, Gebruder Borntraeger, 535 p.

Staub, J. R., and Cohen, A. D., 1978, Kaolinite enrichment beneath coals: A modern analog, Snuggedy swamp, South Carolina: Journal of Sedimentary Petrology, v. 48, p. 203–210.

Teichmuller, M., 1989, The genesis of coal from the viewpoint of coal petrology: International Journal of Coal Geology, v. 12, p. 1–89.

Thiessen, R., 1930, Splint coal: Transactions of the American Institute of Mining and Metallurgical Engineers, v. 88, p. 644–672.

Thiessen, R., and Sprunk, G. C., 1936, The origin of the finely divided opaque matter in splint coals: Fuel, v. 15, p. 304–315.

Wagner, R. H., 1989, A Late Stephanian forest swamp with *Sporangiostrobus* fossilized by volcanic ash fall in the Puertollano Basin, central Spain: International Journal of Coal Geology, v. 12, p. 523–552.

White, D., 1913, Environmental conditions of deposition of coal, *in* White, D., and Thiessen, R., The origin of coal: U.S. Bureau of Mines Bulletin 38, p. 68–79.

Willard, D. A., 1989, Palynological analysis of the Springfield coal of the Illinois basin and paleoecological implications: Geological Society of America Abstracts with Program, v. 21, no. 6, p. 53.

Winston, R. B., 1988, Paleoecology of Middle Pennsylvanian-age peat-swamp plants in Herrin coal, Kentucky, U.S.A.: International Journal of Coal Geology, v. 10, p. 203–238.

Winston, R. B., 1989, Identification of plant megafossils in Pennsylvanian age coal: Review of Paleobotany and Palynology, v. 57, p. 265–276.

MANUSCRIPT ACCEPTED BY THE SOCIETY JANUARY 14, 1993

Geological Society of America
Special Paper 286
1993

Vegetational patterns in the Springfield Coal (Middle Pennsylvanian, Illinois Basin): Comparison of miospore and coal-ball records

Debra A. Willard*
Department of Plant Biology, University of Illinois at Urbana-Champaign, 505 South Goodwin Avenue, Urbana, Illinois 61801

ABSTRACT

Coal-ball peats and miospore floras were sampled quantitatively in profiles from the upper Middle Pennsylvanian Springfield Coal of the Illinois Basin. Coal profiles for miospore analysis were sampled at 13 sites, forming two transects across the Galatia paleochannel. Miospore assemblages near paleochannels differ from those near the coal margin. Near the Galatia paleochannel, four species of tree-fern spores (*Laevigatosporites globosus, L. minimus, Punctatosporites minutus, Thymospora pseudothiessenii*) share dominance throughout the profile, and *Lycospora* is subdominant. In profiles near the coal margin, *T. pseudothiessenii* dominates the lower three-fourths of the seam, and *Laevigatosporites globosus* dominates the upper one-fourth of the seam. *Lycospora* is at its most abundant in the lower one-fourth of the coal, and *Anacanthotriletes spinosus* is abundant in the middle of the seam.

Three coal-ball profiles were collected in conjunction with miospore profiles to compare species abundance in the two records. Lycopods are the dominant biovolume producers in coal-ball floras, and tree ferns usually rank second. This differs from the miospore floras, in which tree-fern miospores are dominant over those of lycopods. Disparities between the two records were evaluated with R-values, ratios of percent abundance of species in the miospore record to that in the coal-ball record. In the Springfield Coal, tree ferns are 2 to 3 times and lepidodendrid lycopods 0.5 to 0.75 times as abundant in the miospore record as in the peat. *Sigillaria* and *Diaphorodendron*, however, are much more poorly represented by spores and have R-values less than 0.2. Although R-values are too variable among zones in profiles to accurately reconstruct the peat, they provide an estimate of how over- or underrepresented species are in the miospore record and should be considered when estimating vegetational biomass from percent miospore abundance.

INTRODUCTION

Studies of the composition, structure, and ecology of Pennsylvanian-age coal-swamp communities traditionally have relied on two primary types of preservation, anatomically preserved peat (coal balls) and spores preserved in coal. Historically, the two records have focused on different aspects of the flora.

Coal balls have been used extensively to interpret the biology of the plants and to determine how reproductive adaptations and anatomical modifications are related to the growth environment of the plant. Data from coal balls also have been used to infer broad-scale paleoecological changes throughout the paleotropical belt and to pinpoint more local ecological parameters that controlled the distribution and abundance of species within a coal swamp. The spore record from coal, on the other hand, initially was applied to biostratigraphic problems. Spores and pollen are so durable and abundant that they typically are found in all but

*Present address: U.S. Geological Survey, 970 National Center, Reston, Virginia 22092.

Willard, D. A., 1993, Vegetational patterns in the Springfield Coal (Middle Pennsylvanian, Illinois Basin): Comparison of miospore and coal-ball records, *in* Cobb, J. C., and Cecil, C. B., eds., Modern and Ancient Coal-Forming Environments: Boulder, Colorado, Geological Society of America Special Paper 286.

the highest rank coals and, potentially, form the most extensive paleobotanical data base. In an ideal study, both coal balls and spores from coal would be incorporated along with other data to provide as complete a picture of the vegetation as possible. Because coal balls have a limited distribution and are absent from many coals, however, such integrated studies often are impossible, leaving spores and pollen as the only record of the vegetation. Therefore, the spore record has been used extensively to estimate original vegetational composition and abundance in an effort to understand better the paleoecology of the coal swamp (Habib, 1966; Smith and Butterworth, 1967; Mahaffy, 1985, 1988). To date, however, the spore record has been used without correction for potential biases, such as those related to preservation or to differences in reproductive biology of source plants. Comparison of the spore and coal-ball records provides a means to calibrate spore abundance with a different, possibly more reliable, record of source vegetation and should help establish guidelines for the use of the miospore record in paleoecological studies of Pennsylvanian-age coals.

The Springfield Coal of the Illinois Basin was chosen for study because of its great lateral extent and thickness, its highly diverse spore flora, and the relatively abundant occurrences of coal balls (Smith and Stall, 1975; Phillips et al., 1985). Contemporaneous paleochannel systems and major topographic features are well-established (Eggert, 1982; Eggert et al., 1983), and general trends in spore abundance have been identified (Phillips and Peppers, 1984). These spore distribution patterns are based on channel samples of the coal, in which the entire seam is represented by one time-averaged sample. In this study, incremental full-seam profiles were collected in ten mines in the Springfield Coal to determine vertical and lateral changes in spore abundance. Where possible, coal-ball profiles were collected along with coal profiles to ascertain the similarity of the two records and to determine constraints on the use of spore profiles in paleoecological studies. This study also was designed to establish the amount of variation in the spore flora of one large coal swamp, to determine how these changes related to topographic features and depositional environments, and to compare the Springfield miospore flora with those of other Middle Pennsylvanian coals.

COLLECTING SITES

The Springfield coal swamp in the Illinois Basin minimally extended over an area of about 30,000 mi^2 (77,700 km^2), approximately the size of South Carolina. It includes total in-place reserves estimated at 71.7 billion short tons, second in the Illinois Basin only to the Herrin Coal (Smith and Stall, 1975; Eggert et al., 1983). The coal usually is more than 1.3 m thick in western Kentucky, southwestern Indiana, and southeastern and central Illinois but is less than 1 m thick in south-central Illinois. Local variation in thickness results from both structural and depositional controls; the thickest coal (4 m) occurs near the Galatia and Leslie Cemetery paleochannels, and the thinnest coal is west of the DuQuoin monocline (Wier, 1973; Smith and Stall,

1975; Hopkins et al., 1979; Eggert et al., 1983). Stratigraphically, the Springfield (No. 5) Coal Member of Illinois is in the Carbondale Formation of the Desmoinesian Series, corresponding to the upper Westphalian D of European terminology (Fig. 1) (Phillips et al., 1985). The Springfield Coal also has been correlated approximately with the Middle Kittanning and Princess (No. 7) Coals of the Appalachian Basin (Kosanke, 1973; Peppers, 1984). West of the Illinois Basin, the Springfield Coal is correlative with the Summit Coal of Iowa and Missouri (Landis, 1965; Ravn, 1986).

Ten collection sites were selected to form two intersecting transects (Fig. 2). Both transects cross the Galatia paleochannel, maximizing the range of depositional and structural conditions sampled. Sites yielding both coal and coal balls are of primary interest in this study, because they allow direct comparison of mega- and microfossil records. Coal balls were present and collected in three mines in this study: Island Creek Providence Mine No. 1 in Hopkins County, Kentucky, Peabody Lynnville Mine's Eby Pit in Pike County, Indiana, and Peabody Martwick Mine in Muhlenberg County, Kentucky. The Lynnville Mine profiles were collected in coal split by overbank deposits from the Leslie Cemetery paleochannel. The Kentucky mines are in a more basinal facies of the coal (Ward, 1977); coal is thicker in the Providence Mine (1.8 m) than in the Martwick Mine (1.5 m). Coal balls have been collected from at least 20 other mines in the Springfield Coal (Phillips et al., 1985), and their data are used for comparison with that generated in the present study.

COLLECTION METHODS

Two or three full-seam coal profiles were collected in each mine. At least one sampled profile represented the typical coal condition for each mine, and selection of other profile locations was based on criteria such as proximity to coal-ball masses, presence of certain roof floras, and presence of shale splits in the coal, among other features. Coal intervals sampled (hereafter called zones) were not of equal thickness but were based primarily on the thicknesses of coal that could be removed as intact blocks (usually 5–10 cm). If megascopic differences in coal character (e.g., fusain bands) were detected in the laboratory, coal zones were split again before maceration.

Coal-ball zones were separated by coal partings. From each zone, as many coal balls as possible were collected, and, where present, coal was removed between coal balls for miospore analysis. Where very little coal existed between coal balls, the coal and parts of adjacent coal balls were macerated.

MIOSPORE PREPARATION AND SAMPLING

Miospores (spores and pollen smaller than 210 μm in diameter [Guennel, 1952]) were the focus of this investigation and were processed using the procedures of Kosanke (1950). The coal was crushed into fragments 3–5 mm in diameter and mixed before treatment with Schulze's solution (1 part concentrated HNO_3: 1 part saturated solution $KClO_3$), neutralization with distilled water, density separation with $ZnCl_2$, and clearing of spores with

5% KOH. The residue was stained with Safranin O and mounted in glycerin jelly. Slides are stored in the slide collections at the Paleobotanical Research Center at the University of Illinois at Urbana-Champaign under accession numbers 21,800–21,976. Six slides from each sample were scanned at 200× to determine which species were present. To determine relative abundance of species, 300 miospores per sample usually were counted from each maceration, using a magnification of 400×. Also, slides prepared by Mahaffy (1988) from three other sites in the Springfield Coal (Burning Star Mine No. 4, Kennedy Pit, and Spur Mine) were quantified.

COAL-BALL ANALYSIS

Coal balls were cut, etched, and peeled according to the techniques of Joy et al. (1956) and Phillips et al (1976) before quantification. As outlined by Phillips et al. (1977), the center peel from each coal ball was placed under a clear plastic sheet with a square centimeter grid system. For each square centimeter of peel surface, the type of organ, its preservational state, and its taxonomic affinity were recorded. Data are reported here as normalized values with all unidentifiable tissues omitted. Coal balls are stored at the Paleobotanical Research Center at the University of Illinois at Urbana-Champaign and bear accession numbers 36,418–36,619, 36,859–36,953, and 37,200–37,303.

DATA MANIPULATION

The nature of data collection along transects necessitates an analysis of changes along gradients, in this case, presumed gradients from the swamp margin to a paleochannel system. Indirect gradient analyses, such as multivariate ordination techniques, serve to reduce the dimensionality of data (Gauch, 1983; Whittaker, 1975). In these analyses, the community variation relative to underlying gradients is summarized in a more understandable form (Peet et al., 1988). Using average abundance of miospore species in each spore profile, I analyzed the data using nonmetric multidimensional scaling (NMDS); the analysis was performed by calculating Euclidean distances with PROC PROX and NMDS with PROC ALSCAL on the mainframe version 5 of

SAS (SAS Institute, Inc., 1986). The ordination was used primarily as an exploratory tool to understand community variation and to relate this variation to possible underlying gradients. Minimum spanning trees were calculated manually from the distance matrix. Such trees link each sample to its nearest neighbor and can establish whether samples that appear close in two-dimensional space are equally close in higher dimensionality space (Digby and Kempton, 1987).

METHODS RELATING QUANTITATIVE MIOSPORE AND PEAT DATA

The pollen record is used routinely by Quaternary paleoecologists to make inferences about vegetational composition and abundance. Several techniques are used to correct for biases from differential pollen production by source plants and from variation in pollen influx to depositional sites. The most appropriate method for Pennsylvanian-age communities involves calculation of conversion factors (R-values) that normalize the spore record against the megafossil record. In research on modern and Quaternary plant communities, R-values are calculated as the ratio of the abundance of pollen of a species to the abundance of that species in the vegetation, measured as the percentage of total basal area in the forest (see Davis, 1963). This method works well in Quaternary studies because many Quaternary taxa, forests, and depositional environments can be analogized directly to modern examples. For Pennsylvanian floras, dominated by lower-vascular plants with no modern analogues, R-values are calculated as a comparison of spore abundance and aerial biovolume of species in the peat (after Davis, 1963):

$$R_a = \frac{\text{species } a \text{ spore \% abundance}}{\text{species } a \text{ \% aerial biovolume}}$$

In calculations for Pennsylvanian-age coals, abundance of vegetation is expressed as percent biovolume determined from quantitative analysis of coal balls. This is assumed to be the best estimate of relative abundance (biomass) of species in the original vegetation. For R-values, biovolume was reported on a root-free basis with two exceptions. Half of *Psaronius* outer roots were

		WESTERN INTERIOR COAL PROVINCE		MIDCONTINENT SERIES	ILLINOIS FM.	EASTERN INTERIOR COAL PROVINCE				APPALACHIAN COAL REGION	
		MISSOURI	IOWA			ILLINOIS	INDIANA	W. KENTUCKY		WEST VIRGINIA	PENNSYLVANIA E. KENTUCKY
MIDDLE PENNSYLVANIAN	MARMATON			DESMOINESIAN	CARBONDALE	DANVILLE	COAL VII	NO. 13	ALLEGHENIAN	LOWER FREEPORT	LOWER FREEPORT PRINCESS NO. 8
		MULBERRY LEXINGTON	MYSTIC			JAMESTOWN HERRIN BRIAR HILL SPRINGFIELD SUMMUM	HYMERA VI HERRIN COAL VA COAL V	NO. 12 NO. 11 NO. 10 NO. 9		U. KITTANNING M. KITTANNING	U. KITTANNING M. KITTANNING PRINCESS NO. 7
		SUMMIT MULKY BEVIER CROWEBURG	SUMMIT MULKY BEVIER WHITEBREAST			SHAWNEETOWN COLCHESTER	SURVANT (IV) COAL IIIA	NO. 8		L. KITTANNING	L. KITTANNING

Figure 1. Stratigraphic relationships of Carbondale Formation coals of the Illinois Basin and correlated coals in the Western Interior and Appalachian Coal regions (modified from Phillips et al., 1985).

considered as aerial (see Phillips and DiMichele, 1981). Half of *Paralycopodites* periderm identified as transitional between root and stem also was counted as aerial material. To account for local differences in R-values, they were calculated for all zones at each coal-ball site; all R-values then were considered together to establish narrower ranges for the ratios.

WHOLE-SEAM AVERAGE ABUNDANCE OF MIOSPORE TAXA

The amount of lateral and vertical variation within the miospore flora can be evaluated through comparison of average abundance of spore taxa. Looking first at major plant groups, tree-fern spores dominate most profiles (47%–69%), and lycopod miospores typically are subdominant (18%–44%) (Fig. 2). Miospores of other plant groups usually are comparatively minor components of spore floras. Only profiles near the Leslie Cemetery paleochannel exhibit strong dominance of lycopod miospores (54%–62%) with those of tree ferns subdominant (28%–32%).

Lateral variation in miospore abundance correlates with paleogeographic location within the Springfield coal swamp and proximity to paleochannel systems. Miospore floras are more diverse near paleochannels where the variety and abundance of tree-fern spores is greater. *Lycospora,* produced by the tree lyco-

pods *Lepidophloios, Lepidodendron,* and *Paralycopodites* (Courvoisier and Phillips, 1975; Willard, 1988, 1989), is most abundant near paleochannels (18%–50%) and least abundant (although still common: 9%–18%) near the coal margin. Conversely, *Thymospora,* produced by *Psaronius* tree ferns (Moore, 1946; Potonie, 1962; Millay, 1979), is most abundant near the margin of the coal (20%–40%) and least abundant near paleochannels (0.6%–20%).

Abundances of other less common spore taxa also appear to

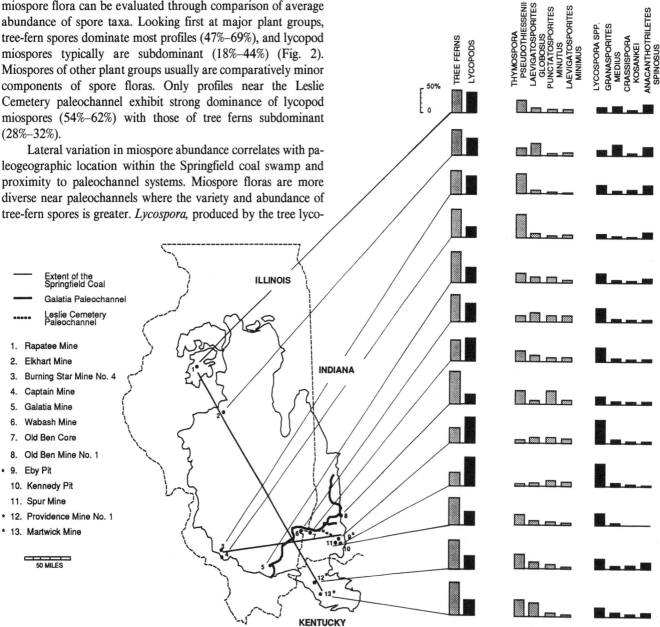

Figure 2. Location of sample sites and percent abundance of miospore taxa on a whole-seam basis from profiles collected in the Springfield Coal of the Illinois Basin. The Galatia paleochannel is indicated by a heavy line, and the Leslie Cemetery paleochannel is indicated by a heavy dashed line. Thin straight lines indicate the intersecting transects formed by miospore samples. Asterisks indicate sites where both coal and coal balls were collected.

be related to location within the Springfield Coal. For example, the abundance of *Anacanthotriletes spinosus,* produced by an unknown but putatively herbaceous lycopod (Baxter, 1971), resembles that of *Thymospora* in its decreased abundance near paleochannels (Fig. 2), whereas the abundance patterns of *Laevigatosporites minimus* relative to paleochannels are more similar to those of *Lycospora.* Miospore profiles collected in northern localities of the coal tend to have greater abundances of *Granasporites medius, Crassispora kosankei,* and *Anacanthotriletes spinosus* than other localities within the coal.

PATTERNS OF MIOSPORE ABUNDANCE IN PROFILES

Two patterns of miospore abundance are found in incremental profiles from the Springfield Coal. Profiles collected near the coal margin share the following characteristics: greatest abundance of *Lycospora* in the lower quarter of the seam; greatest abundance of *Anacanthotriletes spinosus* in the middle half; greatest abundance of *Laevigatosporites globosus* in the upper quarter of the seam; abundant *Thymospora* throughout the seam but least common in the upper quarter (Fig. 3). This pattern is evident both in the full-seam profile and in coal collected between coal-ball zones in Providence Mine No. 1 in Kentucky (Figs. 4 and 5). The similarities between these two profiles imply that vegetation within the coal-ball mass is representative of that throughout the area.

The other pattern of miospore abundance differs strongly from that found near the coal margin and is illustrated by a profile collected within 1 km of the Galatia paleochannel in Amax Coal Company Wabash Mine (Fig. 6); this profile is representative of those collected near the Galatia paleochannel, which existed contemporaneously with peat accumulation. *Lycospora* remains most abundant in the lower part of the seam; *Anacanthotriletes spinosus* is rare throughout the entire profile; and four species of tree-fern spores (*Laevigatosporites globosus, L. minimus, Punctatosporites minutus,* and *Thymospora pseudothiesse-*

nii) are present in roughly equal abundance throughout the profile.

ABUNDANCE OF TAXA IN COAL-BALL PROFILES

The most complete coal-ball profile is that from the Providence Mine No. 1 in Kentucky. This profile was 2.2 m thick, and coal balls were concentrated in the lower half and scattered throughout the upper half of the seam. The Providence Mine is used as the representative coal-ball profile because those in the Martwick Mine and Eby Pit yielded much smaller samples. Each flora was dominated by lycopods (61%–75%), and ferns, primarily *Psaronius* tree ferns, usually are subdominant (Table 1). Pteridosperms and sphenopsids are about equally abundant, and cordaites are rare. Eby Pit, near the Leslie Cemetery paleochannel, differs from the Kentucky Mines in its subdominance of pteridosperms (16%) instead of ferns.

Lycopods dominate most zones of the Providence Mine assemblage; ferns dominate only in zones 6 and 7 (Fig. 7). Similar patterns are evident in the other two profiles. A common pattern in generic abundance begins with dominance of *Lepidodendron* and *Lepidophloios* in the basal zone, followed upward by successive peak abundances of *Diaphorodendron, Paralycopodites, Sigillaria,* and *Psaronius. Diaphorodendron* and *Sigillaria* dominate the next few zones, ending with *Lepidodendron* and *Lepidophloios* (Fig. 8). It should be noted that the top zone does not necessarily correspond to the top of the coal but to the uppermost coal-ball zone collected, so the abundance of *Lepidodendron* and *Lepidophloios* in the upper zones does not imply that they reflect the final stand of the swamp, simply the last stand collected.

ORDINATION OF MIOSPORE DATA

Average abundance of miospore taxa for each profile plus my counts from slides from three profiles prepared by Mahaffy (1988) were analyzed using nonmetric multidimensional scaling (Fig. 9). The profiles cluster into three groups: profiles collected near the coal margins, profiles collected near the Galatia paleo-

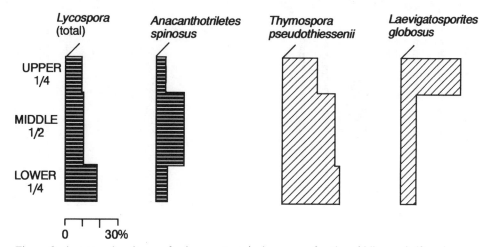

Figure 3. Average abundance of miospore taxa in lower one-fourth, middle one-half, and upper one-fourth of profiles collected near the margin of the Springfield Coal.

channel, and profiles collected near the Leslie Cemetery paleo-channel. The first axis, which separates the three groups, places those samples with high percentages of *Thymospora* and *Anacanthotriletes spinosus* and low percentages of *Lycospora* (those collected near the coal margin) on the left and those with low percentages of *Thymospora* and *A. spinosus* and high percentages of *Lycospora* (those collected near the Leslie Cemetery paleo-channel) on the right (Fig. 9). Profiles collected near the Galatia paleochannel have intermediate abundances of all three taxa, and these profiles are located centrally in the plot of ordination scores. The second axis separates samples based on relative abundance of other tree-fern spores such as *Laevigatosporites globosus* and *Punctatosporites minutus*.

COMPARISON OF SPECIES ABUNDANCE IN PEAT AND MIOSPORE FLORAS

Differences in R-values exist, on a whole-seam basis, between Eby Pit and the two Kentucky mines, particularly for lycopods and sphenopsids (Table 2). In the more typical Kentucky mines, lycopods are about half as abundant in the miospore flora as in the peat, ferns are 2 to 4 times as abundant, and sphenopsids are 1 to 2 times as abundant. Cordaite R-values are extremely variable (0–140). Similar variation exists among major genera in the different mines. The general qualitative pattern of over- or underrepresentation is consistent from mine to mine, but the quantitative expression or degree of representation varies. For example, R-values for *Psaronius* range from 2.9 to 4.3 in the three mines. Among zones of a profile, R-values for individual species and for major plant groups are even more variable, with R-values for *Lycospora*-producing lycopods ranging from 0.2 to

6.1 and values for *Psaronius* tree ferns ranging from 2.0 to 49.6 in zones of the Providence Mine profiles.

To test the accuracy of biovolume estimates from the corrected miospore record, average R-values calculated from the Providence Mine were used with the miospore record from the Martwick Mine to compare resulting abundance estimates with quantitative data derived from coal balls. On an averaged, whole-seam basis, biovolume estimated from the spore record is within 3% of biovolume determined from coal balls (Table 3). On a zone to zone basis, however, the estimates often differed strongly from the actual biovolumes. For example, one zone in which lycopods comprised 49% of the peat biovolume in coal balls was estimated to have 75% lycopods. The estimates uniformly predict the dominant plant group correctly, but are inconsistent in correct predictions of subdominant groups. For example, the subdominance of medullosan pteridosperms in zone 3 is not estimated because of their absence from the miospore record.

REPRESENTATION OF SPECIES IN THE MIOSPORE AND COAL-BALL RECORDS

The coal-ball and spore records preserved different features of Pennsylvanian coal-swamp vegetation; coal balls preserved the peat that accumulated in the swamp, and the spore record preserved the spore and pollen rain of the swamp. Coal-ball floras provide a direct record of plant remains that accumulated as peat, a record of vegetational assemblages, and a general overview of vertical patterns of assemblages in coal-ball profiles. Information derived from coal-swamp vegetation usually is at the local level (10^0–10^2 m) but can be expanded with repeated sampling over broad areas. Spore floras provide a continuous record of the spore and pollen rain throughout the life of a swamp. The spore record

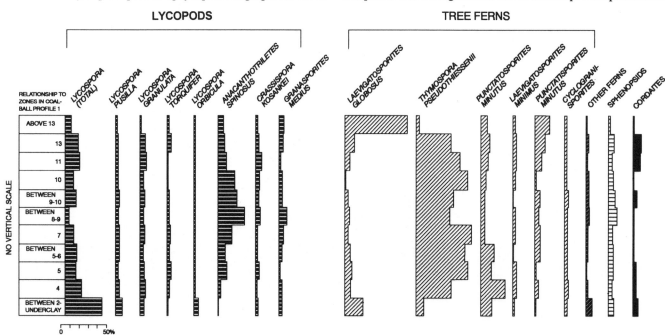

Figure 4. Percent abundance of major miospore species from coal-ball mass (coal-ball profile 1) in the Springfield Coal, Island Creek Coal Company Providence Mine No. 1, Hopkins County, Kentucky.

samples vegetation on a more regional scale (10^0–10^5 m) than coal balls (Traverse, 1988), and lateral changes in vegetational abundance can be detected and related to topographic features of the swamp. The spore record also is very sensitive to appearance and disappearance of species and changes in diversity.

The coal-ball and spore records are subject to different biases, which must be considered in their comparison. The primary bias of the coal-ball record is preservational; much of the original biomass is not preserved during peat accumulation. A rough estimate of the degree of peat degradation is provided by calculation of shoot/root (S/R) ratios for coal-ball profiles. Lycopod trees, which dominated most upper Middle Pennsylvanian coal swamps, would have had S/R ratios of about 4 if found complete, and S/R ratios for lycopods preserved in coal balls from the Springfield and Herrin Coals typically range from 1–2.5 (Phillips et al., 1985). Such enrichment of root material indicates substantial degradation of aerial tissues and reduces the accuracy of estimates of the original standing vegetation. Coal-ball zones with high percentages of root material also affect R-values because of their basis on aerial biovolume of peat. In addition, the full seam thickness usually is not preserved by coal balls (DiMichele and Phillips, 1988), and coal partings that separate coal-ball zones may represent a greater time span than do

coal-ball zones themselves. Thus, coal-ball profiles may provide discontinuous, incomplete records of the history of peat accumulation.

Different biases exist in the spore record. Spores and pollen usually are well-preserved in coals, and the spore flora is thought to record swamp vegetation primarily, with transport a minor bias (Traverse, 1988). Source plants are increasingly well known for the most prominent spore species, and the primary bias in the spore record results from different reproductive strategies among coal-swamp plants. Homosporous tree ferns produced fertile photosynthetic fronds repeatedly throughout their lives and generated enormous numbers of small spores (Lesnikowska, 1989). The heterosporous lycopods include both monocarpic species, which reproduced once near the end of their lives, and polycarpic species, which reproduced repeatedly throughout their lives (DiMichele and Phillips, 1985). Only the microspores of these lycopods are represented in the miospore record, which typically is all that is studied. Pteridosperms were seed plants, but the prepollen of the arboreous medullosan pteridosperms was too large for inclusion in the miospore record, excluding medullosans from miospore studies (Phillips and Peppers, 1984). Pollen of smaller pteridosperms, such as *Callistophyton* and *Heterangium,* is preserved in the miospore record, but the plants are relatively

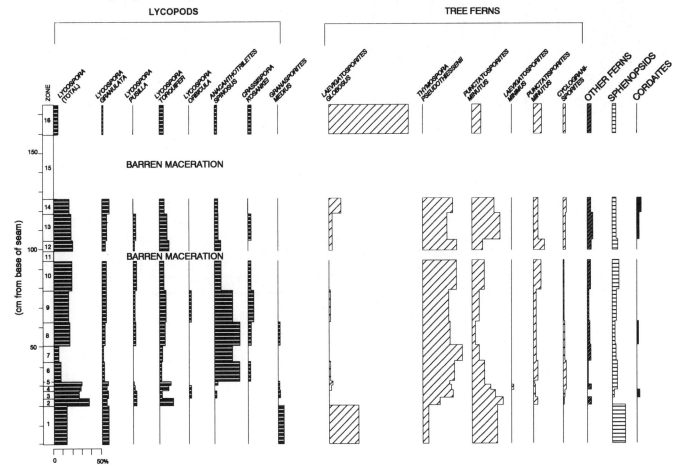

Figure 5. Percent abundance of major miospore species (coal profile 1) in the Springfield Coal, Island Creek Coal Company Providence Mine No. 1, Hopkins County, Kentucky.

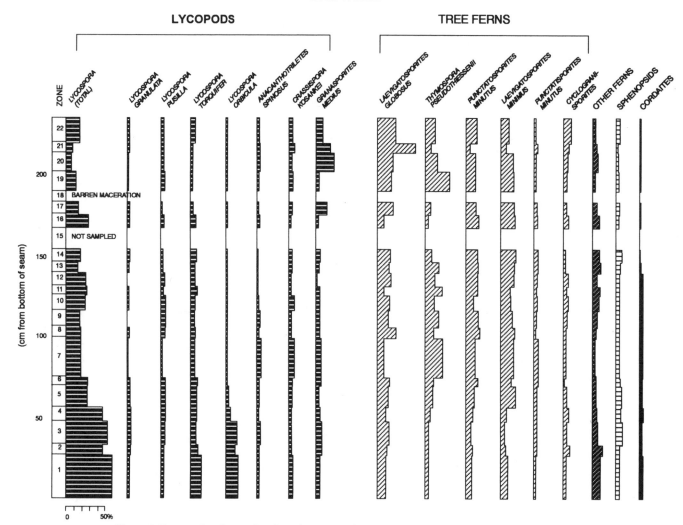

Figure 6. Percent abundance of major miospore species (coal profile 1) in the Springfield Coal, Amax Coal Company Wabash Mine, Wabash County, Illinois.

TABLE 1. AVERAGE PERCENT BIOVOLUME PRODUCED BY MAJOR PLANT GROUPS IN COAL-BALL PROFILES IN THE SPRINGFIELD COAL FROM THE MARTWICK MINE, MUHLENBERG COUNTY, KENTUCKY, PROVIDENCE MINE NO. 1, HOPKINS COUNTY, KENTUCKY, AND LYNNVILLE MINE'S EBY PIT, PIKE COUNTY, INDIANA

Profile	Percent Biovolume				
	Lycopods	Ferns	Pteridosperms	Sphenopsids	Cordaites
Eby Pit Profile 12	71.3	7.7	16.1	4.8	0
Martwick Mine Coal-Ball Profile	74.6	15.7	5.0	4.1	0.3
Providence Mine No. 1 Coal-Ball Profile 1	60.6	27.9	6.1	5.4	0.009

All values were normalized after subtraction of unidentifiable material. Values for the Martwick Mine sample do not include the upper four zones.

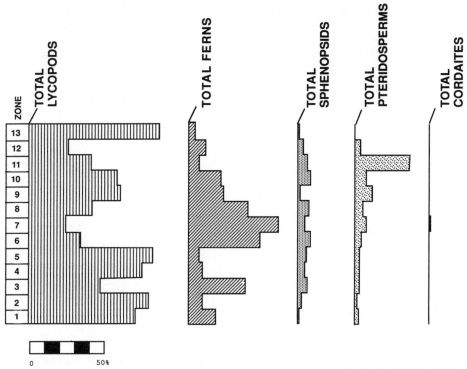

Figure 7. Percent biovolume of major plant groups (coal-ball profile 1) in the Springfield Coal, Island Creek Coal Company Providence Mine No. 1, Hopkins County, Kentucky.

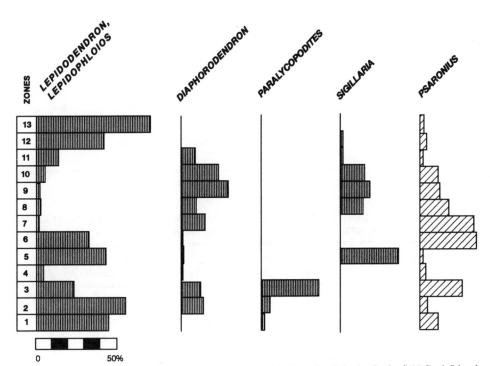

Figure 8. Percent biovolume of arborescent genera (coal-ball profile 1) in the Springfield Coal, Island Creek Coal Company Providence Mine No. 1, Hopkins County, Kentucky. Estimates are made on a root-free basis with two exceptions. One-half of *Psaronius* outer roots and one-half of *Paralycopodites* periderm that was transitional from root to shoot were counted as aerial.

D. A. Willard

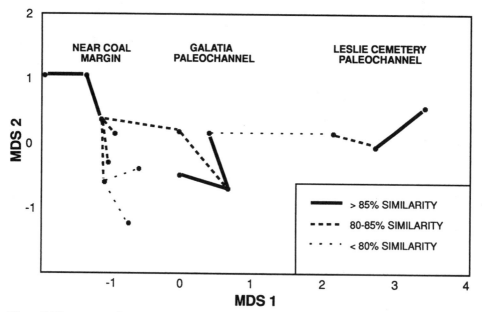

Figure 9. First two axes from nonmetric multidimensional scaling of average miospore abundance from mines in the Springfield Coal, Illinois Basin. A minimum spanning tree is superimposed upon the plot (see text for discussion).

TABLE 2. AVERAGE R-VALUES FOR MAJOR PLANT GROUPS AND GENERA IN EBY PIT PROFILE 12, MARTWICK MINE PROFILE 1 AND PROVIDENCE MINE NO. 1 COAL-BALL PROFILE 1

Mine	Lycopods	Ferns	Sphenopsids	*Lepidophloios, Lepidodendron, Paralycopodites*	*Diaphorodendron*	*Sigillaria*	*Psaronius*
Eby Pit Profile 12	2.12	2.78	6.56	1.65	1.30	0.01	2.90
Martwick Mine Profile 1	0.56	6.30	1.18	1.95	0.18	0.08	4.25
Providence Mine No. 1 Coal-ball Profile 1	0.75	2.58	1.76	0.83	0.11	0.13	3.68

TABLE 3. COMPARISON OF BIOVOLUME OF PLANT TAXA ESTIMATED FROM MIOSPORE DATA WITH ACTUAL BIOVOLUME DETERMINED FROM COAL-BALL PEATS, SPRINGFIELD COAL, MARTWICK MINE, MUHLENBERG COUNTY, KENTUCKY

Zone	Ferns		Lycopods		Sphenopsids		Tree Ferns	
	Estimated	Actual	Estimated	Actual	Estimated	Actual	Estimated	Actual
5	22	24	75	49	2	11	15	14
4	21	7	86	76	8	4	13	6
3	23	8	63	78	16	5	16	5
2	14	12	80	79	0	2	10	8
Whole-Seam Average	19	17	62	65	6	6	11	10.5

rare (typically <5%) in both the coal-ball and miospore records, making calculation of valid R-values difficult. Most biases in the miospore record are of a biological rather than preservational nature, and the main contribution from R-values in this study is the estimation of the degree of over- or underrepresentation of each species in the miospore record (Table 4).

R-VALUES AND APPLICABILITY TO PALEOECOLOGICAL STUDIES

Disparities in species dominance and abundance between the coal-ball and miospore records and intraseam variation in R-values result from the biases discussed above. In this study, R-values provide an index of differential spore representation based on miospore abundance and plant abundance in the peat, not in the standing vegetation. Therefore, zones with extremely poor peat preservation (i.e., high percentage of root material) have aberrant R-values. Similarly, zones with exceptionally well-preserved peat should have R-values approaching their "true" values, those that would result from comparison of spore data with standing vegetation. The great variation in R-values within profiles is the result of differential peat preservation and differences in spore production. The best possible estimates for R-values of Pennsylvanian-age plants can be obtained by concentrating on the best-preserved peats.

The relative over- or underrepresentation of species in the miospore record from the Springfield Coal is established here in very general terms at least, from R-values. Tree ferns are overrepresented markedly in the miospore record, whereas lycopods, as a group, are underrepresented (Table 4). *Lycospora pusilla* and *L. granulata,* produced by *Lepidodendron* and *Lepidophloios,* usually are about 0.75 times as abundant in the spore record as in coal-ball peats, and *L. orbicula,* produced by *Paralycopodites,* appears to be represented about equally well by the miospore and coal-ball records. *Sigillaria* (represented by *Crassispora kosankei*) and *Diaphorodendron* (represented by *Granasporites medius*) are very poorly represented in the miospore record. Miospores of these taxa typically are less than 0.2 times as

abundant in the miospore record as they are in the megafossil record. Similar estimates from other coals have been obtained by Mahaffy (1985, 1988), Peppers (1982), and Willard (1986).

INFERRED TRENDS IN VEGETATIONAL ABUNDANCE FROM MIOSPORE DATA

Lateral trends in vegetational abundance and diversity in the Springfield coal swamp were clarified through comparison of miospore profiles. Three distinct groups were indicated from analysis of miospore abundance: those near the Galatia paleochannel, those near the Leslie Cemetery paleochannel, and those near the margin of the coal (Fig. 9). Among these groups, profiles collected near paleochannel systems are more diverse, with a different suite of tree-fern species and greater abundance of *Lepidophloios* and *Lepidodendron* (*Lycospora* producers) than other sites.

Profiles collected near the coal margin are united primarily by their relatively high percentage of *Anacanthotriletes spinosus* and *Thymospora pseudothiessenii* and lower abundances of *Lycospora* (<20%) (Fig. 9). Of these profiles, those from Captain Mine and Burning Star Mine No. 4 are most similar; both mines are located in the relatively thin coal west of the DuQuoin Monocline and have very high abundances of *Thymospora* and *A. spinosus* but low abundances of lycopod miospores and those of other tree ferns. Profiles collected near the northern coal margin in the Elkhart and Rapatee Mines differ from more southern profiles in their greater abundance of *Diaphorodendron* and *Sigillaria.* Because these plants typically are greatly underrepresented in the miospore flora, the high abundances of their miospores (as much as 28.5%) in these profiles suggest that these taxa were unusually abundant in this region. This hypothesis is supported by the high abundance of *Diaphorodendron scleroticum* in a random sample of coal balls from the Hazeldell Mine in Fulton County, Illinois (W. A. DiMichele, 1989, personal communication).

Miospore profiles near the Leslie Cemetery paleochannel have aberrantly high proportions of *Lycospora* (44%–62%) and

TABLE 4. SUMMARY OF RELATIVE REPRESENTATION OF TAXA IN COAL-SPORE FLORAS

Taxon	Representation in Spore Flora	Typical R-value
Psaronius tree ferns	Overrepresented	2–3
Monocarpic *Lycospora*-producers (*Lepidodendron, Lepidophloios*)	Slightly overrepresented to slightly underrepresented	0.7–1.2
Polycarpic *Lycospora* producers (*Paralycopodites*)	Evenly	*
Diaphorodendron	Underrepresented	0.1–0.2
Sigillaria	Underrepresented	<0.1

R-values were based on whole-seam average abundances of taxa in spore floras and coal-ball peats and represent typical R-values for Middle Pennsylvanian coals.
*R-values have not been calculated for this taxon.

low proportions of tree-fern spores (13%–32%). This record does not necessarily indicate higher than normal abundance of *Lycospora* producers, however. It does indicate that tree ferns were less abundant than elsewhere in the coal, but coal-ball data indicates a greater than normal abundance of medullosan pteridosperms, the subdominant peat producers (Table 1). Because medullosan prepollen is absent from the miospore record, lycopod spores appear more abundant. Preliminary analysis of megaspores from Eby Pit indicates that pteridosperm prepollen is common. The masking of the large medullosan prepollen by *Lycospora* is a common problem when using percentage data based on fixed counts. Also, the exclusive use of miospores in such analyses can hamper detection of important trends and cause data to be misinterpreted.

Using the evidence generated in this study on representation of species by their miospores, palynofloras from other coals of similar age can be reevaluated for comparison with the Springfield Coal. The Princess (No. 7) and Middle Kittanning Coals of the Appalachian Basin are approximately equivalent to the Springfield Coal (Fig. 1) (Kosanke, 1973; Peppers, 1984), but tree-fern spores are more abundant in these Appalachian Basin coals than in the Springfield Coal. Tree-fern species composition also differs, with *Laevigatosporites globosus* the dominant miospore species in the Middle Kittanning Coal (Gray, 1967) instead of *Thymospora pseudothiessenii.* Although this pattern may be interpreted as indicative of greater tree-fern abundance in the Appalachian Basin, consideration of differing reproductive output of different species suggests that a greater abundance of *Diaphorodendron* and *Sigillaria* in the Middle Kittanning and Princess (No. 7) Coals also is possible. If *Diaphorodendron* and/or *Sigillaria* were the dominant lycopod trees in the Appalachian coals, a position held by the more prolific spore producer *Lepidophloios* in the Illinois Basin, tree-fern spores would appear to be more abundant in the miospore record (less "diluted" by *Lycospora* and would mask the presence of *Diaphorodendron* and *Sigillaria.* Indeed, *Sigillaria* has been reported to be abundant in coal balls from the approximately equivalent Derringer Corners locality in the Appalachian Basin (Feng, 1989), making the latter interpretation seem more likely. Further work incorporating miospores, megaspores, and megafossils would be necessary either to confirm or deny the hypothesis.

THE SPRINGFIELD PEAT SWAMP—
DOMED OR PLANAR?

Recently, numerous studies have attempted to analogize modern peats with ancient coal swamps. The search for modern analogues for ancient peat swamps has focused attention on two types of peat deposits, domed (ombrotrophic) and planar (rheotrophic) peats. Domed peats have been hypothesized as the best model to explain vertical patterns observed in lithotype, ash and sulfur content, and preservation in many Middle Pennsylvanian coals of the Appalachian Basin. These patterns include decreased sulfur and ash, poorer preservation, and decreased abundance of woody material going upward in the seam (Cecil et al., 1985;

Esterle and Ferm, 1986; Esterle et al., 1989; Eble, 1988). In modern domed peat deposits, similar patterns are present and reflect the shift from ground-water fed (rheotrophic) peats at the base to rain-water fed (domed) peats at the top. This shift also results in decreasing fertility and more stressful conditions from the base to the top of the peat (Anderson, 1983). Palynological analyses of domed peats reflect the vegetational succession tied to these changes (Anderson, 1983), and analyses of Pennsylvanian coals have used the miospore record as confirmation of the doming hypotheses (Calder, this volume; Eble and Grady, this volume). In a Middle Pennsylvanian domed peat, it is hypothesized that tree ferns would increase in abundance upward in the coal. This hypothesis is based on the relatively great tolerance of tree ferns to a variety of edaphic conditions. Tree ferns were homosporous, were not tied to standing water for reproduction, and would have been more likely to survive in the limiting, stressful conditions than the heterosporous lycopods, whose reproduction was tied more closely to consistently higher water regimes (Phillips, 1979; DiMichele and Phillips, 1985). If this hypothesis of Pennsylvanian plant "succession" in domed peats was correct, the petrographic trend would be a dulling upward sequence corresponding to a change in peat composition from a periderm-rich, lycopod-dominated peat at the base to primarily parenchymatous, tree-fern–dominated peat near the top (Esterle and Ferm, 1986; Eble and Grady, this volume).

The most convincing cases for characterizing Pennsylvanian-age peats as domed or planar are those that incorporate data from coal petrology, geochemistry, and paleobotany (see Winston, 1988; Calder, this volume; Eble and Grady, this volume). This analysis of the Springfield Coal was not designed for such a determination, but consideration of paleobotanical data in conjunction with earlier petrographic research suggests that the Springfield coal swamp probably was a planar peat. Concurrent analyses of petrography and geochemistry are necessary to make such a statement with confidence, but this study does provide valuable insight into the possible use (and misuse) of the miospore record in such analyses.

In extrapolating plant biomass from the miospore record, several items must be considered. First, the representation of species in the miospore record ranges from nearly zero (medullosan pteridosperms) to more than twice the abundance of the plants in coal balls (tree ferns). Second, lycopod genera are not equal in their relative microspore output. *Lepidophloios* and *Lepidodendron* range from under- to overrepresented by their *Lycospora* microspores, but *Sigillaria* and *Diaphorodendron* produced relatively few microspores and are poorly represented in the microspore record. It is not unreasonable to assume that similar variation exists among tree-fern species. Finally, miospore data typically are presented as percentages, not as absolute values, such as miospores per gram. Because percentages are used, the problem of closure arises; that is, the percentages must add up to 100, so a decrease in abundance of one species necessitates an increase in abundance of others. For example, *Lycospora* typically decreases in abundance upward from the base of the Springfield Coal, and tree ferns typically increase in abundance upward (Figs. 4–6). The assumption from miospore data that *Lepidodendron* and *Lepidophloios* abundance decreased while *Psaronius* tree fern abundance increased upward is contrary to patterns shown by coal-ball data. A more likely interpretation (and one more consistent with megafossil evidence) is that *Lepidodendron* and *Lepidophloios* did become less abundant but that *Sigillaria* and *Diaphorodendron* were more prominent in later floras. Similar patterns have been documented in studies of the Herrin Coal (DiMichele and Phillips, 1988). Thus, the increased abundance of tree-fern miospores upward may reflect a change from dominance of *Lycospora* producers to a greater abundance of *Sigillaria* and *Diaphorodendron,* not an increased abundance of *Psaronius* tree ferns. To reduce the chances of misinterpretation of such miospore data in future studies, it would be helpful to report results as absolute miospore frequency; determination of the number of miospores per gram of coal also could be useful in assessing relative peat preservation. Because most studies of Pennsylvanian coals probably will continue to use percentage data, however, this data on relative representation of species in the miospore record serves as a caution in making paleoecological interpretations based solely on percent abundance of miospores.

CONCLUSIONS

Coal-ball and miospore data reflect different aspects of Pennsylvanian coal-swamp vegetation, and both types of data should be incorporated in paleoecological analyses whenever possible. Coal-ball peats may provide the most accurate picture available of quantitative abundances of coal-swamp plants in vegetational assemblages. Analysis of coal-ball profiles reveals patterns of vertical change in assemblages through the life of the swamp. Studies of coal balls also provide a great deal of information on the anatomy, morphology, and reproductive biology of coal-swamp plants as well as a foundation for analysis of community structure, on a local level at least. Miospores record the spore and pollen rain in the forest and yield a better broad-scale, qualitative impression of vegetational patterns. Because miospores are present so abundantly in coal, they are useful for establishing general spatial and temporal patterns in vegetational abundance over broad areas. The inequities in miospore production among different species, however, makes it necessary to calibrate the miospore record with some measure of vegetational abundance before estimating species abundances from miospore data. Coal balls provide the best baseline for such calibration. By interpreting the miospore record in conjunction with peat biovolume, inferences about relative miospore output of the original vegetation can be made, as can estimates of the relative under- or overrepresentation of individual taxa in the miospore record. Continuing comparisons of the two records in other coals should improve accuracy of R-values and allow detection of changes in relative representation of taxa by miospores both within one coal and between different coals. Miospores also are useful in detection of changes in relative abundance of source plants and changes in diversity. Disturbance events are chronicled by certain taxa with distinct, known miospores, and the approximate time of vegetational recovery from disturbance can be estimated.

Comparisons of miospore abundance and peat biovolume indicate that inferences drawn from each data set about the original coal-swamp vegetation are broadly similar, when the biases of each record are considered. Because of the differences in miospore representation of the prominent coal-swamp species indicated by this study, it is necessary that R-values be considered whenever vegetational abundance is estimated from percent miospore abundance. Failure to consider these differences in reproductive strategy can result in estimates that contradict the more direct megafossil evidence of abundance. Further refinement of R-values through analyses of other coals and through comparison of mio- and megaspore abundance with vegetational biovolume from well-preserved coal balls should increase the accuracy of the R-values and, therefore, improve estimates of vegetational abundance from miospore data. As additional data are acquired from megafossils, it also may be possible to describe idealized trees of major species: total biomass, timing of reproduction, total number of fructifications produced, and total number or micro- and megaspores produced. If such models can be realistically described, it may become possible to predict accurately original standing biomass in Pennsylvanian-age coal swamps.

ACKNOWLEDGMENTS

I thank R. M. Bateman, W. A. DiMichele, and T. L. Phillips for discussion and informal review of the manuscript and T. L. Phillips for providing quantitative data from coal balls. R. J. Burnham and G. W. Rothwell also provided helpful comments. Carol Kubitz, scientific illustrator, School of Life Sciences, University of Illinois, and Salvatore Johnson rendered the figures. This research was partially supported by a research grant from the Geological Society of America and represents work submitted in partial fulfillment of the doctoral degree at the University of Illinois at Urbana-Champaign.

REFERENCES CITED

Anderson, J.A.R., 1983, The tropical peat swamps of western Malesia, *in* Gore, A.J.P., ed., Mires: Swamp, bog, fen and moor: Amsterdam, Elsevier, Ecosystems of the world, v. 4B, p. 181–199.

Baxter, R. W., 1971, *Carinostrobus foresmani,* a new lycopod cone genus from the Middle Pennsylvanian of Kansas: Palaeontographica, v. 134B, p. 124–130.

Cecil, C. B., Stanton, R. W., Neuzil, S. G., Dulong, F. T., Ruppert, L. F., and Pierce, B. S., 1985, Paleoclimate controls on Late Paleozoic sedimentation and peat formation in the central Appalachian basin (U.S.A.): International Journal of Coal Geology, v. 5, p. 195–230.

Courvoisier, J. M., and Phillips, T. L., 1975, Correlation of spores from Pennsylvanian coal-ball fructifications with dispersed spores: Micropaleontology, v. 21, p. 45–59.

Davis, M. B., 1963, On the theory of pollen analysis: American Journal of Science, v. 261, p. 897–912.

Digby, P.G.N., and Kempton, R. A., 1987, Multivariate analysis of ecological communities: London, Chapman and Hall, 206 p.

DiMichele, W. A., and Phillips, T. L., 1985, Arborescent lycopod reproduction and paleoecology in a coal-swamp environment of late Middle Pennsylvanian age (Herrin Coal, Illinois, U.S.A.): Review of Palaeobotany and

Palynology, v. 44, p. 1–26.

DiMichele, W. A., and Phillips, T. L., 1988, Paleoecology of the Middle Pennsylvanian-age Herrin Coal Swamp (Illinois) near a contemporaneous river system, the Walshville paleochannel: Review of Palaeobotany and Palynology, v. 56, p. 151–176.

Eble, C. F., 1988, Palynology and paleoecology of a Middle Pennsylvanian coal bed from the central Appalachian basin [Ph.D. thesis]: Morgantown, West Virginia, West Virginia University, 495 p.

Eggert, D. L., 1982, A fluvial channel contemporaneous with deposition of the Springfield Coal Member (V), Petersburg Formation, Northern Warrick County, Indiana: Indiana Geological Survey Special Report, v. 28, 20 p.

Eggert, D. L., Chou, C. L., Maples, C. G., Peppers, R. A., Phillips, T. L., and Rexroad, C. B., 1983, Origin and economic geology of the Springfield Coal Member in the Illinois Basin (Field Trip 9): 1983 Field Trips in Midwestern Geology, v. 2, p. 121–146.

Esterle, J. S., and Ferm, J. C., 1986, Relationship between petrographic and chemical properties and coal seam geometry, Hance Seam, Breathitt Formation, southeastern Kentucky: International Journal of Coal Geology, v. 6, p. 199–214.

Esterle, J. S., Ferm, J. C., and Tie, Y.-L., 1989, A test for the analogy of tropical domed peat deposits to "dulling up" sequences in coal beds—Preliminary results: Organic Geochemistry, v. 14, p. 333–342.

Feng, B. C., 1989, Paleoecology of an upper Middle Pennsylvanian coal swamp from western Pennsylvania, U.S.A.: Review of Palaeobotany and Palynology, v. 57, p. 299–312.

Gauch, H. G., Jr., 1983, Multivariate analysis in community ecology: Cambridge, Cambridge University Press, 298 p.

Gray, L. R., 1967, Palynology of four Allegheny coals, northern Appalachian coal field: Palaeontographica 121B, p. 65–86.

Guennel, G. K., 1952, Fossil spores of the Alleghenian coals in Indiana: Indiana Geological Survey Report of Progress, v. 4, 40 p.

Habib, D., 1966, Distribution of spore and pollen assemblages in the Lower Kittanning Coal of western Pennsylvania: Palaeontology, v. 9, p. 629–666.

Hopkins, M. E., Nance, R. B., and Treworgy, C. G., 1979, Mining geology of Illinois coal deposits, in Palmer, J. E., and Dutcher, R. R., eds., Depositional and structural history of the Pennsylvanian system of the Illinois Basin: Part 2. Invited Papers: Ninth International Congress of Carboniferous Stratigraphy and Geology, Field Trip 9, p. 142–150.

Joy, K. W., Willis, A. J., and Lacey, W. S., 1956, A rapid cellulose peel technique in paleobotany: Annals of Botany, v. 20, p. 635–637.

Kosanke, R. M., 1950, Pennsylvanian spores of Illinois and their use in correlation: Illinois State Geological Survey Bulletin 74, 128 p.

Kosanke, R. A., 1973, Palynological studies of the coals of the Princess Reserve District in northeastern Kentucky: U.S. Geological Survey Professional Paper 839, 22 p.

Landis, E. R., 1965, Coal resources of Iowa: Iowa Geological Survey Technical Paper 4, 141 p.

Lesnikowska, A. D., 1989, Anatomically preserved Marattiales from coal swamps of the Desmoinesian and Missourian of the midcontinent United States: Systematics, ecology, and evolution [Ph.D. thesis]: Urbana-Champaign, Illinois, University of Illinois, 227 p.

Mahaffy, J. M., 1985, Profile patterns of coal and peat palynology in the Herrin (No. 6) Coal Member, Carbondale Formation, Middle Pennsylvanian of southern Illinois, in Dutro, J. T., Jr., and Pfefferkorn, H. W., eds., Paleontology, Paleoecology, Paleogeography: Ninth International Congress of Carboniferous Stratigraphy and Geology, Compte Rendu, v. 5, p. 25–34.

Mahaffy, J. M., 1988. Vegetational history of the Springfield Coal (Middle Pennsylvanian of Illinois) and distribution patterns of a tree-fern miospore, *Thymospora pseudothiessenii,* based on miospore profiles: International Journal of Coal Geology, v. 10, p. 239–260.

Millay, M. A., 1979, Studies of Paleozoic marattialeans: A monograph of the American species of *Scolecopteris:* Palaeontographica, v. 169B, p. 1–69.

Moore, L. R., 1946, On the spores of some Carboniferous plants: Their development: Geological Society of London Quarterly Journal, v. 102, p. 251–298.

Peet, R. K., Knox, R. G., Case, J. S., and Allen, R. B., 1988, Putting things in order: The advantages of detrended correspondence analysis: The American Naturalist, v. 131, p. 924–934.

Peppers, R. A., 1982, Palynology of the unnamed coal in the Staunton Formation, Maple Grove Mine, in Eggert, D. L., and Phillips, T. L., eds., Environments of deposition—Coal balls, cuticular shale, and gray-shale floras in Fountain and Parke Counties, Indiana: Indiana Geological Survey Special Report, v. 30, p. 27–33.

Peppers, R. A., 1984, Comparison of miospore assemblages in the Pennsylvanian System of the Illinois Basin with those in the Upper Carboniferous of Western Europe, in Sutherland, P., and Manger, W. L., eds., Biostratigraphy: Ninth International Congress of Carboniferous Stratigraphy and Geology, Compte Rendu, v. 2, p. 483–502.

Phillips, T. L., 1979, Reproduction of heterosporous arborescent lycopods in the Mississippian-Pennsylvanian of Euramerica: Review of Palaeobotany and Palynology, v. 27, p. 239–289.

Phillips, T. L., and DiMichele, W. A., 1981, Paleoecology of the Middle Pennsylvanian age coal swamps in southern Illinois: Herrin Coal Member at Sahara Mine No. 6, in Niklas, K. J., ed., Paleobotany, paleoecology, and evolution: New York, Praeger, v. 1, p. 231–284.

Phillips, T. L., and Peppers, R. A., 1984, Changing patterns of Pennsylvanian coal-swamp vegetation and implications of climatic control on coal occurrence: International Journal of Coal Geology, v. 3, p. 205–255.

Phillips, T. L., Avcin, M. J., and Berggren, D., 1976, Fossil peat from the Illinois Basin: A guide to the study of coal balls of Pennsylvanian age: Illinois State Geological Survey Educational Series 11, 39 p.

Phillips, T. L., Kunz, A. B., and Mickish, D. J., 1977, Paleobotany of permineralized peat (coal balls) from the Herrin (No. 6) Coal Member of the Illinois Basin, in Given, P. N., and Cohen, A. D., eds., Interdisciplinary studies of peat and coal origins: Geological Society of America Microform Publications 7, p. 18–49.

Phillips, T. L., Peppers, R. A., and DiMichele, W. A., 1985, Stratigraphic and interregional changes in Pennsylvanian coal-swamp vegetation: Environmental inferences: International Journal of Coal Geology, v. 5, p. 43–109.

Potonie, R., 1962, Synopsis der Sporae in Situ: Beihefte zum Geologischen Jahrbuch, v. 52, 204 p.

Ravn, R. L., 1986, Palynostratigraphy of the Lower and Middle Pennsylvanian coals of Iowa: Iowa Geological Survey Technical Paper 7, 245 p.

SAS Institute, Inc., 1986, SUGI supplemental library user's guide, version 5 edition: Cary, North Carolina, SAS Institute, 662 p.

Smith, A.H.V., and Butterworth, M. A., 1967, Miospores in the coal seams of the Carboniferous of Great Britain: Special Papers in Palaeontology, v. 1, 324 p.

Smith, W. H., and Stall, J. B., 1975, Coal and water resources for coal conversion in Illinois: Illinois State Water Survey and Illinois State Geological Survey Cooperative Resources Report 4, 79 p.

Traverse, A., 1988, Paleopalynology: London, Unwin Hyman, 600 p.

Ward, C. R., 1977, Mineral matter in the Springfield-Harrisburg (No. 5) in Member in the Illinois Basin: Illinois State Geological Survey Circular, v. 498, 35 p.

Whittaker, R. H., 1975, Communities and ecosystems: New York, Macmillan, 385 p.

Wier, C. E., 1973, Coal resources of Indiana: Indiana Geological Survey Bulletin 42-I, 40 p.

Willard, D. A., 1986, Comparison of megaspores, miospores, and coal balls in analysis of coal-swamp vegetation [abs.]: American Journal of Botany, v. 73, p. 715.

Willard, D. A., 1988, *Lycospora* from Pennsylvanian-age *Lepidostrobus* compressions of Euramerica [abs.]: American Journal of Botany, v. 75, p. 119–120.

Willard, D. A., 1989, Source plants for Carboniferous microspores: *Lycospora* from permineralized *Lepidostrobus:* American Journal of Botany, v. 76, p. 820–827.

Winston, R. B., 1988, Paleoecology of Middle Pennsylvanian-age peat-swamp plants in Herrin Coal, Kentucky, U.S.A.: International Journal of Coal Geology, v. 10, p. 203–238.

MANUSCRIPT ACCEPTED BY THE SOCIETY JANUARY 14, 1993

Geological Society of America
Special Paper 286
1993

The evolution of a ground-water–influenced (Westphalian B) peat-forming ecosystem in a piedmont setting: The No. 3 seam, Springhill coalfield, Cumberland Basin, Nova Scotia

J. H. Calder
Coal Section, Nova Scotia Department of Natural Resources, P.O. Box 698, Halifax, Nova Scotia B3J 2T9 Canada

ABSTRACT

The basis of modern fresh-water peatland (mire) classification, namely ground-water influence and source of ionic input, has been adopted in this study for ancient mire analysis. Trends that result from a modern mire's evolution from a rheotrophic (ground-water influenced) planar to ombrotrophic (solely rain-fed), raised status, under decreasing influence of ground water, include decreasing pH levels, nutrient/ionic supply, ash content, species diversity, and ratio of arboreous to herbaceous vegetation. These attributes are inferred to give rise to the following upward trends within a coal seam: enhanced preservation and reduced biochemical gelification within similar tissues; decreasing abundance of liptinite macerals of aquatic affinity, sulfur (especially pyritic) content, and waterborne ash; and decreasing floral diversity. Reversals in these trends may signal change in the trophic status of the ancestral mire (e.g., deflation). The identification of such trends relies heavily upon the description of vitrinite in terms of relative gelification. The significance of Eh and the historical use of inertinite in paleo-mire analysis is questioned.

The ancestral mire of the Westphalian B No. 3 seam of the Springhill coalfield, Cumberland Basin, Nova Scotia, formed between a piedmont of coalesced, retreating alluvial fans and the medial reaches of a basin-axis channel belt. The lithologically distinct piedmont, inner mire, and riverine zones of the seam reflect this geomorphic setting. Modeling of a maceral-based index of ground-water influence (strongly gelified tissues and mineral matter versus well preserved tissues) led to the deduction that the paleomire originated as a rheotrophic, and presumably planar, ecosystem that evolved progressively toward a less ground-water–influenced (mesotrophic) state, and possibly to an ombrotrophic, weakly domed system within the inner mire. This maceral-based method suggests a succession of mire types from swamp to fen (and questionably to bog) representing the classic hydroseral succession that forms by the autogenic process of terrestrialization. Contrary to the maceral-based evidence of progressive, albeit weak, raising of the mire surface, ash, sulfur, and miospore diversity increase, and lithotypes become duller upward within the upper third of the seam in the inner zone, suggesting that the mire may have ultimately reverted to a more ground-water–influenced state. A decrease in pH, inferred from an upward increase in tissue structure and decrease in gelification, accompanied inner mire development; elsewhere conditions were less acidic.

The paleomire flora was dominated throughout by the arboreous lycopsids *Lepidodendron hickii* and *Anabathra* (cf. *Paralycopodites*), confirming the rheotrophic nature of the ecosystem and the prevalence of flooded conditions. Floral succession of these arboreous lycopsids is evident within the inner mire. Groundwater discharge from alluvial fans at the piedmont margin favored conditions for the colonization of the forest flora. The feedback mechanism of lateral or upslope paludification was aided by the rapid, noncompetitive growth strategy of the arboreous lycopsids. At the riverine margin, autogenic evolution of the ecosystem was stymied by allogenic fluvial processes and by differential compaction about entombed multistory sandstone bodies. Lithotype trends record a general, but similar history of mire development. The ultimate demise of the mire is ascribed to allogenic change, potentially involving precession-induced climate change in concert with basin subsidence and sediment supply.

INTRODUCTION

The debate over the analogy of raised peatlands (mires) as precursors of coal seams is fundamental to interpretations of the origin of coal. Apart from being one of the most interesting problems to be solved in the future (Teichmüller, 1989), it is of fundamental importance not only from an academic standpoint but also in the development of exploration strategies, which draw heavily upon interpreted controls on the deposition of ancient peat, hence coal. In addition, the nature of the developing, precursor peatlands influences the type and distribution of sulfur and other mineral matter within a seam, the paleoflora that contributed to the accumulating biomass, and the genesis of derived coal macerals.

The debate, in essence, is whether raised (domed) mires and, in particular, tropical bog forests are widely applicable as an analogue for ancient coal seams. The concept was introduced in an important study by Smith (1962) who drew an analogy between modern raised peatlands of tropical Malaysia and Indonesia and the Westphalian coal seams of Yorkshire, United Kingdom. The debate has developed as geologists investigating coal and coal-bearing strata readily embraced the concept of the raised mire and applied it to the ancient coal seams, at times in absence of objective, definitive criteria. Such conclusions were supported by the views of peat researchers, such as Moore (1987): "Such [low ash, wood-derived] material can only have developed in an ecosystem of a tropical 'forest' type in which the abundant supply of water was entirely by precipitation, i.e., ombrogenous." Whereas such tropical bog forests clearly are suitable precursors of low-ash coals, there is troubling contradictory evidence from many coals (Teichmüller, 1989), one of the most obvious problems being the intercalations of flood-derived siliciclastic partings within many coal seams (Teichmüller, 1982). As observed by Moore (1989, p. 90), "There has been an unfortunate lack of information exchange between coal geologists and ecologists concerned with modern peatlands, to the detriment of both areas of study."

This paper concerns the validity of methods employed in assessing ancient peat-forming systems, particularly from the standpoint of raised (ombrotrophic) versus planar (rheotrophic) mires. The fundamental thesis of this paper is that the tools for the analysis of ancient peat deposits (coal) must be selected or developed from those employed in the classification of modern peatlands. The three avenues of approach in the interpretation of ancient peat-forming systems that will be addressed in this paper are lithotype, maceral, and miospore analysis. The case study used herein is the No. 3 seam of the Springhill coalfield, Cumberland Basin, Nova Scotia (Fig. 1), of Westphalian B age.

CLASSIFICATION OF PEAT-FORMING ECOSYSTEMS

In this study, the term mire is employed for any fresh-water wetland system in which peat accumulates (Moore, 1989). The nomenclature and main definitive criteria employed in the classification of the mire types—bog, fen, and swamp (Gore, 1983; Moore, 1987)—are summarized in Table 1. No inference is drawn in this study to plant composition that typifies modern examples of bog, fen, and swamp.

The principal attributes by which mires are characterized are (1) ground-water influence (Kulczynski, 1949; Moore and Bellamy, 1974; Grosse-Brauckmann, 1979; Gore, 1983; Etherington, 1983), (2) nutrient/ionic supply (Kulczynski, 1949; Moore, 1987; Gore, 1983), (3) ash content (Gore, 1983; Moore, 1987), and (4) vegetation type (Gore, 1983; Moore, 1987). The generalized developmental trends of major mire types with respect to these attributes is depicted in Figure 2. Hydrologic character, including ground-water influence and source of ionic input, is both a pragmatic and ecologically meaningful basis on which to classify mires (Moore, 1987). Similarly, the two primary variables in the development of peat-forming plant communities are water and ionic supply (Kulczynski, 1949, and Tallis, 1983), which are intrinsically related. Changes either self-induced (autogenic) or externally induced (allogenic) in these environmental variables result chiefly from modifications to hydrology or surface topography of the mire (Tallis, 1983). A fundamental distinction of mires, therefore, is between *ombrotrophic* mires, which are solely rain fed, and *rheotrophic* mires, which receive recharge from both ground water and rainfall. Rheotrophic mires are usually minerotrophic and eutrophic (nutrient rich), whereas ombrotrophic mires are usually oligotrophic (nutrient poor). Ombrotrophic, rain-fed mires are invariably raised and domed (ombrogenous); the geomorphic terms *domed* and *planar*, commonly used in this volume, are loosely analogous to the hydrologic terms ombrotrophic and rheotrophic, respectively. A term useful in interpreting the temporal evolution of a mire, whether

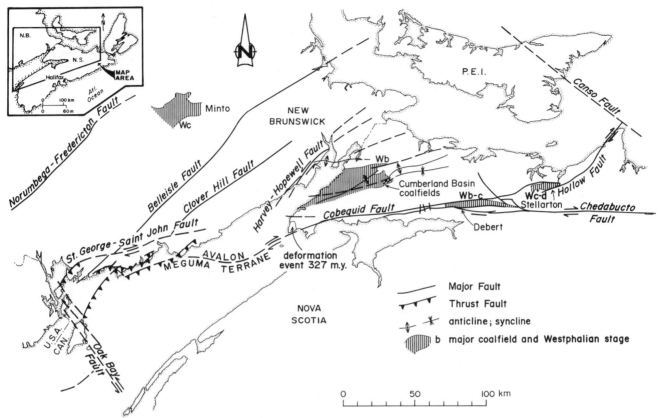

Figure 1. Tectonic setting of Westphalian (Pennsylvanian) coalfields of the southern Maritimes Basin of eastern Canada (after Calder et al., 1991).

modern or ancient, is *mesotrophic,* which refers to a mire tending toward ombrotrophic conditions. A fundamentally important point for geologists concerned with the interpretation of ancient mire development is that many domed ombrotrophic mires *evolve* from a rheotrophic, planar state. The tropical, raised forest mires of Indonesia, for example, originated as ground-water–influenced mangrove soils (Polak, 1975; Anderson, 1964; Cameron et al., 1989).

Rheotrophic and planar mires generally exhibit a higher ash content than do domed, ombrotrophic mires (Moore, 1987, Fig. 6). The reader is referred to Gore (1983) and Moore (1987, 1989) for elaboration of the processes and classification of peat-forming ecosystems.

THE RECOGNITION OF RHEOTROPHIC/PLANAR VERSUS OMBROTROPHIC/DOMED PALEOMIRES

The fundamental templates for the recognition of ancient peatland types are those that characterize modern peat-forming systems (Fig. 2). A mire tending toward a raised (ombrotrophic) status (cf. doming) should be marked by the following upward trends: (1) decreasing ground-water influence; hence (2) decreasing nutrient supply and ionic input, which can be expressed as decreasing fertility (Anderson, 1983); (3) decreasing content of

TABLE 1. WETLAND ECOSYSTEMS AND HYDROLOGICAL TYPES OF MIRES

Wetland Ecosystems			
Peat Forming (Mires)			Non–Peat Forming
Rain Fed (Ombrotrophic)	Flow Fed (Rheotrophic)		
	Dry Season Water Table		
	Below Surface	Above Surface	Near Surface
Bog	Fen*	Swamp†	Marsh§
(bog forest)		(floating swamp, swamp forest)	

Based upon Moore (1987).
Sensu stricto as originally proposed by Tansley (1939). Moore and Bellamy (1974) and Gore (1983) applied fen in the broader sense to all rheotrophic mires.
†"Swamp" in the United States commonly refers to forested wetlands (Penfound, 1952) without regard to accumulation of peat. Those swamp forests which are peat forming are similar to Scandinavian "carr."
§As originally proposed by Tansley (1911). In United States, marsh and swamp are commonly differentiated by presence (swamp) or absence (marsh) of trees (Penfound, 1952).

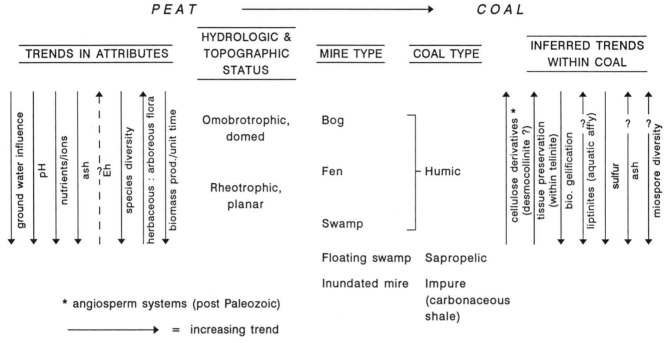

Figure 2. Generalized schematic trends of attributes of mire development for principal rheotrophic to ombrotrophic peat/coal-forming mire types (modified from Calder et al., 1991).

waterborne ash; and (4) a tendency toward more acidic conditions (decreasing pH). Deflation of the mire surface may result in opposite trends. The search for supportive or contrary evidence of these trends within the descendant coal seam is discussed at some length in this paper.

In an earlier paper, Calder et al. (1991) discussed criteria for the recognition of rheotrophic and ombrotrophic conditions in coal seams, especially as they applied to maceral-based methods of analysis. They concluded that the long-standing use of inertinite as a relative measure of Eh level is intrinsically flawed in determination of paleomire types. The major reasons given were (1) Eh level is little used as a definitive criterion of modern mire types; and (2) fusinite, a common inertinite maceral, when of the variety pyrofusinite owes its origin to wildfire, which is an event related to short-term climatic rather than geomorphic conditions (Cope and Chaloner, 1985; Collinson and Scott, 1987, Teichmüller, 1989; Scott, 1989). In addition, such fire-derived fusinite is particularly prone to transport by wind or consequent runoff (Scott, 1989), which further lessens its usefulness as a sensitive paleomire indicator. Moreover, as pointed out by Esterle et al. (1989), recent studies of modern tropical forest mires (Esterle et al., 1989, Grady et al., 1989) have not revealed the upward increasing trend of inertinite precursors cited by numerous workers (Esterle and Ferm, 1986; Littke, 1987; Warwick and Flores, 1987; Eble and Grady, this volume) as being indicative of a seam derived from a paleomire that evolved to a raised (domed) status. Some inert macerals may develop during diagenesis, which

may account for their abundance in coal relative to peat and lignite (Teichmüller, 1989).

One criterion directly transferable from the modern to the ancient is the temporal (upward) decrease in (waterborne) ash content in mesotrophic mires. This trend may be exaggerated by the leaching of inorganic matter (Cohen, 1974, 1989; Cohen et al., 1987 Kosters et al., 1987) as the pH level of the mire decreases with raising of the surface relative to ground-water level. Although the introduction of flood-derived detritus into the mire is the result of an event (flood) as is wildfire, a mire that receives such input is by definition rheotrophic (ground-water influenced). The occurrence of a parting within a seam does not mean that the seam developed under the influence of ground water throughout its entire history, however. Caution must be exercised as well in the interpretation of mires that were sites of volcanic ashfall, resulting in the formation of kaolinitic tonstein.

Nutrient and ionic supply and pH level can be deduced indirectly by considering their effects upon vegetation, maceral, and mineral formation. Decreasing fertility of the tropical bog forests of Indonesia and Malaysia results in the development of areally zoned plant communities (Anderson, 1983) characterized by (1) floral change (only one species of tree spans all six communities, (2) a reduction in species diversity, (3) an increase in stems per unit area, and (4) a decrease in average size of a species (stunting). The first two of the changes described by Anderson can be measured directly in the ancient record through palynomorph analysis. The effects of an increase in the number of stems and stunting of trees (Anderson, 1983) on the development of coal are less straightforward. Stunting and oxidation of vegetation in response to raising of the mire surface has been invoked to explain dulling-upward sequences in coals (Smith, 1962; Esterle and Ferm, 1986; Pierce et al., 1991, among others. In a test of this hypothesis, however, Esterle et al. (1989) found that the modern tropical bog forests of Southeast Asia would give rise to a brightening upward lithotype sequence. An effect that has not been considered is the increase in number of stems or roots, hence outer bark, per unit area that accompanies stunting (Anderson, 1983). There are fundamental botanical problems with the analogy between Carboniferous coals and Indonesian peats, however. Carboniferous peats were composed largely of lycopsid bark; they contained almost no wood. Indonesian peats, on the other hand, are wood rich. Furthermore, lycopsid bark was probably quite unlike that of angiosperms, especially in its chemistry and sheet-like character (DiMichele and Phillips, 1993).

Changes in ionic supply and pH level within the developing mire have a profound effect not only on growing vegetation but also upon macerals and mineral matter. The development of vitrinite macerals in particular is strongly affected by these two variables (Teichmüller, 1982, 1989). A modern tropical domed peat from the Baram River region of Sarawak (Esterle et al., 1989) is characterized by a macroscopic succession from sapric through hemic to fibric types (ASTM, 1987) with a decrease upward and peripherally in maximum size of woody fragments. Microscopically, plant tissues are better preserved upward, whereas the poorly structured "matrix" decreases. Cellulose-derived matter (red textinite, attrinite, densinite) has been found to increase upward within the domed mires of Malaysia and Indonesia (Esterle et al., 1989; Grady et al., 1989), a reflection of floral change upward and inward to herbaceous vegetation (Anderson, 1964; Cameron et al., 1989; Esterle et al., 1989) in response to infertility arising from lessened ground-water influence and nutrient supply.

Biochemical gelification of lignin-derived humins is promoted by the presence of water (particularly when oxygenated) and alkaline ions (especially Ca) and results in progressive loss of cell structure in the vitrinite group (Teichmüller, 1982, 1989). A paleomire tending from planar to domed would therefore be expected to reveal a vertical (temporal) decrease in the effects of biochemical gelification. Such a trend may be largely independent of the kind of tissue entering the system, but the degree of gelification is probably in part tissue dependent. For example, lycopsid periderm would have been more resistant both to gelification and physical degradation than, for example, lycopsid parenchyma or medullosan tissues (DiMichele and Phillips, 1993). Because successional changes in vegetation could contribute variably resistant tissues during the life of the mire, it may prove useful to isolate the effects of gelification on similar tissues throughout the coal bed. An increase in cellulose-derived macerals, presumably attrital desmocollinite, may be expected if a trend

toward stunted, herbaceous vegetation occurred as it does in modern mires (Fig. 2). The spectrum of cellular preservation in etched telinite (telinite, telocollinite, and gelocollinite, unetched) may be especially telling.

CASE STUDY: NO. 3 SEAM, SPRINGHILL COALFIELD, NOVA SCOTIA

The No. 3 seam of the Springhill coalfield, Cumberland Basin, Nova Scotia, provides a case study of the trophic evolution of a Carboniferous mire. In this study, the previously described maceral record (Calder et al., 1991) will be reconsidered in light of lithotype and palynomorph data.

GEOLOGIC SETTING

The Springhill coalfield is situated within the Cumberland Basin of northern Nova Scotia, Canada (Fig. 1), one of the component subbasins of the regional Maritimes Basin. The Maritimes Basin during the late Paleozoic formed part of a predominantly transcurrent zone including the Variscan Belt of Europe and North Africa, which joined the compressional Appalachian and Ural orogens (Arthaud and Matte, 1977). The basin occupied a tropical, nearly equatorial paleolatitude during the middle Westphalian/Pennsylvanian (Scotese et al., 1979). The paleoclimate of the Euramerican coal province during the Westphalian B has been described as being seasonal and relatively drier than the Westphalian A (Phillips et al., 1985), but an ever-wet tropical climate was deduced for the central Appalachian Basin (Cecil et al., 1985). The criteria of Cecil et al. (1985), however, suggest a degree of seasonality for the Cumberland Basin at this time.

During the Westphalian A–B, the Cumberland Basin was a major depocenter of the Maritimes Basin, as witnessed by greater than 4 km of basin-fill deposited during an approximately 4 m.y. (Lippolt and Hess, 1985) period. The coal-bearing strata of the Springhill coalfield (Fig. 3) are assigned to the Springhill Mines Formation (Ryan et al., 1991) of the Cumberland Group (Bell, 1944), part of an overall fining- and ultimately reddening-upward basin-fill sequence of continental origin. The No. 3 seam occurs within a lithofacies assemblage (V, Fig. 3) of nonmarine cyclothems comprising variable but generally equal proportions of gray, sideritic mudrock and pale gray, multistory sandstone, with thick (≤4.3 m) humic coals. Decimeter-thick, lenticular cone-in-cone limestone occurs rarely within the mudrock.

During deposition of this major coal-bearing assemblage, peat-forming ecosystems 4 to 9 km in width flourished along a margin of retreating coalesced alluvial fans derived from the Cobequid highland massif to the south. The northern margin of the peatlands was bordered toward the basin axis by the medial reaches of a river belt 1–2 km wide (Calder, 1986). The interaction of the mires and neighboring geomorphic systems resulted in a distinct areal zonation of the thickest and most areally extensive

J. H. Calder

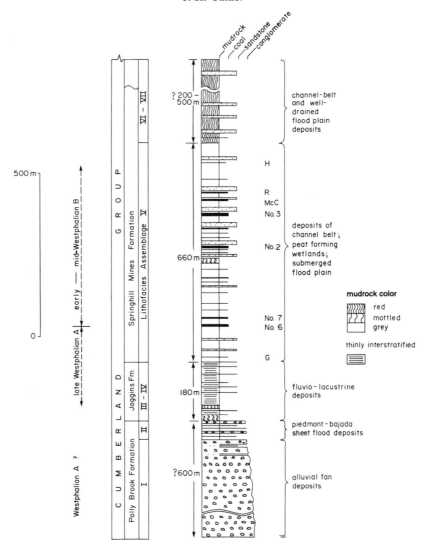

Figure 3. Basin-fill sequence of the Cumberland Group, Springhill coalfield, Cumberland Basin (modified from Calder et al., 1991).

seams, exemplified by the No. 3 seam. Three zones are apparent, defined by the lithology and geometry of the seams and siliciclastic splits (Fig. 4): (1) a *piedmont zone* to the south, bordering the basin margin; (2) an *inner zone*; and (3) a *riverine zone* to the north, bordering the basin axis. Hacquebard et al. (1967) recognized the preferentially mineable inner zone and the rapid lithologic change, which they termed lithification, that typifies the piedmont zone, although these areal terms were not used. Similar areal zonation and basin-fill sequences characterize intermontane coal basins such as those of the Mesozoic (Fuxin Basin) of China (Li et al., 1984), and the Tertiary (Powder River Basin) of the United States (Obernyer, 1978), which are drained by through-flowing basin axis fluvial systems (Flores, 1989).

LITHOTYPE ANALYSIS

More than 30 macroscopic seam profiles were recorded for the No. 3 seam, but only those profiles chosen for comprehensive

(lithotype, maceral, and palynomorph) analysis are considered in detail in this section (whole core from drillholes SH85, piedmont zone; SH81, inner zone; SH72, riverine zone).

Areal variation in lithotype abundance

The macroscopic petrography of the No. 3 seam reveals marked lateral (areal) variation in lithotype composition (Table 2). The middle, or inner, zone of the No. 3 seam is characterized by abundant clarain (42%) and duroclarain (31%) and a paucity of impure coal lithotypes (shaly coal, coaly shale: 7.9%). Within the riverine zone to the north, clarain decreases to 16% whereas durain and impure coal increase to 24% and 37%, respectively. In comparison, the piedmont zone shows the greatest proportion of impure coal (51%). Although there is little difference in duroclarain between the piedmont (34%) and inner (31%) zones, clarain decreases sharply within the former (7%). Vitrain, as macrosopi-

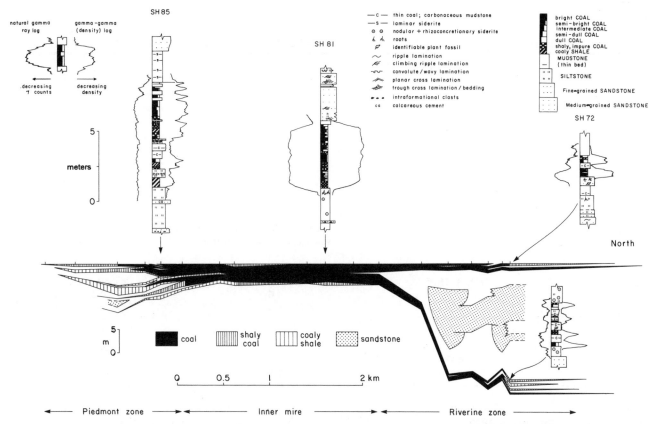

Figure 4. Basin margin to basin axis cross-section of the No. 3 seam, Springhill coalfield, with major depositional zones indicated. Profiles are of drillcore used in lithotype, maceral, and miospore analysis (after Calder et al., 1991).

cally recorded bands, comprises 3% or less of the seam in any zone. Within the three measured sections, fusain is most abundant in the inner zone (5%) but is locally highly abundant in the piedmont zone, comprising more than 13% of the No. 3 seam in drillhole SH27 (Fig. 4). The high percentage of fusain in this intersection can be attributed in part to a solitary 6-cm thick bed composed of erratically oriented cubes of fusain that bear a remarkable resemblance to charcoal.

Vertical sequences (trends) of lithotypes

The macroscopic columns (Fig. 4 and some of the other figures that follow) exhibit an overall upward change in lithotypes. In the piedmont zone (SH85) the basal portion of the seam is composed predominantly of impure lithotypes, whereas the upper part of the seam is dominated by brighter lithotypes (chiefly clarodurain and duroclarain), which increases in abundance upward and thus confer an overall brightening-upward trend. A similar vertical trend is apparent in the inner zone, but with a slight reversal toward duller lithotypes in the upper third of the seam. The inner zone exhibits a greater proportion of brighter lithotypes throughout and virtual absence of siliciclastic partings. The riverine zone to the north shows variable trends with no clear, overall motif to the vertical arrangement of lithotypes.

Preferred sequence of lithotypes: Markov chain analysis

In order to investigate whether lithotypes exhibit a preferred vertical sequence, the macroscopic data from the three zones were subjected to Markov chain analysis (Harbaugh and Bonham-Carter, 1970). The preferred pathways of upward transition from one lithotype to another are presented for each zone in Figure 5. In all three cases, two major populations of lithotypes are apparent based on their preference for preceding lithotypes. These two groups are (1) clarodurain/duroclarain/clarain/ fusain and (2) mudrock/impure coal/durain/vitrain. Upward transitions within the second of these lithotype groups are rather varied from zone to zone (Fig. 5). Hence, this group describes an interrelationship among this lithotype population rather than a vertical succession. The inner zone shows less variation or stronger preference in pathways than either marginal zone and the least number of cross transitions between lithotypes of the two groups.

INTERPRETATION OF LITHOTYPE DISTRIBUTION

Two lithotype populations, rather than one linear sequence (cf. Tasch, 1960), were identified through Markov chain analysis. These reflect two broad environments of peat accumulation: the

J. H. Calder

TABLE 2. SUMMARY MACROSCOPIC CHARACTERISTICS OF AREAL ZONES, NO. 3 SEAM, SPRINGHILL COALFIELD

	Piedmont Zone (South, basin margin)	Inner Mire	Riverine Zone (North, basin axis)
Gross Geometry	Abrupt lateral change dm-scale, commonly ill-sorted mudrock and graded, pebbly sandstone partings becoming numerous over distance of 300 to 500 m, giving rise to marked increase in seam thickness	Relatively free of seam splits	Major splits (as much as 22 m thick) about entombed multistory sandstone bodies, with further splitting of major leaves
Trend of Lithotypes	Upward brightening of lithotypes	Overall upward brightening of lithotypes, with slight reversal in trend at top	Variable trends in lithotype brightness
Lithotype Participation (%)	Drillhole SH85	Drillhole SH81	Drillhole SH72
Vitrain	3.0	1.6	0.9
Vitroclarain*	1.5	4.6	—
Clarain	7.2	42.4	16.3
Duroclarain[†]/clarodurain	33.8	31.3	21.5
Durain	3.2	6.9	24.1
Fusain	0.5[§]	5.3	0.4
Impure coal	50.8	7.9	36.9

*Bright clarain
[†]Dull clarain
[§]Locally as much as 13.5% (drillhole SH27)

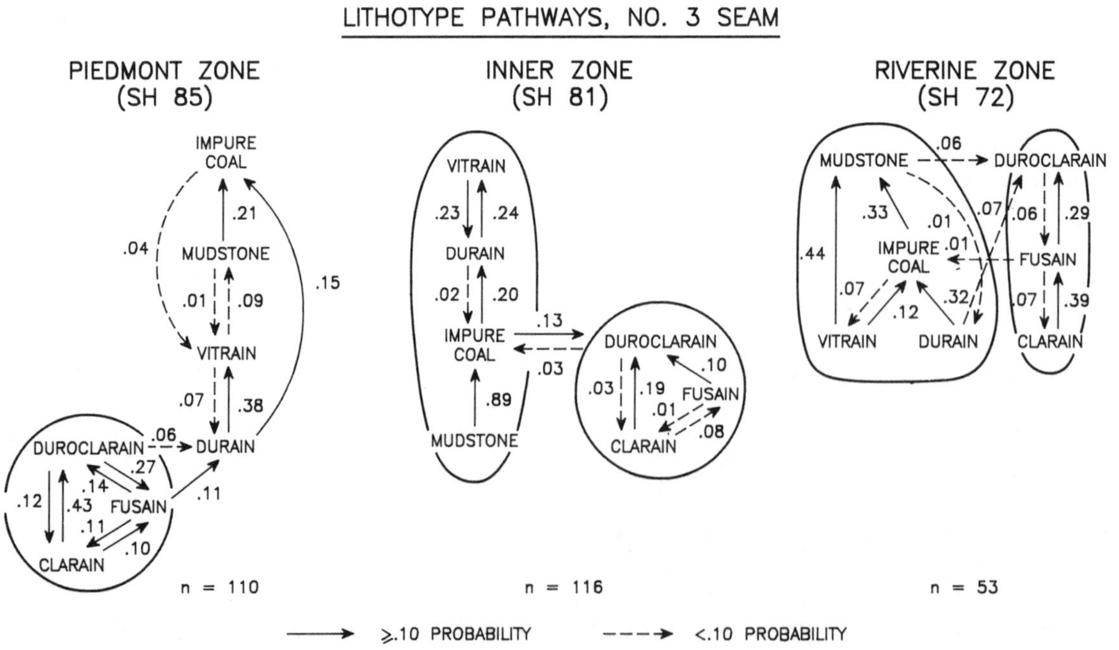

Figure 5. Preferred vertical succession pathways of lithotypes from the three areal zones of the No. 3 seam, as deduced from Markov chain analyses.

"dry" population (fusain, clarain, duroclarain, clarodurain) represents the established mire, whereas the "wet" population (mudrock, impure coal, durain, vitrain) reflects the heavily flooded, inundated mire. Fusain and clarain are closely associated as preceding states in the Markov chain analysis. Fusain, representing fossil charcoal, is believed to be the record of wildfire (Cope and Chaloner, 1985; Scott, 1989), which is essentially an "event." Such fire horizons may be more valid as an indicator of short-term climatic change than of topography within the mire. Vitrain, the brightest lithotype, occurs within the "wet" population of lithotypes in the No. 3 seam. This seeming paradox is attributed to the incorporation in mud of thick, discrete vitrain layers or bands representing bark sheets of arboreous lycopsids, which favored wet habitats.

According to the summary of lithotype abundance by zone (Table 2) and the interpretation of lithotypes (Table 3), both margins developed under generally higher ground-water levels than did the inner zone. The riverine zone to the north witnessed predominantly wetter (limnotelmatic) conditions with the deposition of durain and impure coal. Minor pathways indicated in the Markov chain analyses between "wet" and "dry" communities (Fig. 5) indicate fluctuating ground-water level, but at a century scale given the average lithotype thickness (4.8 cm) and assuming peat accumulation ratios of 2.3 mm yr as in modern tropical forest mires (Anderson and Muller, 1975) and a coal to peat compaction ratio of 7 to 1 (averaged by Ryer and Langer, 1980). Although this zone was proximal to the major river channel belt, which formed under strongly fluctuating discharge, the temporal scale of flood deposits preserved within the fluvial and mire deposits may have differed. Systematic upward changes on a small (laminar) scale were poorly developed, largely because of the interruption of autogenic peat-forming processes by random, allogenic events and perhaps because of the interference of auto-

genic peat evolution by differential compaction around entombed channel-belt sandstone bodies. The stronger pathways of the inner mire suggest that peat accumulation there was steadier, and less interrupted by episodic events. The piedmont zone to the south shows a strongly bimodal composition of duroclarain/clarodurain and impure coal. The general rarity of crossover between the wet and dry lithotype populations of the piedmont zone (Fig. 5) may indicate that these represent temporally distinct habitats.

In order to obtain a true picture of seam mire evolution, it is necessary to inspect the entire seam column for the overall upward sequence of lithotype development. The overall upward-brightening tendency of lithotypes within seam columns of the inner mire and piedmont zones (Figs. 4, 6, and 7) indicate that there was a progressive evolution of the mire throughout much of its history and over much of the peatland area. From the vertical succession of lithotypes evident in the macroscopic seam columns (Figs. 6–8), it can be deduced that the seam originated as a flood-prone mire. As the ecosystem evolved, successive organic accumulates record a relative drop in ground-water level upward and toward the piedmont as the mire became established and accreted vertically and laterally, suggesting that the mire was mesotrophic. A reversal toward duller lithotypes in the uppermost part of the inner mire column (SH81), however, suggests a rise in ground-water level relative to the mire surface. The northern margin of the mire was throughout its existence the site of episodic riverine inundation, which interfered with long-term evolution of the mire.

In summary, four main inferences can be drawn from the lithotype analysis of the No. 3 seam: (1) two broad environments are reflected by the lithotype population—(i) established mire, and (ii) flooded, inundated mire; (2) systematic upward change was interrupted by allogenic random events; (3) steadier, uninter-

TABLE 3. COMPARISON OF PALEOENVIRONMENTAL INTERPRETATIONS OF LITHOTYPES BY TASCH AND CALDER

Tasch		Calder		
Lithotypes	Relative Groundwater Level	Lithotypes	Relative Water Level higher →	Mire Condition
Fusain	Dry ↑	Fusain	— — — - - - - -	Wildfires
Bright coal and bright banded coal (vitrain and vitroclarain) Banded coal (clarain)		Clarain	——————	Established mire
Dull banded coal and dull coal (clarodurain and durain)		Duroclarain Clarodurain	————	Transitional - - - - - - - - - - - - - -
Carbonaceous shale	↓ Wet	Discrete vitrain[*] Durain Impure coal	- - - - - - - - - - - - ————	Flooded, inundated mire

Tasch (1960); Calder (this study, based on Markov chain analysis).
[*]Within mudrock.

rupted peat-forming processes are indicated by the stronger pathways of the inner mire; and (4) the overall trend of lithotypes evident in the macroscopic seam columns indicates a progressive, mesotrophic development of the mire throughout much of its history, particularly in the inner mire and piedmont zone, with a reversal late in its life.

MACERAL-BASED PALEOMIRE ANALYSIS

Factors that must be taken into account in the maceral-based paleoenvironmental interpretation of a coal seam include (1) the botanical precursors of the maceral; (2) transport of macerals within the paleomire; (3) diagenetic history, including physiochemical change; and (4) secondary macerals whose precursors are unknown. The trophic status of the mire profoundly affects not only the contributing flora but the preservation and transformation of the accumulating peat. This is particularly significant within the vitrinite maceral group, which is especially susceptible to diagenetic transformation because of the chemistry of the contributing plant matter.

Maceral abundance profiles: No. 3 seam

The maceral analysis of the No. 3 seam has been described by Calder et al. (1991); lithotype and maceral abundance of the three seam sections are depicted in Figs. 5, 6, and 7. Calder et al. (1991) found, in summary, (1) predominance of the vitrinite maceral group in all three areal zones; (2) predominance of gelified, poorly structured vitrinite (unstructured telocollinite and gelocollinite) over well preserved forms (telinite and weakly structured telocollinite) throughout, except for the upper half of the seam within the inner mire; (3) abundant liptinite (in particular sporinite and locally alginite) at the seam base; (4) distinct vertical zonation of both macerals and ash content within the inner mire; and (5) petrographically distinct benches (cf. "leaves" of Nova Scotian mining terminology) in the riverine zone, including a paucity of structured telinite, telocollinite, and inertinite within the lower bench in comparison with the upper bench.

The record of ground-water influence in peat and coal beds

The effects of two of the four templates for modern peatland classification, namely ground-water influence and nutrient/ionic supply, are reflected in the distribution of the vitrinite maceral types, which record varying degrees of biochemical gelification and, arguably, contribution of lignin-rich versus cellulose-rich tissues (Teichmüller, 1989). Biochemical gelification, influenced strongly by mire water chemistry, and the resistance of tissues (cellulose versus lignin) to both physical and biochemical degradation are two main factors in the degradation of peat.

The degree of biochemical gelification of lignin-derived humins is largely a function of ground-water influence, for it is promoted by the presence of water, O_2, and alkaline ions. Biochemical gelification and loss of tissue structure is enhanced by

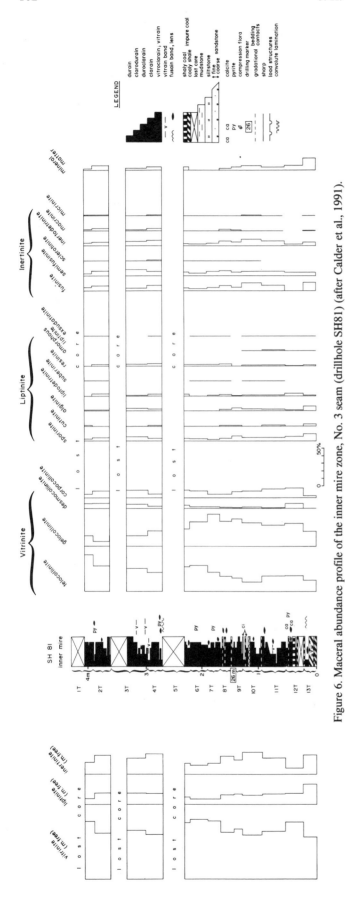

Figure 6. Maceral abundance profile of the inner mire zone, No. 3 seam (drillhole SH81) (after Calder et al., 1991).

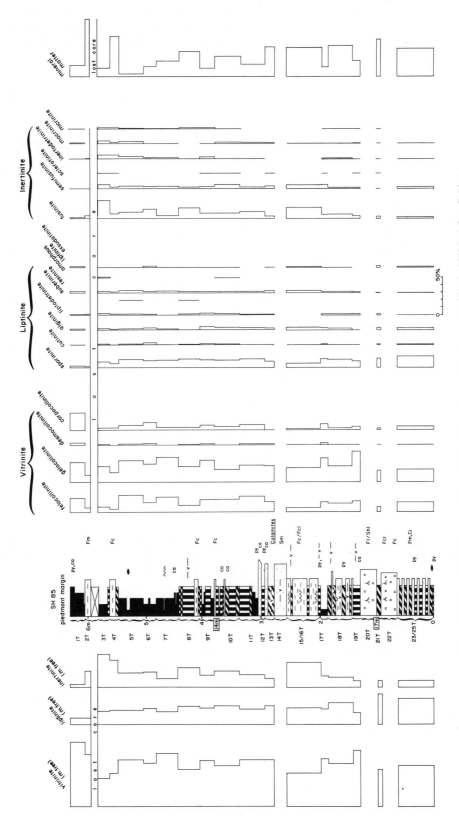

Figure 7. Maceral abundance profiles of the piedmont zone, No. 3 seam (drillhole SH85) (after Calder et al., 1991).

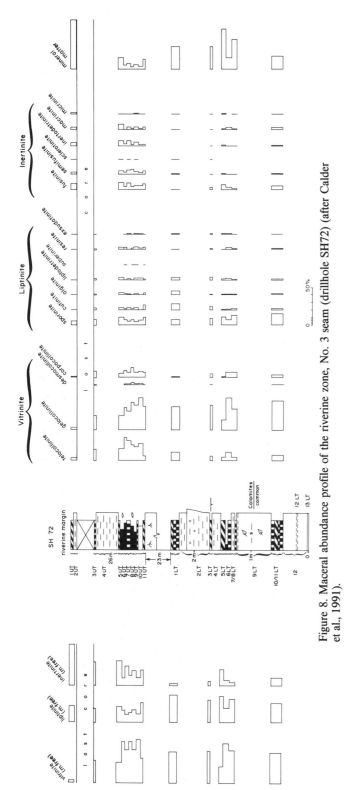

Figure 8. Maceral abundance profile of the riverine zone, No. 3 seam (drillhole SH72) (after Calder et al., 1991).

bacterial activity, fostered by the increase in pH level of mire waters arising from the incorporation of alkaline ions (Teichmüller, 1982). Strongly gelified, unstructured vitrinite (especially gelocollinite unetched; poorly preserved klinite etched) derived from lignin would therefore be expected to decrease upward in a rising mire as ground-water influence diminishes. Calder et al. (1991) therefore stressed the need in studies of mire genesis to subdivide the vitrinite macerals and, in particular, to differentiate maceral types that reflect the gradual loss of tissue structure that accompanies progressive gelification (Koch, 1969; Timofeev and Bogoliubova, 1964).

Cellulose-rich tissues that are derived at least in part from herbaceous vegetation, are more prone to physical degradation (fragmentation) than are woody, lignin-rich tissues. The attritus forms humodetrinite and desmocollinite in coal (Styan and Bustin, 1983; Teichmüller, 1989). The occurrence of cellulose-rich, herbaceous vegetation in modern tropical mires is linked to decreasing fertility and nutrient supply that accompanies rising of the mire surface and diminishing ground-water influence (Anderson, 1983). Cellulose-derived macerals, precursors of desmocollinite (Teichmüller, 1989), have been found to increase upward within domed tropical peats of Indonesia and Malaysia (Esterle et al., 1989; Grady et al., 1989).

The physical and chemical pathways of degradation may conflict: assuming a similar tendency to herbaceous vegetation within ancient ecosystems, desmocollinite would increase upward within a rising mire, whereas strongly gelified vitrinite (e.g., gelocollinite) theoretically would be expected to decrease upward. It is fundamental, therefore, to specify which aspect of degradation is being considered in order to more clearly comprehend the genesis of ancient mires. For this reason, it may be advisable to consider desmocollinite separately from other maceral types in a measure of ground-water influence. Furthermore, modern herbaceous mires may not serve as close analogues to Carboniferous ecosystems. Truly herbaceous plants were rare in Carboniferous ecosystems (W. A. DiMichele, 1990, written communication). It is necessary to resolve the paleoecology of plants less resistant to degradation than, for example, the lignin-rich arboreous lycopsids and which may or may not have occupied an ecological niche similar to the herbaceous flora of modern tropical mires.

Petrographic analysis of etched blocks from the Tertiary Anderson and Smith coals, Powder River Basin (Moore et al., 1990), confirms these two degradation pathways in angiosperm-dominated peats. Moore et al. (1990) described two genetic maceral groups derived by degradation: (a) structureless degradation products of woody material (their Genetic Group IIA), and (b) particulate huminitic components (their Genetic Group IIB). As hypothesized here, these commonly show opposing trends (Moore et al., 1990, their Figs. 8 and 9).

The validity of gelified tissues, which may possibly form at various stages of peat diagenesis (Teichmüller, 1982), as an indicator of ground-water influence is contingent upon its early formation. Evidence of very early gelification within the No. 3 seam is illustrated, however, in Figure 9, a and b. Restricted oxidation halos at contacts with gelocollinite indicate that this matter was already in a compact gelified state at the time that the adjacent structured tissues (semifusinite) were partially oxidized, perhaps through the process of smouldering ground fire (note presence of pyrofusinite, Fig. 9a). The *abundance* of gelified matter is not an absolute indicator of rheotrophic conditions. Direct comparison between seams must take into account regional differences in ground-water chemistry and mire flora. The latter becomes especially important if the seams differ sufficiently in age for the evolution of plants to have become a factor.

Ash content, a third template of modern peatland classification, is another measure of ground-water influence, assuming that the mineral matter is largely waterborne. Mineral matter can form also through diagenetic and biogenic processes (Cohen, 1990) and can be introduced by means such as volcanic ashfall (e.g., Cecil et al., 1979; Dewison, 1989; Ruppert et al., 1989). Such modes of origin do not seem to be widely considered in modern peat classifications. It should be possible, however, to assess more accurately the degree of ground-water influence by the differentiation of waterborne ash from that of other origins.

Macerals as indicators of vegetation

Diessel (1986) used the ratio of telinite + telocollinite + semifusinite + fusinite to desmocollinite + macrinite + inertodetrinite as a relative measure of woody versus herbaceous vegetation within Permian peat-forming systems of Australia. Named the tissue preservation index (TPI), it addresses change in tissue preservation only indirectly as a function of the type of plant tissue entering the system. The TPI of Diessel (with the removal of macrinite) in contrasting lignin-rich, arboreous and cellulose-rich, presumably herbaceous, tissues may reflect a mire's evolution toward a raised/domed condition (Fig. 2), although such a use was not suggested. A similar ratio was expanded by Calder et al. (1991) in an attempt to contrast a maceral assemblage of lignin-rich forest affinity (telinite + telocollinite + fusinite + semifusinite + suberinite + resinite) with assemblages of presumably herbaceous, cellulose-rich (desmocollinite + inertodetrinite) and marginal aquatic (alginite + sporinite + cutinite + liptodetrinite) affinity, termed the "Vegetation Index" (V.I.). (Note the first report of suberinite from a Paleozoic coal by Calder et al. (1991).

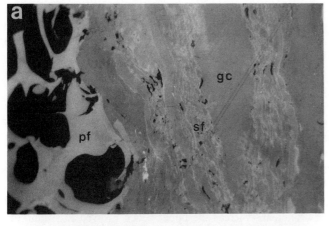
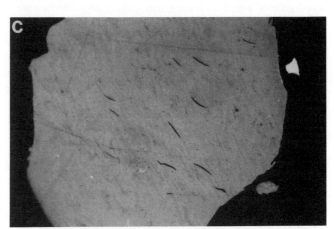
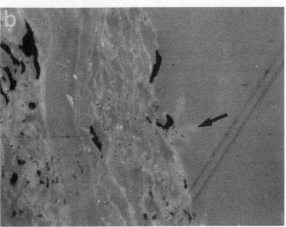
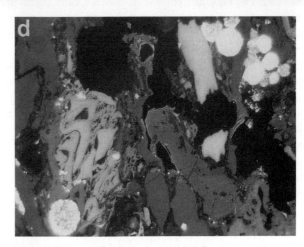

Figure 9. Photomicrographs of the No. 3 seam in reflected light (×500 magnification): (a) pyrofusinite (pf), semifusinite (sf) and gelified collinite (gc), inner mire zone; (b) enlargement of oxidation halos (arrow) at semifusinite/gelocollinite boundary; (c) structured vitrinite (telinite) in the form pseudovitrinite, from inner mire zone; (d) gelocollinite, semifusinite liptinite, mineral matter and framboidal pyrite; basal inner mire (SH81-13T).

Gelocollinite and macrinite were excluded because of less certain botanical affinity (but see Fig. 10d). The accurate reconstruction of vegetational cover, however, exceeds the capability of conventional maceral analysis.

Mire paleoenvironment diagram

The mire paleoenvironment diagram of Calder (1991) is a maceral-based method by which an ancient peat-forming ecosystem can be modeled using parameters similar to those employed in classification of modern mires, as described earlier. The method further develops the concept introduced by Diessel (1982, 1986), but it differs in theory, primarily in the definition of a Groundwater Influence Index (G.W.I.) that abandons the long-standing use of relative Eh levels as deduced through ratios of fusinitized versus vitrinitized tissues. Calder et al. (1991) originally defined the G.W.I. as the ratio of gelocollinite + corpocollinite + desmocollinite (biochemically and physically degraded tissues) + mineral matter to telinite + telocollinite (structured and partially structured tissues), although they discussed both the exclusion of corpocollinite because of uncertain origins and desmocollinite because of possible cellulose affinity. The potential conflict of trends between desmocollinite and lignin-derived gelocollinite arising from potentially different pathways of degradation in peat leads to the conclusion that an index of groundwater influence that includes all vitrinite maceral types is better served by the ratio of gelocollinite + corpocollinite + mineral matter to telinite + telocollinite.

The mire types identified on Figure 10a are the hydrological types: swamp, fen, and bog, with the vegetational subtypes, swamp forest and bog forest. The limnic environment, site of sapropelic accumulation, is also represented. Wetland environments prone to siliciclastic inundation result in the formation of impure coal lithotypes (shaly coal, coaly shale) and carbonaceous mudrock. The high ash content of such deposits places them outside the definition of peat, hence their environments are not truly mire types. The terms "inundated marsh" and "inundated forest" were thus employed by Calder et al. (1991) for coals exhibiting a G.W.I. in excess of 5. The mire types from top to bottom in the mire paleoenvironment diagram (Fig. 10a) represent an evolution from rheotrophic to ombrotrophic conditions and planar to domed morphology accompanied by a change from eutrophic to oligotrophic nutrient status (see also Fig. 2), but index values are somewhat arbitrary, indicating *trends* rather than absolute limits. The G.W.I. value below which ombrotrophic conditions prevailed will be more accurately constrained as other seams are evaluated.

The V.I., described previously, was necessarily used in an attempt to provide the required template of vegetation type in order to achieve a wholly maceral-based method of ancient mire habitat that honored all major templates of modern mire classification. As discussed previously, however, at our present level of understanding of botanical and ecological affinities of macerals, conventional maceral analysis is a poor avenue for reconstruction

of vegetation type, although phyteral analysis (Winston, 1986) holds promise.

PALEOMIRE RECONSTRUCTION OF THE NO. 3 SEAM

According to the maceral-based indices of ground-water influence and vegetation, distinct mire types developed within the three areal zones of the peat-forming system (Fig. 10b), the sole type common to all three zones being the swamp environment. The inner mire of the ancestral No. 3 seam records a systematic evolutionary sequence of mire types throughout its development (Fig. 10c and d), inferred to represent a hydroseral succession (Tallis, 1983). The lower third of the seam developed under highly rheotrophic conditions and inundated marsh/limnic to swamp environments. The mire subsequently felt the effects within the inner mire of the proximity of a basin axis–channel belt which occupied the northern, riverine zone of the mire. The G.W.I. indicates that the surface of the inner mire continued to rise relative to ground-water level with the development of fen, and possibly bog, mire types, following abandonment of the riverine zone by this fluvial system in favor of a more northerly axial position (Fig. 4). Increases in ash, sulfur, and miospore diversity and a trend to duller lithotypes within the upper third of the seam, however, argue against continued doming and may reflect deflation of the mire surface. (See section on processes of mire development.)

Both riverine and piedmont margins of the mire developed under rheotrophic conditions. The piedmont zone records a weakly developed hydroseral succession of mire types but did not evolve beyond the swamp stage; inundated forest characterized this margin during much of its history. The two benches of the No. 3 seam in the riverine zone developed from different mire types: the lower bench accumulated within an inundated marsh environment, whereas the upper bench evolved to swamp and fen. The different petrographic composition and paleomire types of the two benches can be attributed partly to the mesotrophic tendency of the system and the diastem of peat development during fluvial occupation. The contrast is likely caused in part by relief inherited from the different compaction of sediment around the entombed multistory sandstone body. The ancestral mire of the No. 3 seam therefore was predominantly rheotrophic, and

---

Figure 10. (a) Mire paleoenvironment diagram (modified from Calder et al., 1991). Note lower value of Ground-water Influence Index (G.W.I.) assigned to bog than in original version of Calder et al. (1993). (b) Paleomire types of the inner mire, riverine, and piedmont zones, No. 3 seam. (c) Paleomire evolution of the No. 3 seam, inner mire zone, excluding gelocollinite from the Vegetation Index (V.I.). Dashed line indicates inferred actual pathway in section of lost core which comprises a thin mudrock parting. (d) Paleomire evolution of the inner mire including gelocollinite in the V.I. (assumes derivation from lignin, hence arboreous vegetation).

Mire Paleoenvironment Diagram

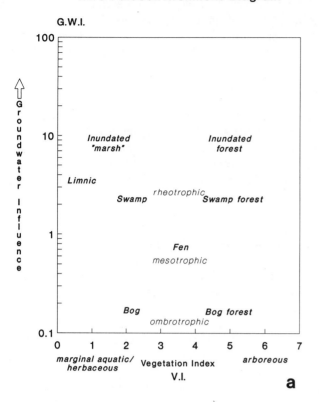

a

Mire Paleoenvironment Diagram

No. 3 Seam

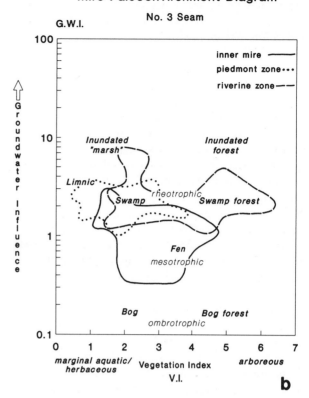

b

Mire Paleoenvironment Diagram
Inner Mire

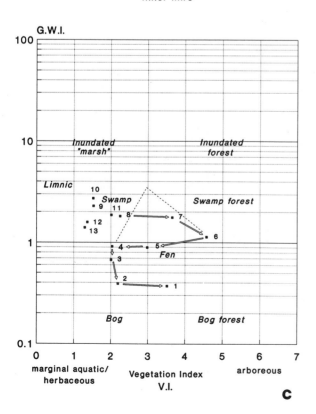

c

Mire Paleoenvironment Diagram
Inner Mire

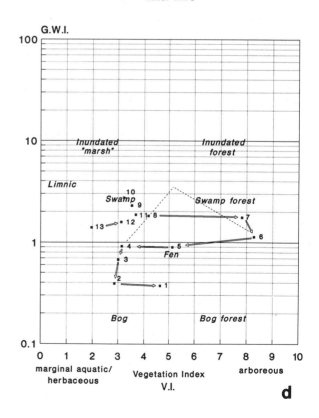

d

exhibited a mesotrophic tendency that ultimately reverted to more ground-water influenced conditions.

Ash and Sulfur distribution

Coal seams of the Springhill coalfield exhibit a typical areal distribution of ash and sulfur content, with the margins enriched in both components relative to the inner areas. The No. 3 seam is typical of the Springhill coals in that the inner or central region has a total sulfur content (by weight) of 1.6%, which increases to greater than 3% (total seam) at seam margins (Table 4). The participation of sulfate, pyritic, and organic sulfur differs between the inner mire and relatively sulfur-enriched margins. Within the inner mire, organic sulfur is the greatest contributor and sulfates contribute the least, whereas in both marginal zones pyritic sulfur is the most abundant form. The ash content of the seam increases sharply and steadily away from the inner zone (10.1% by weight, SH81), especially toward the piedmont margin. Within the inner zone, the content of ash and sulfur is greatest at the seam base and lower near mid-seam (Fig. 11).

Upwardly decreasing pH level within the inner mire, inferred from the increasing ratio of well-preserved tissues (telinite, telocollinite) to strongly gelified tissues (structureless collinite, gelocollinite), accompanied the rising of the mire surface and diminishing influence of ground water. At the piedmont and riverine margins and during the early formation of the inner mire, the supply of alkaline ions contributed from basin margin fault-fed springs (Calder, 1991) or eroding Early Carboniferous limestones (Bell, 1927; Gibling et al., 1989) should have permitted sulfate-reducing bacteria to thrive under conditions of elevated pH (Casagrande, 1987; Cecil et al., 1979; Altschuler et al., 1983). The enrichment of sulfur (particularly pyrite) in areas of rheotrophic influence would be further enhanced by the flocculation of ferrous-iron–supplying clays carried into the mire during flooding. Rapid flocculation occurs when clays are introduced to humic acids (Staub and Cohen, 1979; Cecil et al., 1979); as the

mire evolved, the inner reaches became more acidic and less hospitable to bacteria, while becoming evermore insulated from the introduction of ferrous iron because of increasing concentration of humic acids.

The resulting distribution of ash and sulfur within the brightening upward, lower two-thirds of the inner mire of the No. 3 seam is similar to that of the forest-swamp peat of the Baram River area of tropical Sarawak (Fig. 11) described by Esterle et al. (1989) and Cameron et al. (1989). The latter authors described a minerotrophic (rheotrophic) to ombrotrophic evolution for these tropical mires. The higher ash content and less pronounced upward decline in sulfur content of the No. 3 seam throughout, however, suggests that the paleomire was more strongly influenced by ground water than are the tropical Indonesia mires.

PALYNOMORPH ANALYSIS

In this section, palynomorphs within the No. 3 seam will be investigated to determine whether the miospore record reflects the interpreted temporal and zonal development of the paleomire. Two of the attributes described by Anderson (1983), i.e., floral change and a reduction in species richness temporally (vertically) and areally (toward the inner mire) can be investigated through miospore analyses, thus providing a basis to assess the degree of autogenic, catenary development of the No. 3 paleomire. Further insight into conditions within the mire can be provided by considering the paleoecology of the parent flora of the miospores.

The floral type and parent plant for miospores identified in this study (Table 5) are given in Smith (1962), Smith and Butterworth (1967), Phillips et al (1985), Bartram (1987), Mahaffy (1988), and Willard (1989a, 1989b). The miospore taxa are grouped by floral type: arboreous lycopsids; herbaceous lycopsids; sphenopsids; seed ferns (Pteridosperms), ferns—tree ferns (Marattiales), and small ferns (Filicinae); and cordaitean gymnosperms.

TABLE 4. SULFUR FORMS* IN PIEDMONT, INNER, AND RIVERINE ZONES OF THE NO. 3 SEAM, EXCLUDING INORGANIC PARTINGS

Zone	Sample	Sulfate Sulfur (%)	Pyritic Sulfur (%)	Organic Sulfur (%)	Total Sulfur (%)
Piedmont	SH85-1T to 13T (excluding 2T, 4T)	0.35	2.58	0.42	3.35
	SH85-17T to 19T	0.32	2.37	0.59	3.28
	SH85-23T to 25T	0.38	1.75	0.33	2.46
Inner	SH81-1T to 13T	0.08	0.59	0.95	1.62
Riverine	SH72-1UT to 4UT	0.36	1.48	0.33	2.17
	SH72-5UT to 11UT	0.32	2.39	0.90	3.61
	SH72-1LT to 13LT (excluding 2, 4, 9, 12LT)	0.49	3.82	0.17	4.48

*ASTM procedure D-2392.

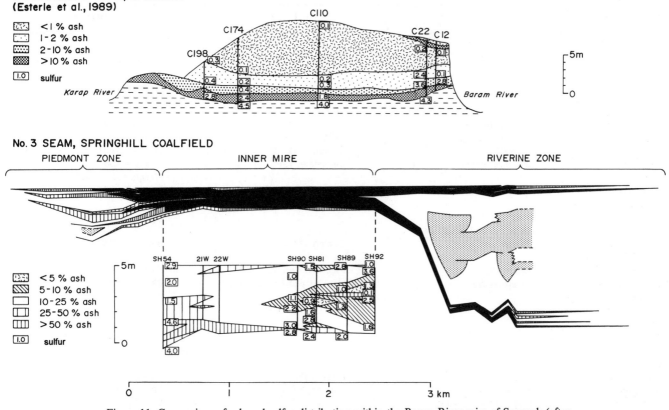

Figure 11. Comparison of ash and sulfur distribution within the Baram River mire of Sarawak (after Esterle et al., 1989) and the inner mire, No. 3 seam.

Methods

All samples submitted for maceral analysis were halved and tendered to G. Dolby and Associates, Calgary, for detailed palynological determinations (Dolby, 1988). Samples were first treated with hydrofluoric acid to remove silicates and subsequently with Schultze solution to extricate spores from the coal. Slides were made of the sieved (+10 and +30 micron) fractions from the concentrate. The +10 micron slides were used for the analyses; the +30 micron slides were subsequently checked for larger forms. For each sample, an initial 200 counts were made of the total palynomorphs to obtain population percentage. A secondary count of palynomorphs was then made, discounting the genus *Lycospora,* so as to obtain a record of as many species as possible. Because of low spore abundance in many samples, 200 counts were not always possible, necessitating a conversion of the data from counts to percentage.

Results of palynological analysis

The results of palynomorph analysis (Figs. 12–14) can be summarized as follows:

(1) Miospores of the genus *Lycospora* are predominant in all but a few samples in the three areal zones of the No. 3 seam (Figs. 12, 13, and 14); the two most abundant miospore species within the No. 3 seam are *Lycospora pusilla* and *L. orbicula.*

(2) The genera *Calamospora* and subdominant *Florinites* are codominant with the *Lycospora* in certain samples associated with fluvial deposits of the riverine to inner zones of the mire (Figs. 12 and 14).

(3) There is a striking paucity of miospores of both herbaceous lycopsid and fern affinity (Table 5).

(4) Species diversity is greater at the piedmont margin (61 species) than in the riverine zone or inner mire sections (44 species each). An overall upward trend of decreasing diversity from 27 through 9 species, followed by a partial recovery to 17 species, is evident within the inner mire (Fig. 12).

(5) Species that occur only in one or two mire zones are low in abundance and generally associated with siliciclastic partings.

(6) The miospore analyses record a marked vertical change in relative proportions of *Lycospora* within the inner mire section ((Fig. 12). Two and possibly three palynomorph assemblages are evident: (a) an assemblage codominated by *Lycospora pusilla* and *L. orbicula* but with *L. orbicula* becoming dominant through time in a large-scale cycle through the lower two-thirds of the seam (SH81-13 to 5T); (b) a second assemblage marked

TABLE 5. PALEOBOTANICAL AFFINITIES OF MIOSPORE GENERA REPORTED FROM THE NO. 3 SEAM

Miospore Genus	Paleobotanical Affinity	Authority	Reference
	Arboreous Lycopsids		
Lycospora	*Lepidodendron*	Chaloner, 1953	Smith, 1962
	Lepidophloios	Andrews and Pannel, 1942	Smith, 1962
	Paralycopodites (Anabathra)	DeMichele, 1980	Bartram, 1987
L. pusilla	*Lepidodendron hickii*	Willard, 1989	Willard, 1989
L. orbicula	*Paralycopodites (Anabathra)*	Willard, pers. commun., 1989	Willard, pers. commun., 1989
L. pellucida	*Lepidophloios harcourtii*	Willard, 1989	Willard, 1989
L. granulata	*Lepidophloios hallii*	Willard, 1989	Willard, 1989
Crassispora	*Sigillaria*	Chaloner, 1953	Bartram, 1987
Granisporites, Apiculatisporis	*Diaphorodendron*	DiMichele, 1985	Bartram, 1987 Phillips et al., 1985
Densosporites cf. *sphaerotriangularis*	*Sporangiostrobus*	Leisman, 1970	Bartram, 1987
	Herbaceous Lycopsids		
Densosporites cf. *loricatus*	*Porostrobus*	Chaloner, 1958	Bartram, 1987
Endosporites	*Chaloneria*	Pigg and Rothwell, 1983	Bartram, 1987
Cirratriradites cf. *saturnii*	*Sellaginellites*	Hoskins and Abbott, 1956	Bartram, 1987
Cingulizonates			Smith and Butterworth, 1967
Cristatisporites			Phillips and others, 1985
Radiizonates			Phillips and others, 1985
	Ferns		
Granulatisporites	?Pteridosperm	Knox, 1938	Smith, 1962
Planisporites	?Pteridosprm	Kidston, 1923-25	Smith, 1962
Laevigatosporites (species <40 microns)	Tree fern (Marattiales)		Mahaffy, 1988
Punctatosporites	Tree fern (Marattiales)	Mamay, 1950; Remy and Remy, 1957	Smith, 1962
Cyclogranisporites	Var. ferns	Remy and Remy, 1955	Smith, 1962
Leiotriletes	Fern (Filicales)	Knox, 1938; Mamy, 1950; Remy and Remy, 1960	Smith, 1962
Lophotriletes	Fern	Remy and Remy, 1957	Smith, 1962
Raistrickia	Fern (Filicales)	Radforth, 1938; Mamay, 1950; Remy and Remy, 1955	Smith, 1962
Savritrisporites	*Senftenbergia* (Filicales)		Smith and Butterworth, 1967
Reticulatisporites	Fern (Filicales, Coenopteridales)		Phillips et al., 1979
	Sphenopsids		
Calamospora	Calamites	Hartung, 1933	Smith, 1962
	Sphenophyll	Remy and Remy, 1955	Smith, 1962
Laevigatosporites (spp. >40 microns)	Calamites	Reed, 1938; Andrews and Mamay, 1951; Remy, 1960	Smith, 1962
Vestispora			Mahaffy, 1988
	Seed Plants (Gymnosperms)		
Florinites	Cordaites	Florin, 1936, 1938-40	Smith, 1962
Schopfipollenites	*Medullosa*		Phillips et al., 1985
Indet. pollen			

*For bibliographic information, see citing reference.

by a sharp decline in *L. orbicula* and concomitant increase in *L. pusilla* (SH81-4 to 3T); and (c) a tentative assemblage marked by the presence of *L. pellucida* but dominated by *L. pusilla* (SH81-2 to 1T).

(7) In marginal zones of the seam (Figs. 13 and 14) further palynomorph assemblages are evident: (d) an assemblage where-in *L. pellucida* was a near equal subdominant with *L. orbicula*; and (e) an assemblage marked by the presence of *L. granulata*, the least abundant species of *Lycospora*, occurring primarily in the basal section of the seam at the riverine margin.

(8) Fungal spores occur in abundance within the lower leaf (bench) of the seam in the riverine zone.

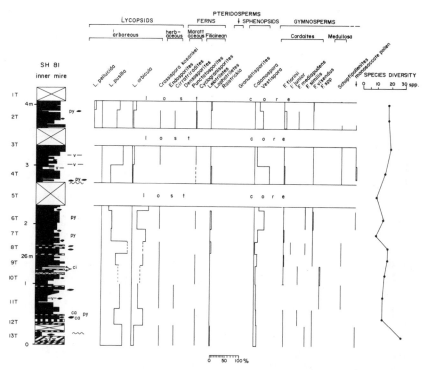

Figure 12. Palynology profile, inner mire, No. 3 seam (SH81).

INTERPRETATION

The use of miospores in paleoecological study is limited by constraints similar to those that limit the use of macerals. The main constraints are (1) macrofloral (i.e., parent plant) affinity; (2) effects of transport; and (3) variable spore production leading to overrepresentation by different species. In the evaluation of mire ecology of the No. 3 seam, an additional factor is the corroded state of some miospores particularly within the piedmont zone (Dolby, 1988), inferred to be due to bacterial activity, high pH and rheotrophic hydrology. Not all spores are equally likely to survive oxidation; therefore, there is a potentially strong taphonomic bias.

Miospore diversity

An upward (temporal) and inward (areal) decrease in species diversity was cited by Anderson (1983) as one of the diagnostic features of the floral zonations (catena) of tropical Malaysian and Indonesian mires. Within the No. 3 seam, miospore species diversity is generally greater at seam margins and within the basal inner mire zone (Figs. 13–15). The increase in species diversity in these areas can be linked to rheotrophic hydrology, because groundwater influenced areas of the mire, which experienced water input from neighboring deposystems should contain a higher proportion of externally derived spores and pollen (Fulton, 1987). These externally derived palynomorphs will occur in greatest abundance within siliciclastic partings

(Marshall and Smith, 1965) introduced to the mire periphery during floods if the partings were impervious enough to protect the spores from decay.

The relationship between increased ground-water flow or rheotrophic conditions and increased miospore diversity is not simply a function of enhanced transport. Increased environmental stress such as waterlogged conditions, relatively low nutrient supply and decreased pH would have had the effect of lowering species diversity within the inner mire. In areas at the mire periphery and beyond, where nutrient supply was greater, species diversity would have been enhanced.

Within the inner mire, species diversity is greatest at the seam base ($n = 27$), decreasing steadily upward ($n = 9$) to mid-seam, but then experiencing a partial recovery (to 17) at the top. The lower trend is the one to be expected from a tropical forested mire such as those of tropical Indonesia (Anderson, 1983). The reversal in the upper half is problematic, since maceral data suggest a continued rising of the mire surface. This, and other trends that appear to contradict evidence of doming, are discussed further in the sections on processes of mire development. The significance of these trends in species diversity must be weighed against the absence of replicate sample analysis and the very low abundance of many species.

General plant community composition

Macrofloral affinities of miospores (Table 5) can be used to provide a reconstruction of the peat-forming plant communities (Scott, 1978), although there are inherent biases due to modes of

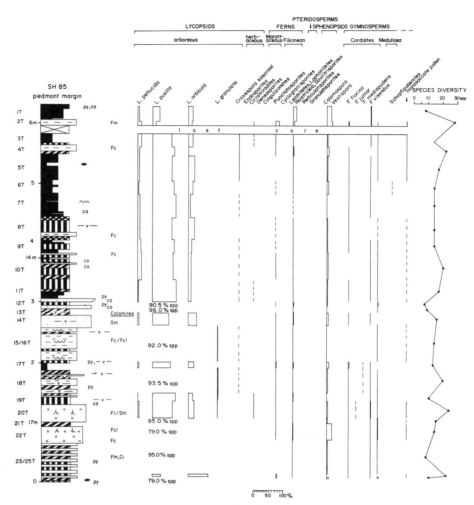

Figure 13. Palynology profile, piedmont zone, No. 3 seam (SH85).

dispersal and preservation (Phillips et al., 1985). The miospore genus *Lycospora* is dominant in all but a few samples from the three areal zones of the No. 3 seam. Even though *Lycospora*-producing lycopsids are one of the most overrepresented in the miospore record (Phillips et al., 1985), the overwhelming abundance of *Lycospora* (as great as 98.5% in an individual sample from the No. 3 seam) nonetheless suggests that the arboreous lycopsids were the dominant plants throughout the mire. *Sigillaria,* an arboreous lycopsid represented by the miospore *Crassispora kosankei,* was a relatively minor tree not only in the ancestral No. 3 seam but apparently in general throughout the Westphalian (Early and Middle Pennsylvanian) mires of Euramerica (Phillips et al., 1985), although exceptions clearly exist (e.g., Scott, 1978, his Fig. 16). In contrast to the abundance of the arboreous types, miospores attributed to herbaceous lycopsids are all but absent (Figs. 12–14). Bartram (1987) has demonstrated, however, that phases of herbaceous lycopsids recorded by megaspores can go undetected in miospore studies due to the exclusion of megaspores from finer sieve fractions; large

Medullosan spores may similarly elude detection (DiMichele and Demaris, 1987).

Two floral groups were subdominant and rarely dominant within the mire: sphenopsids (calamites) and the cordaitean gymnosperms. In contrast to other mires, true ferns, including both tree (Marattaceous) and herbaceous (Filicinean) types, and the seed ferns (pteridosperms) were minor constituents. *Medullosa,* represented by the prepollen *Schopfipollenites,* was apparently rare, although it is probably underrepresented in the miospore record (Phillips et al., 1985). The abundance of calamites and cordaites in relation to the rarely occurring ferns (*sensu lato*) and *Medullosa* may have been partly a function of the adaptive evolution of these plants within mires, as noted by Phillips et al. (1985, p. 95): "In general, cordaites and, to a lesser extent, calamites filled significant habitat gaps in the early Middle Pennsylvanian [circa Westphalian B] until *Psaronius* and *Medullosa* expanded as subdominant plants."

The two most abundant trees and/or prolific producers of miospores within the No. 3 mire were *Lepidodendron hickii (Ly-*

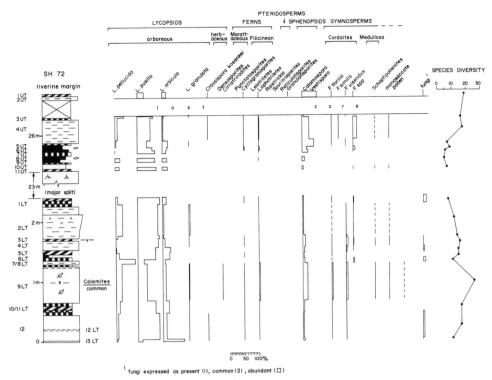

Figure 14. Palynology profile, riverine zone, No. 3 seam (SH72).

cospora pusilla) and *Paralycopodites,* synonomized by Pearson (1986) with *Anabathra* (*L. orbicula*). In the marginal regions of the mire, *Lepidophloios harcourtii* (*L. pellucida*) was a nearly equal subdominant with *Anabathra* and in the lower bench of the seam at the riverine margin was somewhat more abundant than *Anabathra,* assuming equal production and preservation of spores. *Lepidophloios hallii* (*L. granulata*) was the least abundant of the lycopsid trees, occurring primarily in the early stages of mire development in marginal regions. The general composition of the mire forest, until recently, would have been considered atypical of mire floras of the Late Carboniferous (Phillips et al., 1985; DiMichele et al., 1985). Of particular note is the abundance of *Lepidodendron hickii* since *Lepidodendron* is generally not considered to be "centered in coal swamps" (DiMichele et al., 1985). It is becoming apparent, however, that *Anabathra*-dominated floras are less rare than previously supposed, occurring within the Secor coal of Oklahoma, the Hamlin coal of eastern Kentucky, the Katharina seam of Germany (W. A. DiMichele, 1990, written communication) and within zones of the Herrin coal bed (DiMichele and Phillips, 1988) and Low Barnsley seam of Yorkshire (Bartram, 1987).

The paucity of *Lepidophloios hallii,* which was the major tree of some Westphalian D mires (Phillips et al., 1985), may be a function of evolution rather than environment because it did not reach its acme until the Westphalian D when it replaced *L. harcourtii* (DiMichele et al., 1985, their Fig. 8.2), however the general paucity of *L. harcourtii* cannot be similarly attributed to

evolution. *Lepidophloios* is considered to have been the arboreous lycopsid most adapted to wet, aquatic conditions, and its rarity in the No. 3 seam may reflect a less humid or more seasonal climate. Ultimate extinction of *Lepidophloios* has been attributed to a change to a relatively drier, perhaps seasonal, climate (DiMichele et al., 1985; Phillips et al., 1985).

Paleoecology of arboreous lycopsids: Reflection of rheotrophic conditions

The relative abundance of the arboreous lycopsids within the No. 3 paleomire is dissimilar to the general floral composition of the majority of documented Late Carboniferous mires, many of which, however, are of Westphalian C–D age. This implies that the No. 3 seam developed under paleogeographic, hydrological and chemical conditions that were dissimilar to those of the majority of mires which have been studied in the light of paleoecology, although there may be an evolutionary bias as well. Nonetheless, published interpretations of the paleoecology of flora that dominated the ancient No. 3 mires are in accord with their occurrence in a ground-water influenced piedmont-riverine setting.

Lepidodendron hickii, the dominant tree of the ecosystem, is considered to have favored clastic wetlands and "rare nutrient-rich areas, possibly with great fresh-water influx" (DiMichele et al., 1985, p. 236) such as the southern, piedmont margin of the Cumberland Basin margin during the Westphalian B. DiMichele and Phillips (1985, p. 236) attributed its occurrence in Lower and

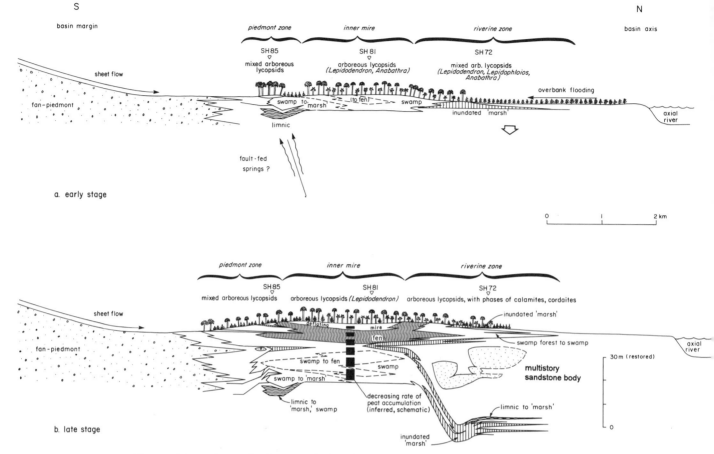

Figure 15. Evolutionary development of paleomire types for the ancestral peat-forming ecosystems of the No. 3 seam, Springhill coalfield: (a) early rheotrophic (eutrophic) stage; (b) late, postmesotrophic (deflationary?) stage. This interpretation reflects reversal in trends of lithotype brightness, ash, sulfur and miospore species abundance at top of seam attributed to deflation rather than continued rising as suggested by maceral record alone. Restored thickness: peat (×7) mud (×2) sand (×1). (Topography is exaggerated.)

Middle Pennsylvanian (Westphalian) peat mires to "unusual conditions, perhaps involving increased clastic supply or seasonal dryness."

Anabathra, an uncommon component of Pennsylvanian peat mires (DiMichele et al., 1985), is thought to have been a "colonizer of disturbed, clastic-rich (transitional) substrates" (DiMichele et al., 1985). Current research being conducted by T. L. Phillips of the University of Illinois and by W. A. DiMichele and D. A. Willard at the Smithsonian Institution similarly indicates that *Anabathra* was ecotonal (an ecotone representing a boundary condition between two floral communities), favoring transitional, flood-prone environments. This theory is supported by the author based on the abundance of *Lycospora orbicula* in the lower half of the No. 3 seam, which developed from a swamp habitat under rheotrophic conditions. It is also consistent with the distribution of *Anabathra* near clastic partings as reported by DiMichele and Phillips (1988). The greater abundance of *L. orbicula* in the inner mire as opposed to the margins and at the high ash base of the inner mire, however, indicates that

Anabathra was less successful than *Lepidodendron hickii* in the marginal mire areas, perhaps being less tolerant than *Lepidodendron* of siliciclastic input. The decline of *Anabathra* (*L. orbicula*) in the development of the upper third of the inner mire may indicate an intolerance to lower pH, water level, or nutrient supply. The occurrence of two 1 cm-thick fusain layers at the boundary between the two palynomorph assemblages raises the possibility that this change may have been promoted by wildfire(s) when the *Anabathra* community was becoming increasingly stressed by the hydrological evolution of the mire.

DiMichele et al. (1985) interpreted *Anabathra* as an element of a "high diversity, drier site assemblage." The author concurs with the association of *Anabathra* (*L. orbicula*) and high species diversity (Figs. 12, 14), but evidence from this study conflicts with the interpretation of relatively dry habitat. Apparently this inference arose from the association of *Anabathra* and fusain (Phillips et al., 1985) and the prevailing concept that fusain represents a drier site, which as discussed earlier, may not be valid given its wildfire origin (Scott, 1989).

Lepidophloios was not a major component of the mire vegetation, but its occurrence in the lower parts of the seam in marginal areas supports the adaptation of this flora to a high water table (Phillips, 1979; DiMichele and Phillips, 1985; DiMichele et al., 1985). It is noteworthy that *Lepidophloios hallii* was found to occur in greater abundance within the Herrin coal of Illinois in areas proximal to a contemporaneous paleochannel (DiMichele and Phillips, 1988). A similar, albeit proportionately lesser, increase in abundance of *L. hallii* occurred during development of the lower leaf of the No. 3 seam in the riverine zone of peat accumulation.

THE NO. 3 SEAM: PROCESSES AND HISTORY OF MIRE DEVELOPMENT

Maceral-based interpretations (Calder et al., 1991) suggest that the protected inner reaches of the ecosystem developed autogenically, resulting in modification to mire hydrology and chemistry. The suggested evolutionary trend to a less ground-water influenced state is typical of many mires, both temperate and tropical. The piedmont (basin margin) and riverine (basin axis) zones developed under differing hybrid processes of peat accumulation arising from their interaction with bordering deposystems (alluvial-fan piedmont and axial-channel belt). Subsequent incorporation of floral paleoecology and trends of ash, sulfur and lithotypes has permitted more accurate reconstruction of mire genesis, which differs somewhat with the maceral-based interpretation. The ultimate demise of the mire must have involved allogenic change that exceeded the adaptive capability of the ecosystem. Climate change linked to the precession signal may have slowed peat accumulation below the rate needed to keep abreast of rapid basin subsidence (Calder, 1991; Calder, 1993).

Inner mire processes

Interpretation of the evolution of the inner mire presents some intriguing dilemmas. The petrographic model of mire development of Calder et al. (1991) suggests a progressive evolution of lessening ground-water influence throughout the life of the mire. More precisely, the upward decrease in gelification, the major attribute measured by the ground-water influence index (G.W.I.), indicates a steady decrease of pH level, O_2, and nutrient supply, inferred to accompany rising of the mire surface. The resulting evolutionary sequence of mire types was originally interpreted (Calder et al., 1991) as swamp → swamp forest → fen, possibly culminating in oligotrophic bog and bog forest. The broader range of evidence considered in this paper, especially pertaining to floral ecology, suggests that the system remained under their influence of ground water and probably never achieved a truly ombrotrophic status.

Although there is compelling evidence of pervasive ground-water influence, the complete evolutionary sequence of mire types is less straightforward and may not have been one progressive continuum of less rheotrophic types as suggested by the

steady decline of gelification. The upper third of the seam column (SH81) exhibits an increase in ash, sulfur and miospore diversity and a concomitant reversion to duller lithotypes, suggesting that the inner mire ultimately reverted to a more rheotrophic state in its latest stage of development. The resolution of these contradictory trends would require that the late, deflating, or stagnating phase of mire development remain acidic.

Upwardly decreasing rates of peat accumulation characterize the tropical forest mires of Southeast Asia as they grow ever less fertile (Anderson and Muller, 1975; Cameron et al., 1989) and certain of these mires are undergoing rapid deflation (T. A. Moore, 1990, personal communication). Let us assume that the surface of the ancient Springhill systems may have actually deflated but also may have accreted at a progressively slower rate in the inner mire region as a result of climate stress (Calder, 1991; 1993) or decreasing fertility. A decrease in rate of biomass accumulation would permit dilution of miospore species indigenous to the mire by externally derived spores (e.g., monosaccate pollen, Fig. 12). Similarly, the steady accumulation of wind and waterborne ash would increase relative to diminishing peat accumulation. Nutrient stress in this mesotrophic system was probably much less pronounced than in modern oligotrophic examples, however. For such a modest change in nutrient supply to affect the rate of peat accumulation would therefore require that the dominant flora (*Lepidodendron* and *Anabathra*) be nutrient sensitive in the extreme.

The probable late, regressive trend of inner mire evolution notwithstanding, the mire first records a mesotrophic tendency from swamp to fen. The succession of mire types within the ancient Springhill peatlands shows a similar tendency to the classic hydroseral succession of temperate zone mires that forms through the process of terrestrialization (*"Verlandung"* in Weber, 1908, Moore, 1987) and to tropical zone mires of Indonesia that form through "vertical paludification" (Cameron et al., 1989; Esterle et al., 1989). They differ in degree, however, especially in that the No. 3 paleomire probably did not evolve to a solely rainfed hydrologic status.

According to the miospore record, floral succession within the inner mire appears to have been weakly developed, which is to be expected in a region with abundant ground water and nutrient supply (Grosse-Brauckmann, 1979; Teichmüller, 1989). Wholesale change in floral groups was rare: arboreous lycopsids in general dominated the mire throughout seam formation time. The apparent domination of mire vegetation by arboreous lycopsids, which required standing-water cover for reproduction (Phillips, 1979), confirms that high ground-water levels prevailed throughout the life of the mire. The only major change in the dominant floral group (arboreous lycopsids to sphenopsids and cordaites) occurred periodically in the upper half of the seam, both in the inner mire (SH81) and riverine zones (SH72).

Floral succession in the inner mire zone of the No. 3 seam is marked by species variation within a single genus or floral group, in particular the arboreous lycosids, as opposed to variation at the scale of floral groups which suggests minimal ecological change.

This succession comprised an initial *Anabathra/Lepidodendron hickii* community, with increasing domination of the mire flora by *Anabathra* before its sharp decline mid-seam. In the succeeding phase, *L. hickii* prevailed as the dominant plant, continuing through the final community, which was marked by the subordinate presence of *Lepidophloios harcourtii.* The domination of the mire flora by the two genera *Anabathra* and *Lepidodendron* is a reflection of the nutrient-rich, inundation-prone piedmont-riverine setting of the mire and supports current paleoecological interpretations of these plants. In the Westphalian B Low Barnsley seam of Yorkshire, Bartram (1987) interpreted *Lepidodendron* and *Anabathra* (*Paralycopodites*)-dominated peat as the early, colonizing and nutrient-rich rheotrophic phases (her 0 and 1) of mire development in a raised bog succession. *Anabathra* may be an indicator of the early rheotrophic, planar stage of peat formation in Westphalian mires elsewhere, given its similar occurrence within the relatively high ash, basal portions of the Westphalian D Herrin (DiMichele and Phillips, 1988) and Springfield (Willard, this volume) coals of the Illinois Basin.

Densosporites, assumed to have been produced from herbaceous lycopods, is uncommon not only within the No. 3 seam (Figs. 13–15) but throughout the Cumberland Basin (G. Dolby, 1989, personal communication). These lycopsids, assumed to have been the climax vegetation of Westphalian B raised mires elsewhere (Smith, 1962; Bartram, 1987; Fulton, 1987; Eble and Grady, this volume) and therefore assumed to be herbaceous by analogy to modern angiosperm-dominated mires, did not develop in this setting in the absence of pronounced, doming of the mire surface possibly due either to prohibitive seasonality (Smith, 1962; Butterworth, 1966) or excessively rapid net subsidence (Fulton, 1987).

Piedmont zone processes

The piedmont zone of the mire initially developed under very high ground-water level, bordering limnic conditions. A weakly developed succession of mire types progressed only to the formation of a swamp, which prevailed for the life of the mire, albeit with episodes of inundated forest in response to piedmont incursions. The piedmont margin shows the weakest development of floral succession, presumably caused by its proximity to water- and nutrient-supplying fans (Grosse-Brauckmann, 1979); *Lepidodentron hickii* was the dominant miospore contributor. The progressive southerly onlap of brighter lithotypes (Fig. 4) indicates that the mire accreted in a progressive, lateral manner toward the distal alluvial fans of the piedmont margin.

The process by which the mire developed in the piedmont zone is fundamental to an understanding of the origin of these and similar basin-margin/piedmont seams. The process of mire development at the piedmont margin, the site of fan-derived ground-water discharge and episodic sheetflood inundation, is analogous to the process of upslope paludification (cf. *Verswampfung* in Weber, 1908) described for the temperate mires of Polesie (Kulczynski, 1949). Here, the ecosystems developed progressively upslope by impeding the drainage of land of slightly higher topography. Peatlands in the Lake Agassiz region, Minnesota, developed in a similar manner (Heinselman, 1963; Tallis, 1983) termed lateral palludification by Cameron et al. (1989). The process was described by Heinselman as follows:

> Upslope growth of the peatland seems to be achieved by damming up the incoming waters from mineral soils. This creates a wet area into which the swamp forest can advance and prepare the way for the bog flora. It is noteworthy that the initial invasion on mineral soils does occur in a wet area but the communities are still forests, not aquatics.

A similarly arboreous (lycopsid) flora characterized the piedmont margin of the No. 3 paleomire. Their dominance in a sheetflood-prone setting was in agreement with their interpreted tolerance to inundation so long as water cover did not become permanent (DiMichele and Demaris, 1987), but their dominance is apparently at odds with the interpreted intolerance of lycopods to inundation by sediment (Gastaldo, 1986). Ground-water recharge to the mires, however, is inferred to have been primarily through seeps and springs at alluvial fan toes (Hunt, 1975; Calder, 1991; 1993), rather than via sediment-carrying surface flow.

In such a habitat prone to inundation, it is suggested that the nutrient-loving plant group(s) with the most successful reproductive and growth strategy would have predominated. The arboreous lycopsids had a heterosporous reproductive strategy (i.e., both microspores and megaspores), which used both wind and water transport for the dispersal of the large quantities of spores generated (Taylor, 1985; Phillips et al., 1985; Collinson and Scott, 1987). They were thus well suited to become reestablished in the aftermath of events such as fan-derived sheetfloods. An illuminating account of the growth and dynamics of a lycopsid forest is given in DiMichele and DeMaris (1987). The rapid growth of the trees, aided by their thick, supportive bark but little wood, and a single episode of mass reproduction (monocarpic) during late stage of growth, permitted them to establish on inundated sites, grow rapidly and once again reproduce. In addition, juvenile lycoposid trees (Fig. 16) were unbranched (in *Lepidodendron aculeatum* until approximately 30 m high), permitting rapid growth without competition between the trees for sunlight (DiMichele and DeMaris, 1987). The arboreous lycopsids, and *Lepidodendron hickii* in particular, were therefore particularly well adapted physiologically to growth in an intrinsically stressful environment such as the piedmont-riverine setting of the southern Cumberland Basin.

Riverine zone processes

Processes of peatland development at the riverine margin were profoundly influenced by channel-belt processes and by differential compaction. The conditions that prevailed in the riverine zone during accumulation of the lower bench are of interest in that they were harbingers of later fluvial occupation. The peat accumulated under highly rheotrophic conditions (Fig. 15) from an arboreous lycopsid-dominated community, which in this area

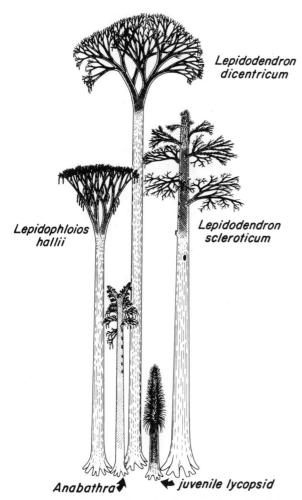

Figure 16. Reconstruction of arboreous lycopsids (after DiMichele and Phillips, 1985).

included *Lepidophloios hallii,* a plant that thrived in flooded regions of high abiotic stress (Phillips, 1979; DiMichele and Phillips, 1985; DiMichele et al., 1985). Rotting indicated by the presence of fungal mycelia (Dolby, 1988), unique to the lower leaf (bench) of the riverine zone, may be an indicator of such stress, but its ecologic significance is not entirely clear. The mire was inundated and was the site of mud deposition, represented by 2–5-m thick mudrock, prior to the site being occupied by the sand-laden axial river, suggesting that an underlying process controlled both mire development and channel-belt migration in this zone. Contributing to this may have been the differential compaction of multistory sandstone bodies and mudrocks. Evidence in support of this hypothesis is found in the sharp contrast in maceral composition of the upper and lower benches, the upper bench indicating peat accumulation under significantly less influence of ground water (swamp and fen), inferring a higher topography for the riverine zone mire above the entombed multistory sandstone body. Another potential control is syndepositional faulting, suggested by the presence in abandoned mine workings

of a fault paralleling the southerly margin of fluvial occupation not only in this, but the underlying No. 1 seam.

The ultimate demise of peat-forming ecosystems is problematic. Cessation of peat accumulation in many marine-influenced settings can be convincingly attributed to combined climate and sea-level change, but for intermontane basins with little or no apparent marine influence the mechanism has been less clear. Steadily decreasing rate of biomass production, hence peat accumulation, may have accompanied climate stress and inferred decrease in mire fertility, reaching a critical value relative to net subsidence of the basin. As a result, the mire may have become progressively susceptible to burial as it fell into disequilibrium with surrounding environmental conditions, notably basin subsidence and sediment supply (Calder, 1991; 1993).

CONCLUSIONS

Evidence of changing ground-water influence is the fundamental parameter for interpreting the genesis of modern and ancient peatlands. Changes in ground-water influence in the No. 3 seam are reflected in the distribution of vitrinite maceral types and especially the variable gelification of similar tissues, the ecology and succession of mire floras, the distribution of detrital mineral matter and vertical trends of lithotypes. Access to the record stored in the vitrinite macerals requires resolution of the progressive gelification of tissues. At the very least, this entails recognition of the maceral telinite and types, telocollinite, gelocollinite, and corpocollinite, although even these terms may be insufficient to fully describe the spectrum of tissue preservation and the products of various pathways of degradation. Etching by oxidative agents should provide important insight into the degree to which cryptic tissue structure has suffered the effects of biochemical gelification (e.g., Warwick and Stanton, 1988; Moore et al., 1990). Reconstructive diagrams of mire type and evolution based on maceral data are useful tools in paleoenvironmental analysis, but it is important to recognize the fact that they are inherently limited in their depiction of the complex evolution of peat-forming ecosystems.

The paleoecology of the mire flora and the record of floral change are fundamentally important to the interpretation of ancient peat-forming ecosystems. Furthermore, paleobotanical data is required to verify processes of degradation which can be profoundly affected by tissue type.

The ancestral paleomire of the No. 3 seam developed as a predominantly rheotrophic system, although there is evidence that it tended toward less ground-water influence as do many peat-forming ecosystems. The predominance of arboreous lycopsids, particularly *Lepidodendron* and *Anabathra,* confirms the pervasive influence of nutrient-rich ground water supplied by seepage at the toes of alluvial fans. Domination of the mire flora by these arboreous lycopsids and the minimal development of floral succession is atypical of many Westphalian mires, where such nutrient-rich, arboreous lycopsid peat forms only a portion of the overall mire succession, usually in the early stages of development.

The inner mire of the No. 3 seam evolved through partial terrestrialization (cf. vertical paludification of Cameron et al., 1989) and may have ultimately reverted (deflated) to a more ground-water influenced condition. Many studies of ancient peat formation tend to focus on "inner mire" zones, rather than on subeconomic marginal zones. Processes of mire development may differ in the inner and peripheral regions. The margins provide insight into the interaction of the peatlands with surrounding deposystems, thus providing a clearer understanding of the ecosystem as a whole, and controls on its development. The feedback mechanism of lateral paludification at the piedmont margin was aided by the dispersed nature and low competence of fan-derived ground waters and by the ecological adaptation and growth strategy of arboreous lycopsids. Peat formation at the riverine margin was profoundly influenced by allogenic processes including differential compaction around multistory sandstone bodies and by channel-belt processes, which in turn may have been locally influenced by contemporaneous faulting.

Insight into the ultimate demise of the peat-forming ecosystems in this intermontane setting, one of the most perplexing aspects of their development, has emerged only recently (Calder, 1991; 1993). Temporal analysis of the Springhill cyclothems suggests that precessional climate change may have critically affected mire hydrology and peat accumulation, forcing the mire into disequilibrium with basin subsidence and sediment supply.

ACKNOWLEDGMENTS

This research was funded by the Coal Section, Nova Scotia Department of Natural Resources and formed part of the author's Ph.D. dissertation, Dalhousie University. Sincere thanks are due Martin Gibling, for his guidance, to Graham Dolby for his careful, detailed palynological analysis, to my colleagues, in particular R. D. Naylor, at N.S.D.M.E. for valued discussion, and to my wife for her support and patience. The scientific content of the manuscript was greatly enhanced by the thought-provoking reviews of W. A. DiMichele and J. C. Hower. Comments by an anonymous reviewer are acknowledged. Thanks are also due the numerous researchers with whom I have held discussions regarding modern peat-forming systems, including J. C. Cobb, A. D. Cohen, C. F. Eble, J. S. Esterle, T. A. Moore and S. G. Neuzil. Typing by J. Cumby, Coal Section, and drafting and photography service by the Cartography Division, N.S.D.N.R., are gratefully acknowledged.

REFERENCES CITED

Altschuler, Z. S., Schnepfe, M. M., Silber, C. C., and Simon, F. O., 1983, Sulfur diagenesis in Everglades peat and origin of pyrite in coal: Science, v. 221, no. 4607, p. 221–227.

American Society for Testing and Materials (ASTM), 1987, Standard classification of peat samples by laboratory testing (D4427-84), in Annual Book of ASTM Standards, Volume 4.08: [city], American Society for Testing and Materials, p. 879–880.

Anderson, J.A.R., 1964, The structure and development of the peat swamps of Sarawak and Brunei: Journal of Tropical Geography, v. 18, p. 7–16.

Anderson, J.A.R., 1983, The tropical swamps of Malesia, in Gore, A.J.P., ed., Mires: Swamp, bog, fen and moor (Regional studies): Amsterdam, Elsevier, Ecosystems of the World, v. 4B, p. 181–199.

Anderson, J.A.R., and Muller, J., 1975, Palynological study of a Holocene peat and a Miocene coal deposit from NW Borneo: Review of Paleobotany and Palynology, v. 19, p. 291–351.

Arthaud, F., and Matte, P., 1977, Late Paleozoic strike-slip faulting in southern Europe and northern Africa: Result of a right-lateral shear zone between the Appalachians and the Urals: Geological Society of America Bulletin, v. 88, p. 1305–1320.

Bartram, K. M., 1987, Lycopod succession in coals: An example from the Low Barnsley seam (Westphalian B), Yorkshire, England, in Scott, A. C., ed., Coal and coal-bearing strata: Recent advances: Geological Society of London Special Publication 32,, p. 187–199.

Bell, W. A., 1927, Outline of Carboniferous stratigraphy and geologic history of the Maritime Provinces of Canada, in Proceedings and Transactions: Royal Society of Canada, Third Series, v. 21, section 4, p. 75–108.

Bell, W. A., 1944, Carboniferous rocks and fossil flora of northern Nova Scotia: Geological Survey of Canada Memoir 238, 277 p.

Butterworth, M. A., 1966, The distribution of Densospores: The Paleobotanist, v. 15, p. 16–28.

Calder, J. H., 1986, Introduction to Cumberland coal basin (Athol Syncline), in

Boehner, R. C., and 6 others, eds., Carboniferous-Jurassic Sedimentation and Tectonics: Minas, Cumberland and Moncton Basins, Nova Scotia and New Brunswick: Basins of Eastern Canada and Worldwide Analogues, Symposium Field Trip Guide: Atlantic Geoscience Society Special Publication 4, p. 63–73.

Calder, J. H., 1991, Controls on Westphalian peat accumulation: The Springhill coal field, Nova Scotia [unpublished Ph.D. thesis]: Halifax, Dalhousie University, 310 p.

Calder, J. H., 1993, The impact of climate change, tectonism and hydrology on the formation of Carboniferous tropical intermontane mires: The Springhill coalfield, Cumberland Basin, Nova Scotia, in Calder, J. H. and Gibling, M. R., eds., The Euramerican Coal Province: Controls on Tropical Peat Accumulation in the Late Paleozoic: Palaeogeography, Palaeoclimatology, Palaeoecology (in press).

Calder, J. H., Gibling, M. R., and Mukhopadhyay, P. K., 1991, Peat formation in a Westphalian B piedmont setting, Cumberland Basin, Nova Scotia: Implications for the maceral-based interpretation of rheotrophic and raised paleomires: Bulletin de la Société Géologique de France, vol. 8, p. 139–154.

Cameron, C. C., Esterle, J. S., and Palmer, C. A., 1989, The geology, botany and chemistry of selected peat-forming environments from temperate and tropical latitudes, in Lyons, P. C., and Alpern, B., eds., Peat and coal; origin, facies and depositional models: International Journal of Coal Geology, v. 12, p. 443–475.

Casagrande, D. J., 1987, Sulphur in peat and coal, in Scott, A. C., ed., Coal and coal-bearing strata: Recent advances: Geological Society of London Special Publication 32, p. 87–105.

Cecil, C. B., Renton, J. J., Stanton, R. W., and Finkelman, R. B., 1979, Mineral matter in coals of the central Appalachian Basin [abs.]: Urbana, Illinois, Ninth International Congress of Carboniferous Stratigraphy and Geology, p. 33.

Cecil, C. B., Stanton, R. W., Neuzil, S. G., Dulong, F. T., Ruppert, L. F., and Pierce, B. S., 1985, Paleoclimatic controls on late Paleozoic sedimentation and peat formation in the central Appalachian Basin (U.S.A.): International Journal of Coal Geology, v. 5, p. 195–230.

Cohen, A. D., 1974, Petrology and paleoecology of holocene peats from the Okefenokee swamp-marsh complex of Georgia: Journal of Sedimentary Petrology, v. 44, p. 716–726.

Cohen, A. D., 1989, Comparison of domed and planar peat deposits in Central America: Geological Society of America Abstracts with Programs, v. 21, no. 6, p. A25–26.

Cohen, A. D., 1990, Distribution of biogenic silica in peats from the Okefenokee Swamp [abs.]: Seventh Annual Meeting of the Society for Organic Petrology Abstracts with Program, p. 10–12.

Cohen, A. D., Spackman, W., and Raymond, R., Jr., 1987, Interpreting the characteristics of coal seams from chemical, physical and petrographic studies of peat deposits, *in* Scott, A. C., ed., Coal and coal-bearing strata: Recent advances: Geological Society of London Special Publication 32, p. 107–125.

Collinson, M. E., and Scott, A. C., 1987, Implications of vegetational change through the geological record on models for coal-forming environments, *in* Scott, A. C., ed., Coal and coal-bearing strata: Recent advances: Geological Society of London Special Publication 32, p. 67–85.

Cope, M. J., and Chaloner, W. G., 1985, Wildfire: An interaction of biological and physical processes, *in* Tiffney, B. H., ed., Geological factors and the evolution of plants: New Haven, Yale University Press, p. 257–277.

Dewison, M. G., 1989, Dispersed kaolinite in the Barnsley Seam coal (U.K.): Evidence for a volcanic origin: International Journal of Coal Geology, v. 11, p. 291–304.

Diessel, C.F.K., 1982, An appraisal of coal facies based on maceral characteristics, *in* Mallett, C. W., ed., Coal Resources—Origin, exploration and utilization in Australia: Symposium proceedings: Geological Society of Australia, Inc., Coal Group, p. 474–483.

Diessel, C.F.K., 1986, On the correlation between coal facies and depositional environments: Advances in the study of the Sydney Basin, *in* Proceedings of the 20th Newcastle Symposium, University of Newcastle, Australia, p. 19–22.

DiMichele, W. A., and Demaris, P. J., 1987, Structure and dynamics of a Pennsylvanian-age *Lepidodendron* forest: Colonizer of a disturbed swamp habitat in the Herrin (No. 6) coal of Illinois: Palaios, v. 2, p. 146–157.

DiMichele, W. A., and Phillips, T. L., 1985, Arborescent lycopod reproduction and paleoecology in a coal-swamp environment of late Middle Pennsylvanian age (Herrin Coal, Illinois, U.S.A.): Review of Palaeobotany and Palynology, v. 44, p. 1–26.

DiMichele, W. A., and Phillips, T. L., 1988, Paleoecology of the Middle Pennsylvanian-age Herrin coal swamp (Illinois) near a contemporaneous river system, the Walshville paleochannel: Review of Paleobotany and Palynology, v. 56, p. 151–176.

DiMichele, W. A., and Phillips, T. L., 1993, Paleobotanical and paleoecological constraints on models of peat formation in the Late Carboniferous of Euramerica, *in* Calder, J. H., and Gibling, M. R., eds., The Euramerican Coal Province: Controls on Tropical Peat Accumulation in the Late Paleozoic: Palaeogeography, Palaeoclimatology, Palaeoecology (in press).

DiMichele, W. A., Phillips, T. L., and Peppers, R. A., 1985, The influence of climate and depositional environment on the distribution and evolution of Pennsylvanian coal-swamp plants, *in* Tiffney, B. H., ed., Geological factors and the evolution of plants: New Haven, Yale University Press, p. 223–256.

Dolby, G., 1988, Palynological analysis of the No. 3 and Rodney seams, Springhill and other samples from the Cumberland Basin, Nova Scotia: Nova Scotia Department of Mines and Energy Report 88-060, 70 p.

Esterle, J., and Ferm, J. C., 1986, Relationship between petrographic and chemical properties of coal seam geometry, Hance seam, Breathitt Formation, southeastern Kentucky: International Journal of Coal Geology, v. 6, p. 199–214.

Esterle, J. S., Ferm, J. C., and Tie, Y. L., 1989, A test for the analogy of tropical domed peat deposits to "dulling-up" sequence in coal beds—Preliminary

results: Journal of Organic Geochemistry, v. 14, p. 333–342.

Etherington, J. R., 1983, Wetland ecology: Studies in biology No. 154: London, Arnold.

Flores, R. M., 1989, Rocky Mountain Tertiary coal-basin models and their applicability to some world basins, *in* Lyons, P. C., and Alpern, B., eds., Peat and coal: Origin, facies and depositional models: International Journal of Coal Geology, v. 12, p. 767–798.

Fulton, I. M., 1987, Genesis of the Warwickshire thick coal: A group of long-residence histols, *in* Scott, A. C., ed., Coal and coal-bearing strata: Recent advances: Geological Society of London Special Publication 32, p. 201–218.

Gastaldo, R. A., 1986, Implications on the paleoecology of autochthonous lycopods in clastic sedimentary environments of the Early Pennsylvanian of Alabama: Palaeogeography, Palaeoclimatology, Palaeoecology, v. 53, p. 191–212.

Gibling, M. R., Zentilli, M., and McCready, R.G.L., 1989, Sulphur-rich coal in the Pennsylvanian of Atlantic Canada: Geologic and isotopic evidence for bedrock evaporites as the sulphur source: International Journal of Coal Geology, v. 11, p. 81–104.

Gore, A.J.P., 1983, editor, Mires: Swamp, bog, fen and moor (General studies): Amsterdam, Elsevier, Ecosystems of the World, v. 4A, 440 p.

Grady, W. C., Eble, C. G., and Neuzil, S. G., 1989, Distribution of petrographic components in a modern domed tropical Indonesian peat: A possible analog for maceral distributions in Middle Pennsylvanian coal beds of the Appalachian Basin [abs.]: Geological Society of America Abstracts with Programs, v. 21, no. 6, p. A25.

Grosse-Brauckmann, G., 1979, Zur Deutung einiger Makrofossil—Vergesellschaftungen unter dem Gesichtspunkt der Torfbildung, *in* Williams, O., and Tüxen, R., eds., Werden und Vergehen von Pfanzengesellschaften, Berliner Internationales Symposium Int. Ver. Vegetationskunde: Vaduz, Liechtenstein, Cramer. J., p. 111–132.

Hacquebard, P. A., Birmingham, T. F., and Donaldson, J. R., 1967, Petrography of Canadian coals in relation to environment of deposition: Ottawa, Department of Energy, Mines and Resources Symposium on Science and Technology of Coal, p. 84–97.

Harbaugh, J. W., and Bonham-Carter, G., 1970, Computer simulation in geology: New York, Wiley-Interscience, 98 p.

Heinselman, M. L., 1963, Forest sites, bog processes, and peatland types in the glacial Lake Agassiz region, Minnesota: Ecological Monograph, v. 33, p. 327–374.

Koch, J., 1969, Mikropetrographische untersuchungen an einigen organischen komponenten jungpleistozäner und holozäner torfe Süddeutschlands und der Schweiz: Geologische Jahrbuch, v. 87, p. 333–360.

Kosters, E. C., and Chmura, G. L., and Bailey, A., 1987, Sedimentary and botanical factors influencing peat accumulation in the Mississippi Delta; Journal of the Geological Society of London, v. 144, p. 423–434.

Kulczynski, S., 1949, Peat bogs of Polesie: Memoir of the Academy of Science Cracovie B., p. 1–356.

Li S., Li B., Yang S., Hueng J., and Li Z., 1984, Sedimentation and tectonic evolution of Late Mesozoic faulted coal basins in north-eastern China, *in* Rahmani, R. A., and Flores, R. M., eds., Sedimentology of coal and coal-bearing sequences: International Association of Sedimentologists Special Publication 7, p. 387–406.

Lippolt, H. J., and Hess, J. C., 1985, Ar^{40}/Ar^{39} dating of sanidines from Upper Carboniferous tonsteins, Compte rendu, dixième Congrès International de Stratigraphie et de Géologie du Carbonifère, Madrid, 1983: v. 4, p. 175–181.

Littke, R., 1987, Petrology and genesis of Upper Carboniferous seams from the Ruhr region, West Germany: International Journal of Coal Geology, v. 7, p. 147–185.

Mahaffy, J. F., 1988, Vegetational history of the Springfield coal (Middle Pennsylvanian of Illinois) and distribution of a tree-fern miospore, *Thymospora pseudothiessenii,* based on miospore profiles: International Journal of Coal Geology, v. 10, p. 239–260.

Marshall, A. E., and Smith, A.H.V., 1965, Assemblages of miospores from some Upper Carboniferous coals and their associated sediments in the Yorkshire

Coalfield: Paleontology, v. 7, p. 656–673.

McCabe, P. J., 1984, Depositional environments of coal and coal-bearing strata, *in* Rahmani, R. A., and Flores, R. M., eds., Sedimentology of coal and coal-bearing strata: International Association of Sedimentologists Special Publication No. 7, p. 13–42.

Moore, P. D., 1987, Ecological and hydrological aspects of peat formation, *in* Scott, A. C., ed., Coal and coal-bearing strata: Recent advances: Geological Society of London Special Publication 32, p. 7–15.

Moore, P. D., 1989, The ecology of peat-forming processes: A review, *in* Lyons, P. C., and Alpern, B., eds., Peat and coal: Origin, facies and depositional models: International Journal of Coal Geology, v. 12, p. 89–103.

Moore, P. D., and Bellamy, D., 1974, Peatlands: London, Elek Science, 221 p.

Moore, T. A., Stanton, R. W., Pocknall, D. T., Flores, R. M., 1990, Maceral and palynomorph facies from two Tertiary peat-forming environments in the Powder River Basin, U.S.A.: International Journal of Coal Geology, v. 15, p. 293–316.

Obernyer, S., 1978, Basin-margin depositional environments of the Wasatch Formation in the Buffalo–Lake de Smet area, Johnson County, Wyoming, *in* Hodgson, H. W., ed., Proceedings, 2nd Symposium on the Geology of Rocky Mountain Coal, Colorado, 1977: U.S. Geological Survey Research Series 4, p. 49–65.

Pearson, H. L., 1986, Structure and taxonomy of the Carboniferous lycopsid *Anabathra:* Bulletin of the British Museum of Natural History, Geology, v. 40, p. 265–292.

Penfound, W. T., 1952, Southern swamp and marshes: The Botanical Review, v. 18, p. 413–446.

Phillips, T. L., 1979, Reproduction of heterosporous arborescent lycopods in the Mississippian-Pennsylvanian of Euramerica: Review of Paleobotany and Palynology, v. 27, p. 239–289.

Phillips, T. L., Peppers, R. A., and DiMichele, W. A., 1985, Stratigraphic and interregional changes in Pennsylvanian coal-swamp vegetation: Environmental influences: International Journal of Coal Geology, v. 5, p. 43–109.

Pierce, B. S., Stanton, R. W., and Eble, C. F., 1991, Facies development in the Lower Freeport coal bed, west-central Pennsylvania, USA; International Journal of Coal Geology, v. 18, p. 17–43.

Polak, B., 1975, Character and occurrence of peat deposits in the Malaysian tropics, *in* Bartstra, G. J., and Casparie, W. W., eds. , eds., Modern Quaternary Research in Southeast Asia; Balkema, Rotterdam, p. 71–81.

Ruppert, L. F. Neuzil, S. G., Cecil, C. B., and Kane, J. S., 1989, Inorganic constituents in a low-ash domed Sumatra peat [abs.]: Geological Society of America Abstracts with Programs, v. 21, no. 6, p. A51.

Ryan, R. J., Boehmer, R. C., and Calder, J. H., 1991, Lithostratigraphic revision of the Upper Carboniferous to Lower Permian strata in the Cumberland Basin, Nova Scotia and the regional implications for the Maritimes Basin in Atlantic Canada: Bulletin of Canadian Petroleum Geology, v. 39, p. 289–314.

Ryer, T. A., and Langer, A. W., 1980, Thickness change involved in the peat-to-coal transition for a bituminous coal of Cretaceous age in central Utah: Journal of Sedimentary Petrology, v. 50, p. 987–992.

Scotese, C. R., Bambach, R. K., Barton, C., van der Voo, R., and Ziegler, A. M., 1979, Paleozoic base maps: Journal of Geology, v. 87, no. 3, p. 217–277.

Scott, A. C., 1978, Sedimentological and ecological control of Westphalian B plant assemblages from West Yorkshire, *in* Proceedings: Yorkshire Geological Society, v. 41, part 4, no. 33, p. 461–508.

Scott, A. C., 1989, Observation on the nature and origin of fusain, *in* Lyons, P. C., and Alpern, B., eds., Peat and coal: Origin, facies and depositional models:

International Journal of Coal Geology, v. 12, p. 443–475.

Smith, A.H.V., 1962, The paleoecology of Carboniferous peats based on the miospores and petrography of bituminous coals, *in* Proceedings: Yorkshire Geological Society, v. 33, p. 423–474.

Smith, A.H.V., and Butterworth, M., 1967, Miospores in the coal seams of the Carboniferous of Great Britain: Special Papers in Paleontology 1, 324 p.

Staub, J. R., and Cohen, A. D., 1979, The Snuggedy Swamp of South Carolina: A back barrier estuarine coal-forming environment: Journal of Sedimentary Petrology, v. 49, p. 133–144.

Styan, W. B., and Bustin, R. M., 1983, Sedimentology of Fraser River delta deposits: A modern analogue for some deltaic coals: International Journal of Coal Geology, v. 3, 101–143.

Tallis, J. M., 1983, Changes in wetland communities, *in* Gore, A.J.P., ed., Mires: Swamp, bog, fen and moor (General Studies): Amsterdam, Elsevier, Ecosystems of the World, v. 4A, p. 311–348.

Tansley, A. G., 1911, Types of British vegetation: Cambridge, Cambridge University Press.

Tansley, A. G., 1939, The British Islands and their vegetation: Cambridge, Cambridge University Press.

Tasch, K. H., 1960, Die Moglichkeiten der Flozgleichstellung unter Zuhifenahme von Flozbildungsdiagrammen: Bergbau-Rundschau, v. 12, p. 153–157.

Taylor, T. N., 1985, Paleobotany—An introduction to fossil plant biology: New York, NcGraw-Hill.

Teichmüller, M., 1982, Origin of the petrographic constituents of coal, *in* Stach, E., Mackowsky, M.-Th., Teichmüller, M., Taylor, G. H., Chandra, D., and Teichmüller, R., eds., Coal petrology, third edition: Berlin and Stuttgart, Gebrüder Borntraeger, p. 219–294.

Teichmüller, M., 1989, The genesis of coal from the viewpoint of coal petrology, *in* Lyons, P. C., and Alpern, B., eds., Peat and coal: Origin, facies and depositional models: International Journal of Coal Geology, v. 12, p. 1–87.

Timofeev, P. P., and Bogoliubova, L. I., 1964, Main factors in the formation of genetic types of humic coals and the specific features of their distribution in different coal-bearing formations of the U.S.S.R., Compte rendu, cinquième Congrès International de Stratigraphie et de Géologie du Carbonifère, Paris, v. 3, p. 1031–1038.

Warwick, P. D., and Flores, R. M., 1987, Evolution of fluvial styles in the Eocene Wasatch Formation; Powder River Basin, Wyoming, *in* Ethridge, F. G., Flores, R. M., and Harvey, M. D., eds., Recent developments in fluvial sedimentology: Society of Economic Paleontologists and Paleontologists and Mineralogists Special Publication 39, p. 303–310.

Warwick, P. D., and Stanton, R. W., 1988, Petrographic characterization of the Wyodak-Anderson coal bed (Paleocene), Powder River Basin, Wyoming, U.S.A.: Organic Geochemistry, v. 12, p. 389.

Weber, C. A., 1908, Aufbau und Vegetation der Moore Nord-Deutschlands: Engler's Botanisches Jahrbuch, v. 40 (Supplement), p. 19–34.

Willard, D. A., 1989a, *Lycospora* from Carboniferous *Lepidostrobus* compressions: American Journal of Botany, v. 76, p. 1429–1440.

Willard, D. A., 1989b, Source plants for Carboniferous microspores: *Lycospora* from permineralized *Lepidostrobus:* American Journal of Botany, v. 76, p. 820–827.

Winston, R. B., 1986, Characteristic features and compaction of plant tissues traced from permineralized peat to coal in Pennsylvanian coals (Desmoinesian) from the Illinois Basin: International Journal of Coal Geology, v. 6, p. 21–41.

MANUSCRIPT ACCEPTED BY THE SOCIETY JANUARY 14, 1993

Geological Society of America
Special Paper 286
1993

Origin of gray-shale lithofacies ("clastic wedges") in U.S. midcontinental coal measures (Pennsylvanian): An alternative explanation

Allen W. Archer
Department of Geology, Kansas State University, Manhattan, Kansas 66506
Erik P. Kvale
Indiana Geological Survey, 611 N. Walnut Grove, Bloomington, Indiana 47405

ABSTRACT

Cyclic laminations in gray shales are noted above many major coal seams in the Illinois Basin. These features, which are termed "tidal rhythmites," indicate significant tidal influence during deposition of the roof strata. Many existing models, however, suggest fluvial dominance and crevasse splay deposition for such gray shales.

Based on sedimentological and paleogeographical considerations, appropriate deposition models require micro- to mesotidal environments, limited wave reworking, and a mud-dominated, tropical setting. Following these requirements, tidally influenced estuarine/deltaic models can be generated for roof strata of many Illinois Basin coals. This model is based upon modern mud-rich tidally influenced estuaries and deltas in near equatorial settings and can be used to explain occurrence of shales that lack obvious marine influence (based upon paleontology) above low-sulfur coals in the Illinois Basin.

INTRODUCTION

Gray-shale lithofacies

The "gray-shale sequence" as defined herein is predominantly clay- and silt-rich shales interbedded with siltstones and generally fine-grained sandstones. In fresh exposures the shale ranges from medium light gray (N6) to dark gray (N3) and weathers to pale yellowish orange (10 YR 8/6) to medium brown (5 YR 4/4). These shales commonly contain well-preserved plant fossils, including upright trees (Shabica, 1970; DiMichele and DeMaris, 1987; Kvale et al., 1989), are generally lacking in marine fossils and commonly contain sideritic concretions containing a variety of terrestrial and marginal marine fossils (Richardson and Johnson, 1969; Baird et al., 1985a, 1985b; Kuecher et al., 1990).

Our use of the term "gray shales" include interbedded mudstones and sandstones that exhibit a variety of primary structures ranging from lenticular to flaser bedding. Well-sorted siltstones that range from massive to laminated also occur in this sequence. Although these deposits are generally nonbioturbated, diverse trace-fossil assemblages do occur locally (see Archer and Maples, 1984). Lenticular sandstones as much as 25 m thick and mudstone are common within the gray-shale sequence. Primary structures associated with thinner sandstones (1–10 m) include massive, relatively structureless sandstones to ripple-laminated sandstones. These sandstones can contain abundant clay partings and flaser- and wavy-bedding. Thicker sandstone bodies (>10 m) may exhibit a variety of large-scale structures including stacked sets of planar and trough cross-stratified sandstones.

Stratigraphic relations

Gray shales and associated coal beds in the Illinois Basin (Fig. 1) can be divided into the Lower Pennsylvanian (Atokan and older) and the Middle Pennsylvanian (Desmoinesian) coals.

Archer, A. W., and Kvale, E. P., 1993, Origin of gray-shale lithofacies ("clastic wedges") in U.S. midcontinental coal measures (Pennsylvanian): An alternative explanation, *in* Cobb, J. C., and Cecil, C. B., eds., Modern and Ancient Coal-Forming Environments: Boulder, Colorado, Geological Society of America Special Paper 286.

Stratigraphic correlations (Fig. 2) of the shales and coal beds used in the following discussion are based on Phillips et al. (1985) and Shaver et al. (1986). Lower Pennsylvanian coal beds are exemplified by the Block Coals of the Brazil Formation in Indiana. The Block Coals are latest Atokan (upper Westphalian C) in age (Eggert and Phillips, 1982). These coals have some of the lowest sulfur and ash values in Indiana. Coals are dull to moderately dull banded coals. They range from 0.2 to 2 m in thickness and tend to be podlike, discontinuous, and of only local economic importance. The gray-shale lithofacies both overlies and underlies (below the underclay) these coals (Fishbaugh et al., 1989; Kvale and Archer, 1990).

Conversely, Middle Pennsylvanian coal beds are correlated over widespread areas and serve as stratigraphic markers in subsurface studies. Examples are the Colchester, Springfield, and Herrin Coals. Overlying gray shales are the Francis Creek, Dykersburg, and Energy Shales respectively. The coal beds are upper Desmoinesian (upper Westphalian D) in age and consist generally of bright-banded coals. They may be as much as 4 m thick. In areas of thick Springfield Coal, the generally bright banded coal may dull upward (Neavel, 1961; Hower and Wild, 1982). The gray-shale lithofacies overlying these coals is only locally developed and have been referred to as "clastic wedges" (Wanless, 1964). Thus instead of gray shales, the most common and areally extensive roof strata includes black, organic-rich shales and limestones. Sulfur and ash values for the coals can be low where thick gray shales are present, but average sulfur and ash values for these coals are considerably higher than those of the lower Pennsylvanian coals.

Past interpretations

Previously, many of the gray-shale sequences have been interpreted as totally nonmarine and indicative primarily of fluvial and lacustrine environments (Franklin, 1939; Gray, 1962; Johnson, 1972; Edwards et al., 1979; Eggert, 1982, 1987; Eggert and Phillips, 1982; Burk et al., 1987). These interpretations are commonly based upon the proximity of the gray-shale lithofacies to coals (which need fresh pore water to form), abundance of upright lycopod trunks, well-preserved plant fossils, presence of insect-wing fossils (Smith, 1871), and the absence of marine macrofossils.

The association of gray shales and low-sulfur coals has been repeatedly documented (Hopkins, 1968; Gluskoter and Hopkins, 1970; Hopkins and Nance, 1970; Edwards et al., 1979; Eggert and Adams, 1979; Eggert, 1982; 1987; Baird et al., 1985a, 1985b, 1986; Burk et al., 1987). Gluskoter and Hopkins (1970) stated low sulfur values (2% or less) were related to relatively rapid deposition of more than 6 m (20 ft) of muds over the peats prior to deposition of sediments containing marine fossils. Al-

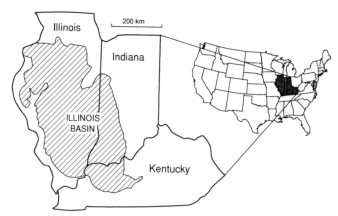

Figure 1. Location map of the Illinois Basin. Hatched area denotes area of preserved Pennsylvanian strata.

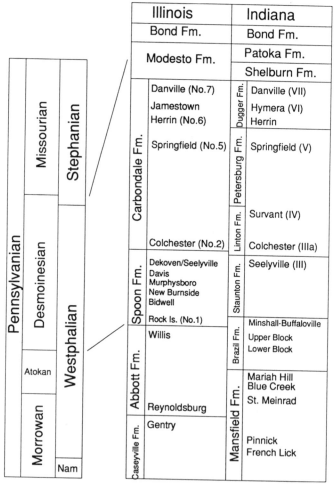

Figure 2. Generalized stratigraphy of the major coals in Indiana and Illinois.

though they did not state that the shales were deposited in fresh water, they suggested that such shales "lack distinct marine affinities (Gluskoter and Hopkins, 1970, p. 91). Many subsequent workers assumed that such shale must have been deposited in fresh water and therefore fluvial/lacustrine in origin (Edwards et al., 1979; Eggert and Adams, 1979; Eggert, 1982, 1987; Burk et al., 1987).

Sandstones that interfinger with the gray-shale lithofacies have been interpreted as fluvial channels formed, at least in part, contemporaneously with coals (Johnson, 1972). For the Springfield and Herrin Coals, these channels have been named the Galatia and Walshville Channels, respectively (Johnson, 1972; Nelson, 1983). Similar gray-shale sequences (Dykersburg Shale and Galatia Channel) overlying the Springfield Coal were considered to be predominantly fresh water and related to crevasse splays (e.g., Eggert, 1982). In addition, gray shales within the Energy Shale of Illinois, which overlies the widespread Herrin Coal, have been interpreted as freshwater crevasse-splay and lacustrine deposits (Edwards et al., 1979; Burk et al., 1987).

Emphasis of this study

We have examined and compared the roof strata associated with Lower and Middle Pennsylvanian coal beds. Lower Pennsylvanian gray shales are less laterally restricted than gray shales of the Middle Pennsylvanian. The shales contain similar sedimentary features, however. We will discuss the lithofacies of roof sequences, details of their sedimentology, and briefly discuss possible modern analogues for the facies described. Many of the features noted are consistent with modern tidal structures. Documentation of tidal deposition above coal seams indicates the need to reevaluate traditionally fluvially dominated models used to explain roof-rock sequences.

CHARACTER OF GRAY-SHALE LITHOFACIES

Geometry

Lower Pennsylvanian gray shales are areally extensive although underlying coals are not widely tradeable. Field-based studies have indicated the lateral relations of gray shales and channel-form sandstones for this interval (Fishbaugh et al., 1989). Numerous large-scale strip and underground mines in Middle Pennsylvanian coals allows for more precise delineation of geometries for gray-shale lithofacies across large areas.

One of the more complete delineations of the gray-shale facies was presented by Hopkins (1968). A generalized model depicting the geometry of a clastic wedge, referred to as a "lenticular gray shale," and the relationship of gray-shale roof strata to underlying low-sulfur coal was subsequently presented by

Gluskoter and Hopkins (1970). These Middle Pennsylvanian gray-shale sequences are commonly depicted as having wedge-shaped geometries when viewed in cross section (Fig. 3). In the idealized model, the wedges fine and thin away from this core facies. The wedges can locally punctuate the common repetition of thin, widespread beds of underclays, coals, black shales, and limestones that constitute normal Illinois Basin cyclothems (Wanless, 1964). At the outer margin of a wedge, the gray-shale sequence thins and becomes discontinuous. Lenses or pods of gray shale are overlain by black shale and limestone (Fig. 4). Organic-rich black shales are the dominant roof type for the widespread Middle Pennsylvanian coals and areally comprise about 90% of the roof strata above Middle Pennsylvanian coals in the Illinois Basin (see distribution maps in Hopkins, 1968, and Gluskoter and Hopkins, 1970).

The Middle Pennsylvanian Energy Shale of Illinois is a typical gray-shale wedge. The wedge is as much as 26 m thick along the sandstone-dominated Walshville Channel (Fig. 4). Away from the channel the overall grain size decreases and marine-fossil content increases eastward over an approximate distance of 35 km. In addition to a lateral fining, there is also a pronounced coarsening upward of more proximal (coarser) facies over more distal (finer) facies (Burk et al., 1987).

Sedimentological observations

Only within the past decade have tidal deposits been recognized in Pennsylvanian rocks of the Illinois Basin. Depositional models for epicontinental seaway sedimentation by Shaw (1964) suggested that no tidal activity occurred in cratonic areas. Sediments deposited under tidal influence, however, are commonly noted in cratonic basins (Klein, 1977; Klein and Ryer, 1978). Recent theoretical models suggest significant tidal flux (in the micro- to mesotidal range) is possible within shallow epicontinental seaways (Ericksen and Slingerland, 1990; Slingerland, 1986; Slater, 1985). Epicontinental seaways were common in the Illinois Basin. Previous subsurface and outcrop studies in the Illinois Basin have failed to recognize Pennsylvanian tidal deposits for a number of reasons. Occurrence of abundant plant remains was thought to be indicative of nonmarine conditions. Lack of obvious marine macrofossils was also used as a nonmarine indicator. Features such as large-scale tidal bundles (Visser, 1980; Allen, 1981; Kriesa and Moiola, 1986) and herringbone cross-stratification were not recognized. Subsurface studies have not identified shore-parallel or shore-perpendicular sandstone bodies, such as tidal sand ridges, barrier bars, or chenier ridges, that would suggest significant wave or tidal influence.

Although tidal features had not been widely recognized in the Illinois Basin, analyses of fine-scale sedimentology indicates that tidal processes can be interpreted from the gray-shale lithofa-

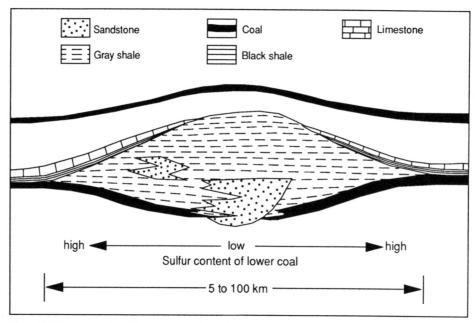

Figure 3. Diagrammatic section transverse through a gray-shale clastic wedge. Modified from Gluskoter and Hopkins (1970, p. 94).

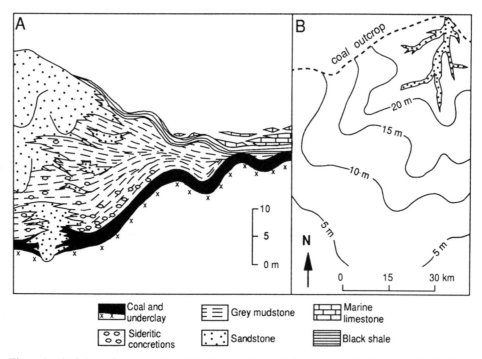

Figure 4. (A) Schematic cross section of the Francis Creek Shale above the Colchester Coal. Modified from Baird et al. (1985a). (B) Hypothetical isopach map of siliciclastic lobe showing distribution of sandstone channels and thicknesses of associated shales. Modified from Baird et al. (1988a).

cies. Within parts of gray-shale sequences, vertical variations in sandstone or mudstone laminae or bed thicknesses are very systematic and can be directly equated to tidal "neap-spring" cycles. In the Illinois Basin, tidal deposits have been documented in a number of Lower and Middle Pennsylvanian, coal-bearing sections (Kuecher, 1983; Baird et al., 1986; Broadhurst, 1988; Archer and Kvale, 1989; Archer et al., 1989; Fishbaugh et al., 1989; Kuecher et al., 1990; Kvale and Archer, 1989a, 1989b, 1990, 1991; Kvale et al., 1989).

The Brazil Formation tidalites are particularly significant because of the traditional interpretation of a nonmarine roof for low-sulfur coal. Kvale and Archer (1989a, 1990) have shown that this assumption may not be valid because rhythmites, interpreted as tidally generated, immediately overlie low-sulfur (<1%) and low-ash (<4.4%) coals. Such rhythmites include parallel-laminated mudstones (Fig. 5A) as well as parallel-bedded sandstone and mudstone couplets (Fig. 5B). Similar features have been noted in the gray shale facies above other major coals of the Illinois Basin. Primary structures in roof strata of low-sulfur parts of the Springfield Coal in Indiana reveal laminae and fine-scale cycles that are related to tidal deposition (Fig. 5C). Our sampling in underground mines in areas of low sulfur Herrin Coal in Illinois has documented the occurrence of apparent tidal rhythmites (Fig. 5D).

ENVIRONMENTAL SETTING

Consideration for modern analogues

Earlier Illinois Basin gray-shale models are based upon reconstructions that invoke single active fluvial channels or channels associated with bird's foot type deltas. We suggest such models are inaccurate. A more complete model should (1) be tidally influenced with a limited tidal range (microtidal to mesotidal), (2) have limited marine wave reworking, (3) be mud dominated, (4) occur within a tropical setting, and (5) be laterally associated with extensive peat accumulations. These features are more commonly associated with estuarine models and tidally influenced muddy deltas in peat-producing coastal areas.

The Orinoco Delta in northeastern South America and the Mahakam Delta on the east-central coast of Kalimantan (Borneo) are examples of deltas that meet these criteria. Based upon the works of van Andel (1967) and Pfefferkorn et al. (1988) for the Orinoco and by Allen et al. (1979) for the Mahakam, important generalities can be summarized. Both systems have significant tidal influence, with tidal ranges of 1 to 3 m at the head of distributaries. Microtidal conditions exist as much as 150 km inland from the delta fronts. Both areas experience low wave energies and exhibit broad, mud-rich tidal flats that form much of the lower delta plain. Such flats have abundant plant debris and minimal marine fauna. Both occur in low latitude tropical climates and are laterally associated with peat-forming environments; however, high quality peats are not forming on active deltaic lobes.

Tidally influenced, mud-rich estuaries also potentially meet the criteria described above. There has been less study of tropical, tidally influenced estuaries than of temperate varieties. Many estuaries exhibit a tripartite subdivision where there is an upper, fluvial-dominated sand-rich part, a middle mud-rich part, and a lower, marine-dominated sand-rich part (Dalrymple et al., 1990). Even within the upper fluvial part of an estuary, tidal influence may extend inland for hundreds of kilometers (Schubel, 1984). A turbidity maximum occurs in the middle part of the estuary and this zone can experience very high rates of mud deposition (Allen, 1971). Clay particles ar commonly aggregated by a variety of processes, including flocculation along the saline/freshwater boundary and various types of biogenic processes. Rates of mud accumulation as high as 10 cm/week have been reported (Dalrymple and Makino, 1989). Many of the tidal rhythmites described from the Illinois Basin may have been deposited in such estuarine settings. Similar estuarine settings have been proposed for coal-bearing strata of the Cretaceous of Alberta (Rahmani, 1988) and Paleocene of Texas (Breyer, 1987; Breyer and McCabe, 1986).

Problems with existing models

Several problems exist with the prevailing Pennsylvanian coal–measures models that rely heavily on nonmarine, fluvially dominated processes. Interbedded sandstones, siltstones, and shales in the gray-shale facies are commonly interpreted as crevasse splays (Burk et al., 1987). Crevasse-splay deposition is contingent upon the development of a significant natural levee system, such as that which characterizes the Mississippian River Delta or the Brahmaputra River. However, levee deposits have never been unequivocally documented from the Pennsylvanian of the Illinois Basin. Interpretations invoking crevasse-splay deposition, which date back to Johnson (1972) and Allgaier and Hopkins (1975), are problematic because "[n]o natural levees of the Walshville channel have been conclusively demonstrated" (Nelson, 1987, p. 12).

On-going studies of modern domed and planar peats present along the coast of Malaysia (e.g., Cecil et al., 1988; Cecil and Englund, 1989) have proposed similarities between these peats and the coal-producing peats that were present during the Pennsylvanian in the eastern U.S. (including the Illinois Basin). Such studies have suggested similar climates and depositional environments for these modern and ancient peats. Currently in

Figure 5. Examples of tidal rhythmites from roof strata immediately overlying low-sulfur coals. Mudstone laminae and flaggy beds within such rhythmites exhibit systematic thinning and thickening that can be related to "neap-spring" tidal cyclicity (see Kvale et al., 1989; Archer and Kvale, 1989; Kvale and Archer, 1990). Arrows denote those laminae deposited during the weakest monthly tides. (A) Parallel-laminated mudstone from the Brazil Formation, Daviess County, Indiana. Note the well-developed thickening and thinning of laminae in the upper part of the photo. Lower laminae preserve only portions of monthly tidal cycles. Also note how the laminae are draped around the siderite nodule near the bar scale. (B) Parallel-bedded sandstone and mudstone couplets, Brazil Formation, Clay County, Indiana. In mine walls rhythmic thickening and thinning of sandstone beds is locally preserved (see Kvale and Archer, 1990). (C) Sandstone and mudstone couplets with some disruption by bioturbation, Dykersburg Shale immediately overlying the Springfield Coal, Pike County, Indiana. (D) Parallel-laminated mudstone, Energy Shale immediately overlying Herrin Coal, Franklin County, Illinois.

Malaysia, the rivers that are present between the peat deposits exhibit very low frequencies of flooding, even during the monsoonal season. Because the amounts of rainfall are constantly high throughout the year (and seasonal flooding is generally absent), channels are capable of confining discharge on a yearly basis. As a result, splays are very rare (James Staub, Southern Illinois University, 1989, personal communication). In fact, splaying is so rare that peat deposits can form right up to the edges of the rivers.

Another problem with fluvial depositional models in the Illinois Basin is channel deposits. Summaries of subsurface studies of some of the gray-shale "channels" associated with Middle Pennsylvanian coals tend to diagrammatically illustrate them as continuous over many kilometers. For example, Eggert and Adams (1979, their Fig. 6), in mapping channels contemporaneous with the Springfield Coal in Indiana, indicate three large channels. The major channel is simply mapped as a zone where

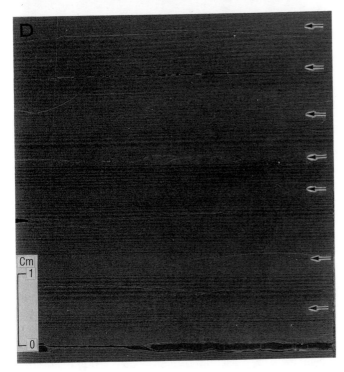

sandstones are replaced down channel to the west by sandy shales, which were interpreted as possible channel-fill deposits (Friedman, 1960). These and other examples illustrate that it is not at all clear that channels, such as those illustrated in Figure 8, actually were connected into an integrated drainage network. Integrated, continuous channel systems would favor a fluvial-dominated depositional model.

Alternative models

Penecontemporaneous coal-forming environments in the Illinois Basin contain facies analogous to the modern Mahakam and Orinoco Deltas and estuaries described previously. Modern deltas have been formed during Holocene sea-level rise; similarly, in areas of lower sediment flux, fluvial systems have been transformed into estuaries. Detailed analysis of a gray-shale sequence has indicated its formation during an ongoing transgression (Baird and Shabica, 1980).

Figure 9A illustrates a transgressive sequence and associated muddy, tidally influenced delta interpreted for the strata above the Springfield Coal in Illinois that can be compared to the clastic wedge model of Figure 3. Rather than the fluvial model for gray-shale deposition the gray shales may have been deposited in a small, mixed fluvio-delta tidal. Alternatively, the shales could easily have been formed flanking a major active estuarine channel. Some of the thick "sandstone" channels of the fluvial models actually consist entirely of ripple-bedded sandstones and interbedded mudstones that can be explained in terms of mixed tidal-fluvial processes within an estuary rather than riverine processes. Much of the gray-shale component flanking the sandstone cores of the shale wedges may be related to the development and lateral shifting of subtidal to intertidal channels and flats falnking an active distributary channel complex or estuary margin (e.g., Dalrymple and Makino, 1989; Dalrymple et al., 1990) (Fig. 9b) rather than splays and floodplains. The pervasive occurrence of laminated mudstones and lenticular-, wavy-, and flaser-bedding, some containing well-defined tidal cycles, supports this interpretation.

In the more distal parts of the gray-shale wedges, pod-like gray shales commonly occur within "swales" or elongate linear depressions (Smith, 1970; Krausse et al., 1979; Burk et al., 1987). Such shales may have been deposited within tidal creeks that developed over the peats during initial transgressions. Water salinities in the channels were likely brackish to nearly fresh because of the high amounts of rainfall. The pH values were likely to be acidic because of runoff from marginal peat swamps. The high pH and low salinity precluding benthic fauna and could explain why bioturbation and marine body fossils are rare to absent. Conversely, aerial parts of terrestrial plants and insects, which could fall directly into muddy tidal creeks from the surrounding forests, could be readily preserved in the laterally accreting tidal channels.

the coal is absent. The second channel subparallels the first and is filled with shale and sandstone. The third is discontinuous and interpreted to be an abandoned meander loop. They suggest that the major channel is part of the larger Galatia Channel system (Fig. 6). Conversely, Droste and others (1986), mapping the same stratigraphic interval, interpreted the rock sequence as a bird's foot delta complex marked by discontinuous sandstone-filled distributary channels (Fig. 7). Discontinuous sandstone-fills are common in many areas. One example, a channel overlying the Herrin Coal, was mapped by Friedman (1960). An axial section of the main channel (Fig. 8) indicates that the channel

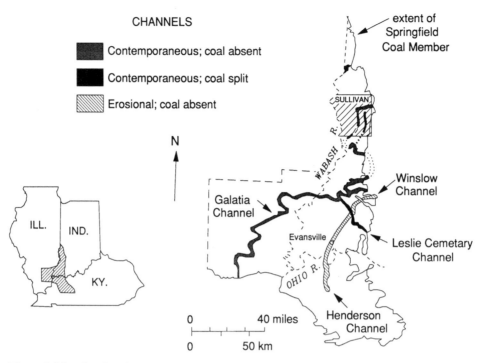

Figure 6. Map showing channels thought to have affected Springfield Coal. Diagram modified from Eggert et al. (1983).

In laminated gray shales, early diagenetic sideritic concretions preserved not only plants but also a host of soft-bodied terrestrial to marine fauna. Such flora and fauna have been extensively studied in the Francis Creek Shale in the Mazon Creek area (see Baird et al., 1985a, 1985b, 1986; Kuecher et al., 1990). Less extensively studied but similar fossiliferous siderites have been reported from dark-gray, laminated shale from above the Lower Block Coal of Indiana (Richardson and Johnson, 1969) and within the Energy Shale of Illinois (Gastaldo, 1977; Jacobson et al., 1979). This combined geochemical-paleontological evidence suggests similar environments were developed, at least locally, within the various gray-shale wedges throughout the stratigraphic interval under discussion.

The occurrence of low-sulfur coals is readily explained within the tidally influenced settings. If the depositional systems were brackish because of high daily rainfall, a relatively rapid accumulation of tidally emplaced muds within a lower delta plain setting would have served to mantle and protect the peats from long-term exposure to seawater sulfates that would have ultimately resulted in high-sulfur coals. Thus our model invokes high rates of brackish-water, tidally controlled deposition as a necessary component for the occurrence of low-sulfur coals.

Hydrodynamics within a delta plain or an estuary are a function of currents generated by an interaction of tides, waves,

and fluvial discharge. For instance, distributary channels on a delta with a significant tidal influence differ from completely fluvially dominated distributary channels (such as the Mississippi) by having open, flaring estuaries at the distributary mouth instead of levee-bounded distributaries. Estuarine deposits will reflect the change from a fluvial dominance to that of a mixed tidal and fluvial influence by becoming wider and muddier down channel where the zone of mixing of salt and fresh water occurs (e.g., Allen, 1971). Seaward transition of fluvially derived sands to mixed sands and muds is used to identify estuary deposits in the rock record (e.g., Rahmani, 1988; Karvonen, 1989) and may explain the sorts of facies relationships observed in the distribution of Pennsylvanian "channel" sandstones.

The degree of marine wave energy also has a profound effect on deltas and estuaries. Wave processes tend to keep muds in suspension and rework the sands into spits, beaches, or chenier ridges. Sandstone bedforms associated with wave processes are uncommon within the coal-bearing strata of the basin, thus it appears wave reworking of the Illinois Basin Pennsylvanian coasts was minimal.

CONCLUSIONS

The occurrence of tidally generated sedimentary structures within gray-shale lithofacies indicates that tidally influenced en-

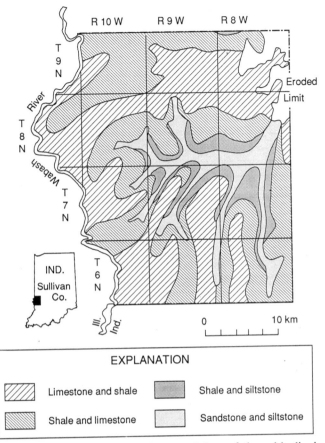

Figure 7. Map showing a bird's foot morphology of channel bodies in Sullivan County, Indiana. Compare with Figure 8. Diagram modified from Droste et al. (1986).

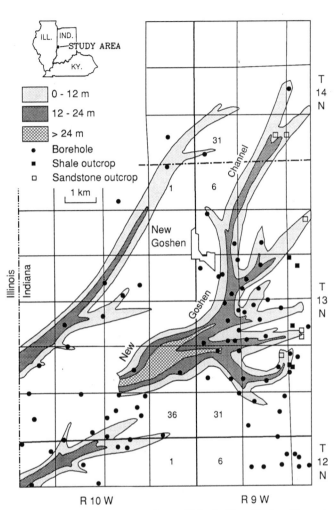

Figure 8. Isopach map of the sandstone fill in the New Goshen Channel and two tributaries. Modified from Friedman (1960).

vironments produced a significant portion of the gray shales above Illinois Basin coals. Many of the major coals may have been formed within coastal settings rather than the commonly depicted upper-delta plain setting. The gray-shale lithofacies was dominant during the Lower Pennsylvanian. Conversely, during the Middle Pennsylvanian, the relationship of the laterally restricted. volumetrically unimportant gray-shale wedges to the widespread black shale facies indicates only limited siliciclastic input to the coast after major peat-forming episodes. The lack of siliciclastics available to mantle the coals during ensuing transgressive phases resulted in high-sulfur coals in areas removed from the areas of gray-shale deposition.

As more details of the tidally influenced environments are understood, enhanced delineation of roof-strata distribution should be possible. Discarding strictly fluvial models and incorporating mixed fluvial and tidal models should help resolve some of the problems that remain regarding the formation, distribution, and quality of Illinois Basin coals.

ACKNOWLEDGMENTS

We wish to thank C. B. Cecil, J. C. Cobb, P. J. DeMaris, W. A. DiMichele, J. B. Droste, R. A. Gastaldo, D. Harper, and W. J. Nelson for stimulating discussions regarding the Pennsylvanian of the Illinois Basin and their possible modern analogues. This paper was substantially improved through critical reviews by Stephen F. Greb, John C. Ferm, and Ralph E. Hunter.

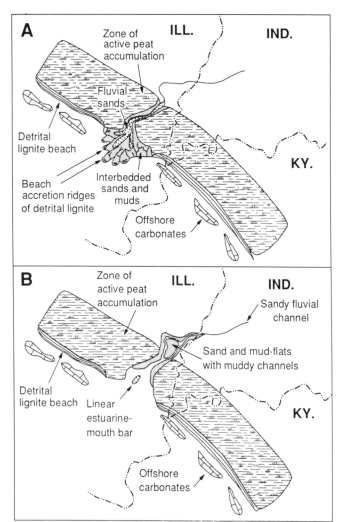

Figure 9. Gray-shale lithofacies depositional models based upon (A) muddy, tidally influenced delta and (B) estuarine system. In the deltaic system, potential lithofacies variations formed by transgression of a mixed fluvial and tidal delta formed immediately following deposition of the peats that were to form the Springfield Coal. In the estuarine model, these same lithofacies are produced within a localized estuarine sequence developed during transgression of a preexisting fluvial system.

REFERENCES CITED

Allen, G. P., 1971, Relationships between grain size parameter distribution and current patterns in the Gironde Estuary (France): Journal of Sedimentary Petrology, v. 41, p. 74–88.

Allen, G. P., Laurier, D., and Thouvenin, J., 1979, Étude sedimentologique du Delta de la Mahakam: Paris, Total, Compagnie Française des Pétroles, Notes et Mémoires, No. 15, 156 p.

Allen, J.R.L., 1981, Lower Cretaceous tides revealed by cross-bedding with mud drapes: Nature, v. 289, p. 579–581.

Allgaier, G. J., and Hopkins, M. E., 1975, Reserves of the Herrin (no. 6) Coal in southeastern Illinois: Illinois State Geological Survey Circular 489, 31 p.

Archer, A. W., and Kvale, E. P., 1989, Seasonal and yearly cycles within tidally laminated sediments: An example from the Pennsylvanian of Indiana, U.S.A., *in* Cobb, J., ed., Geology of the Lower Pennsylvanian in Kentucky, Indiana, and Illinois: Bloomington, Indiana Geological Survey, Illinois Basin

Consortium, no. 1, p. 45–56.

Archer, A. W., and Maples, C. G., 1984, Trace-fossil distribution across a marine to nonmarine gradient in the Pennsylvanian of southwestern Indiana: Journal of Paleontology, v. 58, p. 448–466.

Archer, A. W., Devera, J. A., Kvale, E. P., and Nelson, W. J., 1989, Stop 9—Upper Caseyville/Lower Tradewater roadcut on Interstate 57, southern Illinois, *in* Cecil, C. B., and Eble, C., eds., Carboniferous geology of the eastern United States: Washington, D.C., American Geophysical Union, 28th International Geological Congress, Field Trip Guidebook T143, p. 30–32.

Baird, G. C., and Shabica, C. W., 1980, The Mazon Creek depositional event: Examination of Francis Creek and analogous facies in the mid-continent region, *in* Langenheim, R. L., and Mann, C. J., eds., Middle and Late Pennsylvanian strata on margin of Illinois Basin, Vermilion County, Illinois Vermillion and Parke Counties, Indiana: Danville Illinois, Society of Economic Paleontologists and Mineralogists, 10th Annual Field Conference Guidebook Great Lakes Section, p. 79–92.

Baird, G. C., Shabica, C. W., Anderson, J. L., and Richardson, E. S., Jr., 1985a, Biota of a Pennsylvanian muddy coast: Habitats within the Mazonian delta complex, northeast Illinois: Journal of Paleontology, v. 59, p. 253–281.

Baird, G. C., Sroka, S. D., Shabica, C. W., and Beard, T. L., 1985b, Mazon Creek–type fossil assemblages in the U.S. midcontinent Pennsylvanian: Their recurrent character and palaeoenvironmental significance, *in* Whittington, H. B., and Conway, M. S., eds., Extraordinary fossil biotas; their ecological and evolutionary significance: Royal Society of London Philosophical Transactions, ser. B, v. 311, p. 87–99.

Baird, G. C., Sroka, S. D., Shabica, C. W., and Kuecher, G. J., 1986, Taphonomy of Middle Pennsylvanian Mazon Creek area fossil localities, northeast Illinois: Significance of exceptional fossil preservation in syngenetic concretions: Palaios, v. 1, p. 271–285.

Breyer, J. A., 1987, A tidal origin for coarsening-upward sequences above two Wilcox lignites in East Texas: Journal of the Geological Society of London, v. 144, p. 463–469.

Breyer, J. A., and McCabe, P. J., 1986, Coals associated with tidal sediments in the Wilcox Group (Paleogene), South Texas: Journal of Sedimentary Petrology, v. 56, p. 510–519.

Broadhurst, F. M., 1988, Seasons and tides in the Westphalian, *in* Besly, B. M., ed., Sedimentation in a synorogenic basin complex: The Upper Carboniferous of northwest Europe: London, Blackie and Son, p. 264–272.

Burk, M. K., Deshowitz, M. P., and Utgaard, J. E., 1987, Facies and depositional environments of the Energy Shale Member (Pennsylvanian) and their relationship to low-sulfur coal deposits in southern Illinois, Journal of Sedimentary Petrology, v. 57, p. 1060–1067.

Cecil, C. B., and Englund, K. J., 1989, Origin of coal deposits and associated rocks in the Carboniferous of the Appalachian Basin, *in* Cecil, C. B., and Eble, C., eds., Carboniferous geology of the eastern United States: Washington, D.C., American Geophysical Union, 28th International Geological Congress, Field Trip Guidebook T143, p. 84–88.

Cecil, C. B., Dulong, F. T., Cobb, J. C., Supardi, and Turnbull, P., 1988, Allogenic processes in the central Sumatra basin—A modern analog for the origin of Lower Pennsylvanian coal-bearing strata in the eastern United States: U.S. Geological Survey Circular 1025, p. 8.

Dalrymple, R. W., and Makino, Y., 1989, Description and genesis of tidal bedding in the Cobequid Bay-Salmon River estuary, Bay of Fundy, Canada, *in* Taira, A., and Masuda, F., eds., Sedimentary facies in the active plate margin: Tokyo, Terra Scientific, p. 151–177.

Dalrymple, R. W., Knight, R. J., Zaitlin, B. A., and Middleton, G. V., 1990, Dynamics and facies model of a macrotidal sand-bar complex, Cobiquid Bay–Salmon River Estuary (Bay of Fundy): Sedimentology, v. 37, p. 577–612.

DiMichele, W. A., and DeMaris, P. J., 1987, Structure and dynamics of a Pennsylvanian-age *Lepidodendron* forest: Colonizers of a disturbed swamp habitat in the Herrin (No. 6) coal of Illinois: Palaios, v. 2, p. 146–157.

Droste, J. B., Lane, N. G., and Maples, C. G., 1986, A birdfoot delta in the subsurface Pennsylvanian of Sullivan County, Indiana: Indiana Academy of Sciences Transactions, v. 95 p. 313–317.

Edwards, M. J., Langenheim, R. L., Jr., Nelson, W. J., and Ledvina, C. T., 1979, Lithologic patterns in the Energy Shale Member and the origin of "rolls" in the Herrin (No. 6) Coal Member, Pennsylvanian, in the Orient No. 6 Mine, Jefferson County, Illinois: Journal of Sedimentary Petrology, v. 49, p. 1005–1014.

Eggert, D. L., 1982, A fluvial channel contemporaneous with deposition of the Springfield Coal Member (V), Petersburg Formation, Northern Warrick County, Indiana: Indiana Geological Survey Special Report 28, 20 p.

Eggert, D. L., 1987, Earlier sediment differential compaction in Gibson County, Indiana: International Journal of Coal Geology, v. 8, p. 305–314.

Eggert, D. L., and Adams, S. C., 1979, Distribution of fluvial channel systems contemporaneous with the Springfield Coal Member (Middle Pennsylvanian) in southwestern Indiana: Ninth International Congress on Carboniferous Stratigraphy, Washington, D.C., and Champaign-Urbana, Illinois, Compte Rendu, v. 4, p. 342–348.

Eggert, D. L., and Phillips, T. L., 1982, Environments of deposition—Coal balls, cuticular shale and gray-shale floras in Fountain and Parke Counties, Indiana: Indiana Geological Survey Special Report 30, 43 p.

Eggert, D. L., Chou, C-L., Maples, C. G., Peppers, R. A., Phillips, T. L., and Rexroad, C. B., 1983, Origin and economic geology of the Springfield Coal Member in the Illinois Basin (Field Trip 9), *in* Shaver, R. H., and Sunderman, J. A., eds., Field Trips in Midwestern Geology: Bloomington, Indiana, Geological Society of America, Indiana Geological Survey, and Indiana University, v. 2, p. 121–146.

Eriksen, M. C., and Slingerland, R., 1990, Numerical simulations of tidal and wind-driven circulation in the Cretaceous Interior Seaway of North America: Geological Society of America Bulletin, v. 102, p. 1499–1516.

Fishbaugh, D. A., Kvale, E. P., Archer, A. W., and Gray, H. H., 1989, Association of tidal and fluvial sediments within Lower Pennsylvanian rocks, Turkey Run State Park, Parke County, Indiana: American Association of Petroleum Geologists Eastern Section Meeting and Field Trip, Indiana Geological Survey Guidebook, 46 p.

Franklin, D. W., 1939, Lithologic and stratigraphic study of the Lower Pennsylvanian strata, Orange County, Indiana [Ph.D. thesis]: Urbana, University of Illinois, 48 p.

Friedman, S. A., 1960, Channel-fill sandstones in the Middle Pennsylvanian rocks of Indiana: Indiana Geological Survey Report of Progress 23, 59 p.

Gastaldo, R. A., 1977, A Middle Pennsylvanian nodule flora from Carterville, Illinois, *in* Romans, R. C., ed., Geobotany: New York, Plenum Press, p. 1–308.

Gluskoter, H. J., and Hopkins, M. E., 1970, Distribution of sulfur in Illinois coals, *in* Smith, W. H., et al., eds., Depositional environments in parts of the Carbondale Formation—Western and northern Illinois: Illinois Geological Survey Guidebook Series 8, p. 89–95.

Gray, H. H., 1962, Outcrop features of the Mansfield Formation in south-western Indiana: Indiana Geological Survey Report of Progress 26, 40 p.

Hopkins, M. E., 1968, Harrisburg (No. 5) coal reserves of southeastern Illinois: Illinois Geological Survey Circular 431, 25 p.

Hopkins, M. E., and Nance, R. B., 1970, Sulfur content of the Colchester (No. 2) Coal Member at the Banner Mine, Peoria and Fulton Counties, Illinois, *in* W. H. Smith, et al., eds., Depositional environments in parts of the Carbondale Formation—Western and northern Illinois: Illinois Geological Survey Gudiebook Series 8, p. 96–98.

Hower, J. E., and Wild, G. O., 1982, Petrographic variations in the Springfield (No. 9) Coal in western Kentucky: International Journal of Coal Geology, v. 2, p. 17–30.

Jacobson, R. J., Utgaard, J. E., and Deshowitz, M. P., 1979, Channel fill and associated deposits of the Walshville channel and Anvil Rock channel (Day 6 road log, Stop 12), *in* Palmer, J. E., and Dutcher, R. R., eds., Depositional and structural history of the Pennsylvanian system of the Illinois Basin: Part 1, Road log and description of stops: Urbana, Illinois, Ninth International Carboniferous Congress Guidebook, p. 96–102.

Johnson, D. O., 1972, Stratigraphic analysis of the interval between the Herrin (No. 6) Coal and the Piasa Limestone in southwestern Illinois [Ph.D. thesis]: Urbana, University of Illinois, 105 p.

Karvonen, R. L., 1989, The Ostracode Member east-central Alberta: An example of an estuarine valley fill deposit, *in* Reinson, G. E., ed., Modern and ancient examples of clastic tidal deposits—A Core and peel workshop, *in* Proceedings, Second International Research Symposium on Clastic Tidal Deposits, August 22–25: Canadian Society of Petroleum Geologists, p. 105–116.

Klein, G. DeV., 1977, Tidal circulation model for deposition of clastic sediment in epeiric and mioclinal shelf seas: Sedimentary Geology, v. 18, p. 1–12.

Klein, G. DeV., and Ryer, T. A., 1978, Tidal circulation patterns in Precambrian, Paleozoic, and Cretaceous epeiric and mioclinal shelf seas: Geological Society of America Bulletin, v. 89, p. 1050–1058.

Krausse, H.-F., Damberger, H. H., Nelson, W. J., Hunt, S. R., Ledvina, C. T., Treworgy, C. G., and White, W. A., 1979, Roof strata of the Herrin (No. 6) Coal and associated rock in Illinois—A summary report: Illinois Geological Survey Illinois Minerals Note 72, 54 p.

Kreisa, R. D., and Moiola, R. J., 1986, Sigmoidla tidal bundles and other tide-generated sedimentary structures of the Curtis Formation, Utah: Geological Society of America Bulletin, v. 82, p. 2585–2592.

Kuecher, G. J., 1983, Rhythmic sedimentation and stratigraphy of the middle Pennsylvanian Francis Creek Shale near Braidwood, Illinois [M.S. thesis]: Chicago, Northeastern Illinois University, 143 p.

Kuecher, G. J., Woodland, B. G., and Broadhurst, F. M., 1990, Evidence of deposition from individual tides and of tidal cycles from the Francis Creek Shale (host rock to the Mazon Creek Biota), Westphalian D (Pennsylvanian), northeastern Illinois: Sedimentary Geology, v. 68, p. 211–221.

Kvale, E. P., and Archer, A. W., 1989a, Recognition of tidal processes in mudstone-dominated sediments, Lower Pennsylvanian, Indiana, *in* Cobb, J. C., ed., Geology of the Lower Pennsylvanian in Kentucky, Indiana, and Illinois: Illinois Basin Studies 1: Lexington, Kentucky, Illinois Basin Consortium, p. 29–44.

Kvale, E. P., and Archer, A. W., 1989b, Tidal influence in deposits associated with coals (Pennsylvanian), southwestern Indiana: Geological Society of America Abstracts with Programs, v. 21, p. 26.

Kvale, E. P., and Archer, A. W., 1990, Tidal deposits associated with low-sulfur coals, Brazil Fm. (Lower Pennsylvanian), Indiana: Journal of Sedimentary Petrology, v. 60, p. 563–574.

Kvale, E. P., and Archer, A. W., 1991, Characteristics of two Pennsylvanian-age semidiurnal tidal deposits in the Illinois Basin, U.S.A., Canadian Society Petroleum Geology Memoir 16, p. 179–188.

Kvale, E. P., Archer, A. W., and Johnson, H. R., 1989, Daily, monthly, and yearly tidal cycles within laminated siltstones of the Mansfield Formation (Pennsylvanian) of Indiana: Geology, v. 17, p. 365–368.

Neavel, R. C., 1961, Petrographic and chemical composition of Indiana coals: Indiana Geological Survey Bulletin 22, 81 p.

Nelson, W. J., 1983, Geologic disturbances in Illinois coal seams: Illinois Geological Survey Circular 530, 47 p.

Nelson, W. J., 1987, The Hornsby District of low-sulfur Herrin Coal in central Illinois (Christian, Macoupin, Montgomery, and Sangamon Counties), with contributions by DeMaris, P. J., and Bauer, R. A., Illinois State Geological Survey Circular 540, 40 p.

Pfefferkorn, H. W., Fuchs, K., Christian, H., Hofmann, C., Rabold, J. M., and Wagner, T., 1988, Recent geology and taphonomy of the Orinoco Delta—Overview and field observations: Heidelberg, Heidelberger Geowissenschaftliche Abhandungen, Band 20, p. 21–56.

Phillips, T. L., Peppers, R. A., and DiMichele, W. A., 1985, Stratigraphic and interregional changes in Pennsylvanian coal-swamp vegetation: Environmental inferences: International Journal of Coal Geology, v. 5, p. 43–109.

Rahmani, R. A., 1988, Estuarine tidal channel and nearshore sedimentation of a Late Cretaceous epicontinental sea, Drumheller, Alberta, Canada, *in* de Boer, P. L., van Gelder, A., Nio, S. D., eds., Tide-influenced sedimentary environments and facies: Boston, D. Reidel, p. 433–471.

Richardson, E. S., Jr., and Johnson, R. G., 1969, Mazon Creek fossils: North

American Paleontological Congress Proceeding, v. 1, p. 1222–1235.

Schubel, J. R., 1984, Estuarine circulation and sedimentation: An overview, *in* Haq, B. U., and Milliman, J. D., eds., Marine geology and oceanography of Arabian Sea and coastal Pakistan: New York, Van Nostrand, p. 113–136.

Shabica, C. W., 1970, Depositional environments in the Francis Creek Shale, *in* Smith, W. H., et al., eds., Depositional environments in parts of the Carbondale Formation—Western and northern Illinois: Illinois Geological Survey Guidebook Series 8, p. 43–52.

Shaver, R. H., and 16 others, 1986, Compendium of Paleozoic rock-unit stratigraphy in Indiana—A revision: Indiana Department of Natural Resources, Geological Survey Bulletin 59, 203 p.

Shaw, A. B., 1964, Time in stratigraphy: New York, McGraw-Hill, 365 p.

Slater, R. A., 1985, A numerical model of tides in the Cretaceous seaway of North America: Journal of Geology, v. 93, p. 333–345.

Slingerland, R., 1986, Numerical computation of co-oscillating paleotides in the Catskill epeiric sea of eastern North America: Sedimentology, v. 33,

p. 487–497.

Smith, S. I., 1871, Notice of a fossil insect from the Carboniferous Formation of Indiana: American Journal of Science, ser. 3, v. 1, p. 44–46.

Smith, W. H., 1970, Lithology and distribution of the Francis Creek Shale in Illinois, *in* Smith, W. H., et al., eds., Depositional environments in parts of the Carbondale Formation—Western and northern Illinois: Illinois Geological Survey Guidebook Series 8, p. 89–95.

van Andel, Tj. H., 1967, The Orinoco Delta: Journal of Sedimentary Petrology, v. 37, p. 297–310.

Visser, R., 1980, Neap-spring cycles reflected in Holocene subtidal large-scale bedform deposits: A preliminary note: Geology, v. 8, p. 543–546.

Wanless, H. R., 1964, Local and regional factors in Pennsylvanian cyclic sedimentation, *in* Merriam, D. F., ed., Symposium on cyclic sedimentation: Kansas Geological Survey Bulletin 169, v. 2, p. 593–606.

MANUSCRIPT ACCEPTED BY THE SOCIETY JANUARY 14, 1993

Index

Typeset by WESType Publishing Services, Inc., Boulder, Colorado
Printed in U.S.A. by Johnson Printing, Boulder, Colorado